Discovery and Classification in Astronomy

Astronomical discovery involves more than detecting something previously unseen. The reclassification of Pluto as a dwarf planet in 2006, and the controversy it generated, shows that discovery is a complex and extended process - one comprising various stages of detection, interpretation, and understanding. Ranging from Galileo's observation of Jupiter's satellites, Saturn's rings, and star clusters, to Herschel's nebulae and the modern discovery of quasars and pulsars, Steven J. Dick's comprehensive history analyzes the anatomy of discovery and identifies discovery as the engine of progress in astronomy. The text traces more than 400 years of telescopic observation, exploring how the signal discoveries of new classes of astronomical objects relate to and inform one another, and why controversies such as Pluto's reclassification are commonplace in the field. The volume is complete with a detailed classification system for known classes of astronomical objects, offering students, researchers, and amateur observers a valuable reference and guide.

STEVEN J. DICK served as the NASA Chief Historian from 2003 to 2009 and was the Charles A. Lindbergh Chair in Aerospace History at the National Air and Space Museum from 2011 to 2012. He has worked as both an astronomer and a historian of science, and he has written extensively on astronomy, astrobiology, space exploration, and scientific institutions. Minor planet 6544 Stevendick is named in his honor.

To Adeline, Benjamin, and John
Discovering New Worlds of Their Own

This method of viewing the heavens seems to throw them into a new kind of light. They now are seen to resemble a luxuriant garden, which contains the greatest variety of productions, in different flourishing beds; and one advantage we may at least reap from it is, that we can, as it were, extend the range of our experience to an immense duration. For, to continue the simile I have borrowed from the vegetable kingdom, is it not almost the same thing, whether we live successively to witness the germination, blooming, foliage, fecundity, fading, withering, and corruption of a plant, or whether a vast number of specimens, selected from every stage through which the plant passes in the course of its existence, be brought at once to our view?

William Herschel,
Catalogue of a Second Thousand Nebulae and Clusters of Stars (1789)

Discovery is what science is all about … the "cutting edge" of scientific knowledge.

Norwood Russell Hanson, "An Anatomy of Discovery" (1967)

The real art of discovery consists not in finding new lands, but in seeing with new eyes.

Marcel Proust, The Prisoner, *Remembrance of Things Past* (1923)

Contents

Preface

The title of this volume will bring to mind for some readers Martin Harwit's book *Cosmic Discovery: The Search, Scope, and Heritage of Astronomy,* published more than three decades ago. As an astronomer, Harwit's practical aims in that book were twofold: first, to determine what fields of astronomy have the largest number of potential discoveries and, second, to explore what fields of astronomy might promise the most immediate advances and striking returns. By analyzing just enough history to answer these questions, Harwit hoped to provide answers of use both to students entering the field facing the daunting question of what part of astronomy to study and to policymakers who had only limited resources to dole out for the advancement of astronomy. In this effort at applied history he had only mixed success, but by taking the concept of discovery as a serious object of analysis, Harwit did what only very few astronomers, historians, or philosophers have done before or since. While it will become clear that I differ with Harwit on many issues, not least the crucial question of what constitutes a new "class" of object, I record here my debt to him, and others, for their boldness in attacking a concept so broad as "discovery."

Among philosophers and historians I am especially indebted to the work of Norwood Russell Hanson and Thomas Kuhn. The former, the colorful founder of the pioneering Department of History and Philosophy of Science at Indiana University (where long ago I spent five years as a graduate student), is best known for his book *Patterns of Discovery*, which elaborated the crucial idea of theory-laden observations. His final article, "An Anatomy of Discovery," was submitted to the journal *Philosophy* the day before his death in 1967 when the private plane he was piloting crashed into a hillside. The article remains a seminal exposition, in which Hanson parsed the subject in the manner to which philosophers are peculiarly accustomed. To Thomas Kuhn, best known for his work on scientific "paradigms" and "revolutions," I am indebted for the idea that discovery is an extended process, an idea borne out and elaborated in his book *The Structure of Scientific Revolutions*. Unfortunately, in my view, Kuhn's ideas about the extended structure of discovery were overwhelmed, both in his own work and in its subsequent critiques, by the emphasis on paradigms and revolutions. While most of science does not consist of revolutions, discovery is much more common; one might even say it is the engine of science. Discovery therefore deserves to be brought to the fore as a central concept in that endeavor.

Among historians I am also grateful for the work of Ken Caneva, one of the few to give discovery its due. His important article, "'Discovery' as a Site for the Collective Construction of Scientific Knowledge," has influenced my work and indicates that what is true for astronomy may also be true for the wider scientific world. A few other historians, including Simon Schaffer, have sporadically illuminated the problems of discovery over the last few decades, and many more historians have done work in their specialized areas that relates to the problem.

It is the contention of this book that the concepts of discovery and classification go hand in hand, for the urge to classify discoveries of new objects seems to be an inherent part of the human mind. One cannot have a classification system without classes, and, arguably, one cannot definitively determine a Class (as opposed to another taxonomic level such as a Type or a Family) without a classification system. In natural science, classification has been essential to chemistry (the Periodic Table), physics (the Standard Model), and most especially biology (featuring most recently the dueling Five Kingdom and Three Domain systems). I am indebted to those scientists, historians, and philosophers, cited in the text, who have elaborated the history and philosophy of these systems. In biology I would in particular mention the work of Ernst Mayr as a scientist, historian, and philosopher of biology, who as a systematist himself was attuned to the numerous pitfalls of classification and wrote extensively on the subject. In addition, historian and philosopher Jan Sapp's volume, *The New Foundations of Evolution*, vividly demonstrates how classification can undergo a sea change with new insights and techniques. Astronomers have much to learn from such analysis in other sciences, as was all too painfully evident during the Pluto discussions in 2006 at the International Astronomical Union General Assembly in Prague (see Introduction and Chapter 1).

Perhaps unsurprisingly, when they think of discovery at all as a distinct concept, astronomers continue to be interested in the practical questions originally raised in Harwit's *Cosmic Discovery*. One of their primary objectives, after all, is to make discoveries. Thus the IAU symposium "Accelerating the Rate of

Astronomical Discovery," held at the General Assembly of the International Astronomical Union in Rio de Janeiro in 2009, once again had a very practical goal, as indicated by its title. But there are more than practical issues at stake in discovery, and anyone plumbing its depths will soon find fertile soil not only in its history, but also for the philosophy, sociology, and psychology of science.

I am indebted to the Smithsonian Astrophysical Observatory/NASA Astrophysics Data System (http://adsabs.harvard.edu/) as a tremendous resource for much of the scientific literature used and cited in the notes. The reader can access most of that literature by accessing the ADS Web site and entering author and date. I have also found very useful the compilations and commentaries by Marcia Bartusiak, Kenneth Lang, and Owen Gingerich, as well as the centennial volume of the *Astrophysical Journal*, edited by Helmut Abt, all cited in the bibliography. The *Journal for the History of Astronomy*, edited for forty years by Michael Hoskin at Churchill College, Cambridge, is a vast treasure trove of information that I have also used extensively, and it is available via ADS as well.

Thanks also to those astronomers and historians who read and commented on all or part of the manuscript in its present or earlier forms, including David DeVorkin, Robert Smith, Marc Rothenberg, Martin Harwit, Ken Caneva, Owen Gingerich, Michael Hoskin, the late Don Osterbrock, Michael Crowe, Ken Kellermann, Barbara Becker, Theodore Arabatzis, Virginia Trimble, Jay Pasachoff, Chris Corbally, Richard Gray, Mark Kuchner, Nancy Roman, Woody Sullivan, Brian Mason, Bill Hartkopf, and anonymous reviewers for Cambridge University Press. This is the fifth of my books over the last three decades to emerge from that distinguished Press. My thanks to the Syndics of Cambridge University Press; to my diligent and efficient editor Vince Higgs; and to the copyediting, production, and graphics staffs, including Shari Chappell and Karen Verde.

I owe a considerable debt to the unparalleled astronomy treasures of the U.S. Naval Observatory in Washington, DC, and its ever-helpful librarians Sally Bosken and Gregory Shelton, who have gone above and beyond the call of duty. For imagery, thanks to the Emilio Segrè Visual Archives of the American Institute of Physics; Zolt Levay at the Space Telescope Science Institute for Hubble-related images; Alison Doane, curator of Astronomical Photographs at Harvard College Observatory; Lauren Amundson for images from the Lowell Observatory archives; Hesse Jochen at the Department of Prints and Drawings of the Zentralbibliothek, Zurich; Mary Ann Quinn at the Rockefeller Archive Center; as well as many other individuals and institutions credited in the captions.

Finally, I benefited greatly from a scintillating special session on "Discovery and Classification in Astronomy," held at the 28th General Assembly of the

International Astronomical Union in Beijing, China, in August 2012, in which I participated with Ron Ekers, Martin Harwit, Barry Madore, David DeVorkin, Ken Kellermann, and Ray Norris. The discussions generated there, beginning with an anatomy of the Pluto controversy six years earlier in Prague by the president of the IAU at the time, and extending to all of astronomy, illustrated once again the richness of the subject of discovery.

My hope is that the book spurs provocative discussion about discovery and classification, not only in astronomy but also in the broader realms of science. This volume was supported by NSF STS grant # 1016024 from the Science, Technology and Society Program, under Program Director Frederick Kronz, gratefully acknowledged. The final parts were completed during my time as the Charles A. Lindbergh Chair in Aerospace History at the National Air and Space Museum in Washington, DC. My thanks to that fine institution and its world-class scholars.

Steven J. Dick
Washington, DC
October 2012

Abbreviations

A & A	*Astronomy and Astrophysics*
Abt	*American Astronomical Society Centennial Issue: Selected Fundamental Papers Published This Century in the Astronomical Journal and the Astrophysical Journal*, Helmut Abt, ed., *Astrophysical Journal*, 525 (Chicago: University of Chicago Press, 1999)
AJ	*Astronomical Journal*
AJSA	*American Journal of Science and Arts*
AN	*Astronomische Nachrichten*
ApJ	*Astrophysical Journal*
ARAA	*Annual Reviews of Astronomy and Astrophysics*
ASP	Astronomical Society of the Pacific
BAAS	*Bulletin of the American Astronomical Society*
BAIN	*Bulletin of the Astronomical Institute of the Netherlands*
Bartusiak	*Archives of the Universe: A Treasury of Astronomy's Historic Works of Discovery*, Marcia Bartusiak, ed. (New York: Pantheon Books, 2004)
BEA	*Biographical Encyclopedia of Astronomers*
BJHS	*British Journal for the History of Science*
CMWO	*Contributions of Mt. Wilson Observatory*
DSB	*Dictionary of Scientific Biography*
HCO	Harvard College Observatory
HR Diagram	Hertzsprung-Russell Diagram
HSPS	*Historical Studies in the Physical Sciences*
HSTPR	Hubble Space Telescope Press Release
JAHH	*Journal of Astronomical History and Heritage*

JBAA	*Journal of the British Astronomical Association*
JHA	*Journal for the History of Astronomy*
JHB	*Journal for the History of Biology*
JRASC	*Journal of the Royal Astronomical Society of Canada*
Lang and Gingerich	*A Source Book in Astronomy and Astrophysics, 1900–1975*, Kenneth R. Lang and Owen Gingerich, eds. (Cambridge, MA: Harvard University Press, 1979)
MK System	Morgan-Keenan System (post-1953)
MKK System	Morgan-Keenan Kellman System (pre-1953)
MNRAS	*Monthly Notices of the Royal Astronomical Society*
NDSB	*New Dictionary of Scientific Biography*
OHI	Oral History Interview
PA	*Popular Astronomy*
PASP	*Publications of the Astronomical Society of the Pacific*
PNAS	*Proceedings of the National Academy of Sciences*
PTRSL	*Philosophical Transactions of the Royal Society of London*
QJRAS	*Quarterly Journal of the Royal Astronomical Society*
SciAm	*Scientific American*
SSR	*Space Science Reviews*

Introduction: The Natural History of the Heavens and the Natural History of Discovery

On August 24, 2006, the International Astronomical Union (IAU) - the only institution that counts when it comes to official designations of astronomical bodies - declared that Pluto was not a planet. More specifically, astronomers demoted Pluto from a planet to a dwarf planet, and (to the chagrin of many scientists and the confusion of the general public) declared that a dwarf planet was not a planet all, thus reducing the number of classical planets in the solar system to eight for the first time since 1930 when Pluto was discovered. Pluto's demotion not only meant a rewriting of the textbooks, but also set off a surprisingly intense scientific and public outcry - an interesting cultural phenomenon indicating not only the importance of classification to scientists, but also a deeper investment in astronomy among the general public than one might have thought.[1]

As a longtime member of the IAU, I was among those voting on that fateful day in Prague. Although I had attended every triennial IAU General Assembly since 1988, many of them as an officer in its History of Astronomy Commission, I had never seen the meeting dominated by a single issue as it was on this occasion. Though literally hundreds of sessions were held over the two weeks of the meeting, discussing a broad panoply of astronomical subjects, and though numerous other resolutions were considered and passed at this General Assembly, the resolutions involving Pluto were the center of attention, the subject of numerous sessions, and the topic of the buzz in the hallways. As was tradition, the resolutions were voted on during the last day of the General Assembly, after much discussion the previous two weeks, leaving only 424 delegates to vote out of the thousands who had attended.

The vote unleashed a reaction the likes of which the IAU had never seen on any issue. On the plane returning to Washington, I was astonished that the onboard TV news clips highlighted the IAU vote. Back home all forms of media gave Pluto the attention normally reserved for politics, sports, and Hollywood stars. Nor was the reaction short-lived. A few weeks later the California Assembly, noting the millions of textbooks that would have to be revised and that Pluto shared its name with California's famous Disney dog, passed a resolution that "condemns the International Astronomical Union's decision to strip Pluto of its planetary status for its tremendous impact on the people of California and the state's long term fiscal health." Seven months later the legislature of New Mexico, home to Pluto's discoverer Clyde Tombaugh, passed a more sober bill declaring that "as Pluto passes overhead through New Mexico's excellent night skies, it be declared a planet and that March 13, 2007 be declared 'Pluto Planet Day' at the legislature." Walking the streets of Berkeley early in 2007, the casual stroller down Telegraph Street would have spotted the Mars Vintage Thrift Shop with the curiously melancholy marquee pleading "I Want Pluto To Be A Planet Again." And the debate among scientists continued long after the vote, with some threatening to overturn the decision.

The Pluto affair, in all its fame and infamy, opened a window to what is often seen as a boring subject – the discovery and classification of astronomical objects – part of the much larger problem of discovery and classification in science. Pluto showed that, far from being boring, classification – and classification systems – are an essential part of science. And although astronomy's oldest classification systems for stars and galaxies are barely more than a century old, the endeavor of seeking out new classes of astronomical objects dates, like Linnaean classification in biology, to the eighteenth century, in particular to the work of William Herschel. Quite aside from his discovery of Uranus in 1781 and coining the term "asteroid" for the new class of objects discovered in 1801, Herschel's sweeps of the heavens with his large telescopes revealed a huge number of nebulae, which he perceived to be in different stages of growth and development. Herschel compared himself to a naturalist, and saw himself as revealing the natural history of the heavens. The heavens, he wrote in 1789, "now are seen to resemble a luxuriant garden, which contains the greatest variety of productions." Herschel used the biological analogy repeatedly in his work, and it was picked up by such pioneering naturalists as Alexander von Humboldt in the nineteenth century. One still sees the analogy between astronomy and biology used today, particularly as it pertains to classification.[2]

The number of classes of astronomical objects discovered since Herschel's nebulae has proliferated considerably. Some would seem to be obvious, like comets, asteroids, quasars, pulsars, and spiral galaxies. But just how many classes of astronomical objects are there? It depends, of course, on the definition of "class," for which there is no standard. One of the central themes of this volume is to show how astronomers grappled with the problem of the discovery of new classes of objects, on the one hand a necessary activity prior to any attempt at classification systems, and on the other hand a difficult and confusing task in the absence of a classification system to determine taxonomic levels. In Appendix 1 to this volume, as an exercise in the meaning of "class" and the construction of classification systems discussed in Chapter 8, we propose eighty-two different classes of astronomical objects. I hasten to add that the list cannot be totally definitive, because any such listing depends on a variety of factors, including the principles of any particular classification system and its assignment of taxonomic levels. Yet, I would claim that most astronomers would in fact accept the majority of these classes. Thus, while our analysis in this volume does not depend on the particulars of what I dub in Appendix 1 "Astronomy's Three Kingdoms," the eighty-two classes delineated there provide a useful set of classes for discussion and analysis. The "Three Kingdom" (3K) system also serves as a point of reference for the discussion of classes in the entire volume, whereby the reader can see where a particular class fits in the larger picture. Whatever its shortcomings, Appendix 1 offers the first self-consistent comprehensive classification system for astronomy, one that may prove useful for both scientific and pedagogical purposes.

It is the central thesis of this book that discovery is a complex and extended process, consisting not only of detection, but also of interpretation and various levels of understanding following interpretation. Its microstructure embraces technological, conceptual, and social roles. Moreover, we shall see that discovery and classification in many ways go hand in hand. As was the case with Pluto, they are almost always accompanied by controversy, and sometimes (but not always) by consensus. Thus, discovery, classification, controversy, and consensus are inevitably bound together in a complex conceptual, philosophical, technological, and sociological circle. More often than not, discovery is preceded by a "pre-discovery" phase (see Appendix 2) and always followed by a "post-discovery" phase, which includes its reception among scientists and the public as well as social issues of credit and reward. We develop definitions and describe characteristics of these phases in the course of the book as we analyze various discoveries; those readers wishing to see the definitions up-front may consult the Glossary.

Our entrée into the subject begins with Pluto in Part I, where we see firsthand the complex intertwining of these factors in a concrete case that has stretched over the better part of a century. In Part II we provide narratives and analysis of discovery in turn for selected classes of objects in the realms of the planets, stars, and galaxies. The first two realms (or "Kingdoms" in the parlance of the Three Kingdom system) had been roughly distinguished from antiquity, since one of the most obvious aspects of the night sky was that the planets (Greek for "wanderers") were distinguished from the "fixed stars," the background against which they moved. The third realm was not indisputably proved as separate from the sidereal realm until the work of Edwin Hubble in the 1920s. And the Families and Classes of these realms took centuries to discover. In these chapters we often let the discoverers tell their own stories. This, of course, is not the whole story, as Barbara Becker has recently reminded us in her scientific biography of the pioneer spectroscopist William Huggins. To take scientific narratives at face value, she says, is to fall into "an alluring trap," which may hide regrettable missteps and mask "the complexities and uncertainties that mark the first forays into a new realm of scientific investigation."[3] She was speaking of establishing a new branch of astronomy like astrophysics, but the same cautionary approach holds true for narratives of scientific discovery.

In Part III we examine possible patterns of discovery across astronomy's three realms, and compare them to the varieties of discovery in astronomy and other sciences. It is here that we discuss the macrostructure of discovery, consisting of detection, interpretation, and various levels of understanding; the microstructure of discovery with its conceptual, social, and technical components; the phases of pre-discovery and post-discovery; the varieties of discovery; and how discoveries end. We then turn to classification issues, first discussing the classes themselves and then classification systems, and surprisingly, we find they have both a pre-discovery and post-discovery role, as well as a role in discovery itself. Part IV takes up telescopes as engines of discovery, and theory as a motivator of discovery, finding that the former constitutes a strong, and the latter a weak, driver for discovery, but that theory is most essential in the interpretation and understanding phases. In Part V we see how all of these discoveries were synthesized by the grand discovery of cosmic evolution, which increasingly is being integrated into popular and scientific culture. We conclude with reflections on the meaning of discovery.

This book is both broad and narrow – broad in the sense that it covers discoveries in all fields of astronomy, narrow in the sense that it focuses on the discovery and classification only of classes of astronomical objects. It does not address the subsequent discovery of new members of each class, except briefly in Section 7.3, interesting as that may be. Nor, except by way of comparison in Chapter 7, does it address the discovery of astronomical phenomena such as interstellar magnetic fields, the cosmic microwave background remnant from the Big Bang, and the expansion (and acceleration!) of the universe. To do so not only would have made this a much larger book, it would also at some level have been comparing apples and oranges, though at another level apples and oranges are very interesting to compare. The detailed comparison of the varieties of scientific discovery must await another effort.

Finally, it is not my intention in finding patterns of discovery to be prescriptive, in other words, to tell scientists how to make discoveries. Rather, the effort in this volume at collecting and analyzing discovery narratives, as well as the difficulties of classification, may be seen as *a natural history of discovery*, solidly grounded in history. That does not preclude lessons learned, particularly classification efforts such as Pluto and its fellow dwarf planets, the deluge of exoplanets now being discovered, and the numerous classes, members of classes, and perhaps even new Families or Kingdoms of objects that remain to be discovered in the future. The natural history of discovery in astronomy, properly analyzed as such, is no less fascinating than the natural history of the heavens - or of its historical counterparts in chemistry, physics, and biology.

Part I Entrée

1

The Pluto Affair

A terrific thrill came over me. I switched the shutter back and forth, studying the images. Oh! I had better look at my watch and note the time. This would be a historic discovery.

Clyde Tombaugh, 1980[1]

Discovery is where the scientist touches Nature in its least predictable aspect. It discloses to us the regularities of Nature, but in itself, discovery is fickle, striking at the unexpected moment. This is the view that I must take after my serendipitous discovery of the moon of Pluto.

James W. Christy, 1980[2]

Things in the solar system can equally well be categorized in many different ways. Things with atmospheres. Things with moons. Things with life. Things with liquids. Things that are big. Things that are small. Things that are bright enough to see in the sky ... All of these are perfectly valid categories ... As with birds, your favorite solar system classification will depend on your interests.

Michael Brown, 2010[3]

The story of the discovery of Pluto has been told many times by its discoverer, historians, and the media, but in recent years has become all the more compelling because of the notorious reputation it has acquired following its perceived downgrading in 2006 to "dwarf planet" status. Seen in historical context over the last eight decades since its discovery, this rather small object in our solar system has assumed an outsized importance, precisely because it lies at the outer fringes of our solar system, at the borderline of normally assigned "classes" of objects in terms of its size and mass, and therefore at the border of normality in astronomy, where routine ends and creativity begins. Such borders are precisely what make Pluto interesting, and as such they illuminate in microcosm some of the many issues raised in this volume about the nature of discovery, interpretation, and classification in astronomy.

The discovery of Pluto in 1930 was not widely believed at the time to constitute the discovery of a new class of objects, uncertainties notwithstanding. Rather, it seemed to fit (if uncomfortably, in the eyes of some) into the class of planets; the discovery was, after all, made after a deliberate search for such a planet, focusing on a specific prediction, even if that prediction is now known to be spurious. Nor did the discovery in 1978 of Pluto's first moon, Charon, constitute a new class of objects, but only added a member to the class of satellites first established (beyond our own Moon) by Galileo's discovery of the moons of Jupiter. But by allowing an accurate mass for Pluto to be calculated, Charon began Pluto's decline from planetary status. In the end it was the discovery of a new class of Trans-Neptunian objects beginning in 1992 that precipitated the change in Pluto's class status, resulting in its "demotion" in 2006 by vote of the International Astronomical Union. And, in a linguistic contradiction truly to be regretted, a dwarf planet was ruled not to be a planet.

The story of tiny Pluto involves history, science, culture, negotiation, democracy, politics, and surprisingly strongly held views among the public, teachers, students, and the various disciplines that compose the astronomical community. In short, Pluto is a good entrée into those issues that, as we shall see, have plagued astronomy over the last few centuries, and continue unabated with every new discovery.

1.1 The Discovery of Pluto: Prediction and Observation

The planet Neptune had hardly been discovered in 1846 before predictions were being made of a Trans-Neptunian planet. The historian William G. Hoyt has chronicled these predictions, beginning with a very general one by the French physicist Jacques Babinet in 1848, continuing in the early twentieth century with the predictions of the eccentric astronomers William H. Pickering and Percival Lowell. These predictions culminated with Lowell's presentation of a specific location for the supposed planet, based primarily on the residuals in the motion of Uranus, which had completed much more of its orbit since its discovery than Neptune, and therefore offered more data for the comparison of theory with observation. Lowell's predictions were published in his "Memoir of a Trans-Neptunian Planet" in 1915, the year before his death. Lowell himself even initiated two searches, one from 1905–1907 with a Brashear 5-inch refractor and another in 1911 with a 40-inch reflector, both unsuccessful.[4] It was the third search, begun at the Lowell

Figure 1.1. Clyde Tombaugh, the discoverer of Pluto, at the eyepiece of the 13-inch Lawrence Lowell telescope in 1931. Tombaugh was not, however, the discoverer of the first dwarf planet, a class the International Astronomical Union declared in 2006. Nor was Pluto the first future dwarf planet to be discovered; the minor planet Ceres, also now classed a dwarf planet, was discovered in 1801 by Giuseppe Piazzi (Figure 2.5). Courtesy Lowell Observatory Archives.

Observatory in Flagstaff, Arizona in 1929, that would yield the new planet and become famous in the annals of astronomical history.

Crucial to the decision to undertake a third search were Lowell Observatory Director Vesto M. Slipher, and perhaps even more, Lowell's nephew, the recently appointed Observatory Trustee Roger Lowell Putnam. Putnam, a successful businessman who had studied mathematics at Harvard, was naturally anxious to find his "uncle Percy's" planet, and it was he who obtained the $10,000 in funding for the telescope from Lowell's younger brother, Harvard President A. Lawrence Lowell. After much deliberation, a 13-inch telescope was settled on as the optimal search instrument, its lens shaped by C. A. R. Lundin from the storied firm of Alvan Clark and Sons. The instrument arrived in February, and the search fell to the young Clyde Tombaugh, a Kansas farm boy whom Slipher had hired the previous month (Figure 1.1).[5] After the telescope was assembled and tested, the search was initiated on April 6, 1929, when Tombaugh took his first photographic plate.

Tombaugh had taken about a hundred 14 x 17-inch photographic plates by mid-June, 1929, when Slipher asked him to start examining the images for a

Figure 1.2. Small portion of discovery plates for Pluto, January 23 and 29, 1930, with arrows indicating Pluto, a fifteenth-magnitude object among thousands of others. Pluto was subsequently found to have been photographed as early as 1909, by the Yerkes Observatory, but it was unrecognized – a case of "pre-discovery." The discovery plate, measuring 14 x 17 inches, is now on long-term loan from Lowell Observatory, exhibited in the "Exploring the Planets" gallery at the Smithsonian National Air and Space Museum in Washington, DC. Courtesy Lowell Observatory Archives.

possible planet. This was to be done with an instrument called a "blink comparator," in which two plates of the same region of the sky taken on different dates are compared. During this "blinking" process the astute observer is required to detect the slightest motion over the two observation dates; the distant fixed stars will remain fixed, but closer interlopers will have moved. With about 400,000 objects on some of Tombaugh's plates, which included objects down to 17th magnitude and were invariably taken under slightly different atmospheric conditions, this was no easy task.

As instructed, Tombaugh began blinking plates, interrupted only at each dark-of-the-Moon lunation to photograph more regions. In February 1930, Tombaugh was blinking plates taken six days apart, on January 23 and 29, in the western Gemini region of the sky – the same area where William Herschel had discovered Uranus 149 years earlier. On Tuesday afternoon, February 18, Tombaugh spotted a slightly jumping image. He has given us his reaction at the moment of discovery: "After scanning a few fields to the left, I turned the next field into view. Suddenly, I spied a fifteenth magnitude image popping out and disappearing in the rapidly alternating views. Then I spied another image doing the same thing, about 3 millimeters (or .125 inches) to the left. 'That's it,' I exclaimed to myself." Considering the small amount of motion (Figure 1.2), Tombaugh knew instinctively it was beyond Neptune, a fact borne out by subsequent calculations.[6]

Exercising the caution that is common in such matters, the new object was observed over a period of weeks, and it was almost a month before news of the discovery was released to scientists and the public. On March 13, Slipher sent a telegram reporting the discovery to Harvard College Observatory, indicating the discovery of the new object had resulted from Lowell's predictions, undoubtedly with the Lowell Observatory Trustee in mind: "Systematic search begun years ago supplementing Lowell's investigation for Trans-Neptunian planet has revealed object which since seven weeks has in rate and motion and path consistently conformed to Trans-Neptunian body at approximate distance he assigned. Fifteenth magnitude. Position . . . agreeing with Lowell's predicted longitude." Further details were given in an Observatory Circular also dated March 13, which gave the Trans-Neptunian distance at about 40 to 43 astronomical units, "about Lowell's predicted distance . . . apparently fulfilling Lowell's theoretical findings." The next Circular, dated April 12, again emphasized the connection with Lowell's prediction: "This remarkable Trans-Neptunian planetary body has been found as a direct result of Lowell's work, planning, and convictions, and there appears present justification for referring to it as his Planet X." Although Slipher still called the object "Planet X," he noted that the name "Pluto," suggested by eleven-year-old Venetia Burney from Oxford, England, was being suggested to the American Astronomical Society and the Royal Astronomical Society.[7]

Was the discovery of Pluto, then, a triumph of prediction? Not exactly, even though clearly Slipher initially hailed it as such. Historians agree that the search was indeed motivated by Lowell's 1915 prediction, itself based on the observation that the observed orbital motions of Uranus were slightly irregular compared to theory. Moreover, there is no doubt the new object was found within 6 degrees of Lowell's prediction. Aside from that, however, the connection between theory and observation runs into trouble. Already in his second Circular, Slipher pointed out that the color of the new object was yellowish, more like the terrestrial planets than the bluish giants Uranus and Neptune. Moreover, while Planet X was 100 times brighter than the asteroid Ceres if removed to that distance, it was fourteen times fainter than expected – certainly not bright enough for the ten Earth masses Lowell had predicted to account for the Uranus and Neptune residuals. "It is pertinent to state," Slipher wrote in April 1930, "that if Mars were removed to the distance of 41.3 astronomical units his stellar magnitude would not be greater than Planet X. That is, this new body may be comparable with Mars in size and mass," depending on the albedo, or surface reflectivity, of the new object.

A Mars-sized Pluto meant that it could not account for the observed residuals of Uranus, as the astronomer E. W. Brown realized already in May 1930 when he wrote, "The orbit published by the Lowell Observatory for the newly discovered planet shows definitely that it cannot have any connection with that predicted." This was confirmed by E. C. Bower and others, when in 1931, making use of a number of pre-discovery observations, they refined the orbit close to its modern configuration. This then gave rise to the idea of a tenth planet, and the inevitable searches for a new "Planet X," which proceeded almost to the end of the century. All to no avail, for in 1993, astronomer Myles Standish showed that the unexplained residuals in the motions of Uranus were likely due to the use of incorrect masses for the gas giant planets, primarily Neptune.[8] Nevertheless, as David DeVorkin has written, in the 1930s, some astronomers "remained preoccupied with classifying the object," not only because of its apparently small size but also because its orbit deviated by 17 degrees from the mean of the orbital plane of the other planets. Some even suggested it was the harbinger of a new class of planetoidal objects in the outer solar system.[9] The preoccupation with classification – also playing out at this time with comets and minor planets – demonstrates how closely astronomers identified the classification process with understanding the true nature of an object.

Today, historians and scientists alike generally agree that, while Lowell's predictions motivated the search, the location he predicted was only serendipitously close. "The accuracy of Pickering's and Lowell's predictions was a truly fortunate and astounding coincidence," wrote Gibson Reaves in his review of the whole debate. And writing in his memoir on the fiftieth anniversary of his discovery, Tombaugh himself was clear about his perception of the role of prediction: "The success in finding the new planet was not due to the complex mathematical theory," he wrote, "but to basically simple observational procedure and an enormous amount of painstaking work. Contrary to widespread opinion, the mathematical prediction was of little aid in actually finding the planet, because of earlier negative observational results. Singling out the planet from a sky background, teeming with millions of stars, was like finding a needle somewhere in a large haystack."[10]

The bottom line is that when Tombaugh discovered Pluto in 1930 it was only a point of light, detected among the thousands of background stars by its extremely slow motion. That motion translated to a 248-year orbital period, placing it at the edge of the solar system. That much was certain. For decades the mass, size, and density of Pluto were considered to be similar to the planet Mars. There was no question, therefore, of Pluto being recognized by community consensus as a new class of object at that point, even in the midst of speculation of its true nature. That determination would hinge on a true knowledge of its mass, which remained very uncertain until the discovery of its first moon, Charon, almost fifty years later.

1.2 The Discovery of Charon: Pluto's Mass at Last

On the morning of June 22, 1978, James W. Christy, an astronomer at the U.S. Naval Observatory in Washington, DC, began to measure the position of images of Pluto utilizing the Observatory's Starscan measuring machine. The photographic plates had been taken with the Observatory's 61-inch astrometric reflector, located in Flagstaff, Arizona, some four miles from the spot where Tombaugh had discovered Pluto forty-eight years before. The idea was to improve the accuracy of the orbital elements of Pluto, part of the Observatory's larger mission of improved orbits for the Moon and planets, data essential for generating ephemerides in the *Nautical Almanac*. Christy himself had requested that the images be taken, and accordingly in April and May 1978, the planet was imaged on three nights, twice each night, for a total of six plates.

As Christy settled into his routine that June morning he noticed that the images of Pluto were elongated, showing a faint southerly extension on April 13 and 20, and a faint northerly extension on May 12 (Figure 1.3). Such elongations are not unusual: astronomical images sometimes trail or are imperfect due to atmospheric conditions. In fact, the plates had been marked as defective before being sent from Flagstaff. The same plates, however, showed an elongated planetary image and round stellar images, a cause for puzzlement. Examination under the microscope did nothing to solve this mystery. On a busier day Christy might

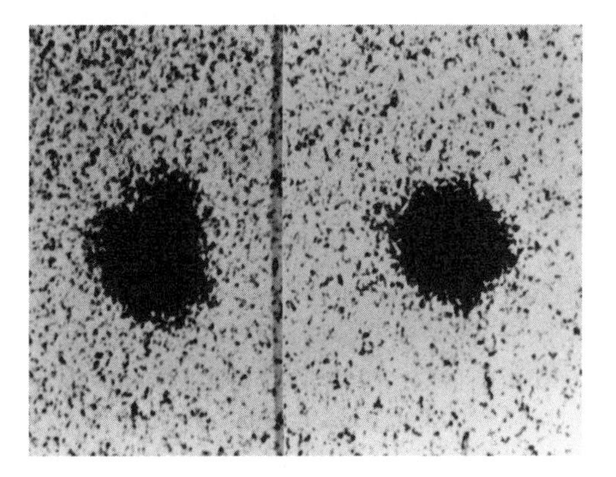

Figure 1.3. Discovery image of Pluto's first moon, Charon, detected on June 22, 1978 (left). The image was taken with the 1.55-meter astrometric reflector at the U.S. Naval Observatory station in Flagstaff, Arizona. The elongation that James W. Christy interpreted as a moon of Pluto had been present on photographic plates taken at least as early as 1965, but it was thought to be spurious. The bulge is absent on the image at right, since the moon has gone behind Pluto. The Hubble Space Telescope later clearly resolved Charon and Pluto, and discovered four more moons. U.S. Naval Observatory.

have let the puzzle pass; in fact, he might not have even attempted to measure the "defective" plates. However, he was about to go on a week's leave, was in a more relaxed mode than usual, and began to consider the possible causes of the elongation. A gigantic eruption on Pluto seemed unlikely to be sustained over a month. The idea of "month" led to the idea of "moon," but at that point Christy "felt a little ridiculous at the thought." He and his colleague F. J. Josties calculated that Pluto's motion was a factor of ten times too small to produce such an elongation over the 1.5-minute exposure. The possibility of a faint background star coincidentally adjacent to Pluto was considered, and an examination of the Palomar Sky Survey actually showed one that was close, but not close enough. At this point the suspicion of a Plutonian satellite became stronger, strong enough that Christy told his colleague and supervisor, Robert S. Harrington.[11]

In order to verify this conjecture, the following day, June 23, Christy examined plates taken with the 61-inch telescope in 1965 and 1971 for the purpose of measuring the diameter of Pluto, and in 1970 for a project to measure the motion of the center of the planet. Plates taken on two nights in 1965 and five nights in 1970 also showed the elongations, and the latter even permitted an estimate of a period of about six days, approximately the period of Pluto's known light curve of 6.387 days. Equating the putative moon's period with the light curve, Christy now informed Harrington not only that Pluto had a moon, but also that its period was 6.387 days! With this information Harrington calculated an ephemeris for the moon's position, while Christy measured the position angles of the elongation on the old and new plates. As Christy recalled, "The agreement between measurement and the 6.387 day period was remarkable. Pluto had a moon, and the moon had an orbit. Twenty-eight hours had elapsed since I had first looked at the Pluto images."[12]

The discovery was duly reported in newspapers around the world, and by July 14, Christy and Harrington had submitted their scientific paper to the *Astronomical Journal*. Christy named the satellite "Charon," after the ferryman in Greek mythology who conveyed the souls of the dead across the River Styx to the underworld ruled by Pluto. (As an added bonus, his wife's name was Charlene.)[13] But the discovery was not without controversy. Although speckle interferometry observations resolved the moon in 1980, "resolving any doubt as to the existence of the satellite" in the opinion of Harrington and Christy, some astronomers refused to accept the moon for several years. Remarkably, one additional mode of proof was the transit of the putative moon in front of Pluto, and its occultation by Pluto, events that occur only during an eight-year period every 124 years, but that were predicted to occur in 1985. The actual observation of these events in February 1985 clinched the reality of Charon to the satisfaction of everyone. The 1990 Hubble Space Telescope image clearly separating the two was icing on the cake, and on October 21, 2005, Hubble astronomers announced the discovery of two small outer moons from images taken in May of that year. These moons were named Hydra and Nix, and were joined by the discovery of a fourth small outer satellite in 2011, and a fifth in 2012.[14]

The importance for our story is that the discovery of Charon in 1978 allowed the definitive mass of Pluto to be determined for the first time. With the simple application of Kepler's laws to two bodies in orbit, the resulting mass was only 1/400th of the Earth and had a diameter of less than 1,500 miles, considerably smaller than our Moon. Plotting determinations of the mass of Pluto over time, Alexander Dressler jokingly predicted the planet would disappear by 1984![15] More seriously, this definitive knowledge of Pluto's small size triggered a low-grade discussion as to its status as a planet, one that would be further fueled by the discovery of new objects beyond Pluto, in the so-called Kuiper Belt.

1.3 Discovery of the Kuiper Belt: Bad News for Pluto

Pluto's status became considerably more precarious in the early 1990s, when astronomers began to discover a variety of objects beyond Neptune, in a zone from 30 to 55 astronomical units from the Sun known as the "Kuiper Belt" to most Americans, or the "Edgeworth-Kuiper Belt" to most Europeans. Such objects had long been hypothesized. Theories of planetary accretion, as well as comets and asteroids known to exist in the inner solar system, indicated that similar bodies should be found in the outer solar system. The technology of the time, however, made them very difficult to detect, as the painstaking discovery of Pluto in 1930 proved. Still, this did not prevent speculation. Both the American astronomer Gerard Kuiper and the Irish astronomer Kenneth Edgeworth noted in the 1940s and 1950s that there was no reason to expect the solar system ended with Neptune or Pluto. Charles Kowal's discovery in 1977 of Chiron, located in an orbit between Saturn and Uranus, added plausibility to the argument that more objects remained to be discovered in the outer solar system. It was the first of the "Centaur" type of minor planets, so called because they exhibit characteristics of both minor planets and comets.[16]

It was advances in technology, in particular charge-coupled device (CCD) detectors, that finally allowed astronomers to image extremely faint objects in the outer solar system even beyond Chiron. With such a CCD at the focal plane of the University of Hawaii's 2.2-meter telescope on Mauna Kea, in 1992 David Jewitt and Jane Luu spotted the first of these, a 22nd magnitude object dubbed 1992 QB1. As we shall detail in Chapter 2, the discovery of many more Trans-Neptunian objects followed, all slightly smaller than Pluto. By early 1999, more than 130 Trans-Neptunian objects were known. It was only a matter of time before one larger than Pluto would be discovered.

The debate over the status of Pluto as a planet began to heat up in 1998 when the International Astronomical Union (IAU), the arbiter in all such matters, got involved. In late 1998 Brian Marsden, director of the IAU's Central Bureau for Astronomical Telegrams and the Minor Planet Center, located at the Harvard-Smithsonian Center for Astrophysics, proposed a dual designation for Pluto as both a planet and a minor planet. "If Pluto were discovered today instead of 70 years ago," Marsden proclaimed, "it would be considered a minor planet and given a minor planet number." But many astronomers objected to the idea of declaring Pluto's dual status as both a planet and a minor planet, including some at Lowell Observatory where Pluto had been discovered. "Marsden's idea makes no sense," said Robert Millis, the Observatory's director. "Why call Pluto fish and fowl? We take a fair degree of pride in this object. It deserves to be considered something more than a minor planet." Planetary scientist S. Alan Stern, the future principal investigator for the Pluto New Horizons mission and future associate administrator for Space Science at NASA, agreed, and hinted at similar problems in the solar system: "I've heard all these fanciful arguments throughout this debate about why Mercury should be tossed out, Mars is really an asteroid, Jupiter is a star, and the moon should really be called a near-Earth object," he said. In Stern's opinion large spherical asteroids such as Ceres should be declared planets. By this definition, he suggested, the number of planets in the solar system was a dozen and climbing. Astronomer Michael A'Hearn, the president of the IAU's Planetary Systems Science Division at the time and chair of its Small Bodies Names Committee, was quoted as saying, "The debates have been remarkably emotional . . . What does it tell me? It tells me that scientists are less rational than I thought they were."[17] To put it another way, it is clear that more than science enters into such debates.

When in 1999 the latest developments of the Pluto debate were reported in an article "Pluto: The Planet that Never Was" in the respected journal *Science*, which concluded "Pluto's career as a planet seems to be ending," it provoked a rebuttal from a group of eight planetary scientists, headed by Bob Millis. Millis, the director of the Lowell Observatory since 1989, had a considerable stake in the debate. Showing a singular lack of predictive ability, they asserted that no committee of the IAU has decided, nor will they decide, that Pluto is not a planet: "Such is not the case, nor is a decision on Pluto's planetary status within the purview of this committee." They argued that not enough is known about Kuiper Belt Objects (KBOs) to insure that Pluto is one of them, beginning with the size of the objects: while Pluto's diameter has been *measured* at just under 2400 kilometers, they emphasized, the size of 1992 QB1 has only been *estimated* at 200 kilometers by its assumed very low reflectivity (analogous to comet nuclei), which is not well known and subject to adjustment. And if the reflectivity is similar to that of Pluto and Charon, they pressed, KBOs would be even smaller, "tiny in comparison with both Pluto and Charon." Even less is known about the physical characteristics of KBOs, they argued, so any conclusions are "premature and arbitrary." In their view, Pluto showed far more similarities to the other planets than to KBOs, including a tenuous atmosphere, a satellite, polar ice caps, and was "by far the dominant known body in its particular realm of the solar system." Moreover, they argued that Pluto's formation process was likely similar to the other planets of the solar system, distinguished only by icier primordial conditions. For all these reasons, they concluded, Pluto should be considered the prototype of a third type of planet, not the prototype of the KBOs.[18] This mini-discussion presaged many of the elements of the coming maelstrom: should planetary status be based on size, on physical properties, on being the dominant body in its region, or on its formation process?

The reaction to the idea of demoting Pluto was so strong that in February 1999, IAU General Secretary Johannes Andersen issued a press release denying that Pluto was being demoted. "Recent news reports have given much attention to what was believed to be an initiative by the International Astronomical Union (IAU) to change the status of Pluto as the ninth planet in the solar system," the statement began. "Unfortunately, some of these reports have been based on incomplete or misleading information regarding the subject of the discussion and the decision making procedures of the Union. The IAU regrets that inaccurate reports appear to have caused widespread public concern." The statement went on to say that, although a Working Group of the IAU Division of Planetary Systems Sciences was conducting a technical debate on Trans-Neptunian objects and ways to classify planets by physical characteristics, neither that group nor any other within the IAU has made a proposal to change the status of Pluto as the ninth planet of the solar system. They did point out, however, that our knowledge of the landscape of the solar system was changing, and that a number of smaller objects with orbits and possibly other properties similar to Pluto had been discovered. Moreover, "It has been proposed to assign Pluto a number in a technical catalogue or list of such Trans-Neptunian Objects (TNOs) so that observations and computations concerning these objects can be conveniently collated. This process was explicitly designed to not change Pluto's status as a planet."[19]

Meanwhile the public component of the debate heated up. One practical manifestation was the debate hosted on May 24, 1999 by Neil deGrasse Tyson, director of the Hayden Planetarium of the American Museum of Natural History in New York City. Among his duties Tyson was project scientist for the new $230 million Rose Center for Earth and Space, and was considering how to treat Pluto in its new exhibits. Making no small plans, and concerned about the expense of exhibits that might have to be changed if Pluto's status was changed, Tyson invited some of the world experts on the subject to a debate. The five astronomers on the panel included Michael A'Hearn, Alan Stern, Brian Marsden, Jane Luu, and David H. Levy, the latter the biographer of Tombaugh and of comet Shoemaker-Levy 9 fame. In his engaging book *The Pluto Files*, Tyson described the outlines of the debate. Luu began by laying out the two sides: some assert that Pluto's tiny size and similarity to other objects beyond Neptune mean it should be classified as a minor planet, while others, "outraged at the idea," believe that demoting Pluto would "dishonor astronomical history and confuse the public." As for herself, "I personally don't care one way or the other. Pluto just goes on the way it is, regardless of what you call it ... in the end, the question goes back to this: should science be a democratic process, or should logic have something to do with it?" Stern argued that the issue should not be a matter of democracy, but of a physical test: if an object was big enough to fuse hydrogen it was a star, if it was too small to be round by virtue of gravity and hydrostatic equilibrium, it was not a planet. Everything in between was a planet. Marsden argued that Pluto was a member of the Kuiper Belt, but was willing to grant it dual status as both a major planet and a minor planet, in accordance with his previous suggestion. A'Hearn also argued for dual classification, but as a planet and a Trans-Neptunian object, since the purpose of classification was to help us understand the objects involved, and Pluto seemed to be both (at least during that part of its orbit when it was beyond Neptune!). Levy argued that science wasn't just for scientists or taxonomists, but for people, and normal people considered Pluto a planet.

The overflow audience of 800 at this New York City event indicated Pluto was, surprisingly, becoming a cultural phenomenon. The debate led Tyson and his curators to demote Pluto from planethood in its exhibits, declaring only eight planets existed in the solar system. In this the American Museum of Natural History, it would turn out, was ahead of its time by seven years.[20] The debate had highlighted an array of dichotomies that would show a stubborn persistence over the coming years: democracy or logic, planet or minor planet, planet or Trans-Neptunian object? Tiny Pluto was becoming a major headache to more and more people.

The growing number of objects discovered beyond Neptune, as well as Marsden's proposal, had their effect, essentially forcing astronomers to consider the definition of a planet. At the triennial meeting of the IAU in the summer of 2000 in Manchester, England, a day-long session was devoted the problem of "The Trans-Neptunian Population," including a paper on "Pluto: A Planet or a Trans-Neptunian Object?" The latter, written by A'Hearn, sensibly suggested a dual classification of Pluto as both a planet and a Trans-Neptunian object.

Taking a broad view of classification, and allowing that analogies are never perfect but can lead to useful insights, A'Hearn turned to biology and presented one of the most thoughtful, dispassionate, and well-considered discussions in the entire Pluto debate. He began with an analogy to the fossil known as archaeopteryx, first discovered in 1861, only two years after Charles Darwin's *Origin of Species* was published. A'Hearn pointed out that it was immediately recognized as unusual, and likely to provide insights into the evolutionary relationships of birds and dinosaurs. Subsequent research showed that archaeopteryx had both distinctive bird-like features including feathers, a wishbone, and hind limbs, and dinosaur-like features, including its teeth and pelvis. For fifty years there was strong disagreement about its proper classification as dinosaur, a bird, or the ancestral bird, with consensus beginning to emerge in the mid-1920s, and details still being argued even today.

Similarly, A'Hearn argued, Pluto has both planetary characteristics (spherical shape, tenuous atmosphere), and Trans-Neptunian Object characteristics (size, dynamics). The important point, he emphasized, is that "Pluto should be looked at from both sides, since, like archaeopteryx, it is on or near the boundary between two classes and thus can shed unique light on the relationship between the groups." Pluto could help us understand how icy planets work and how TNOs developed. The classification of Pluto, he pointed out, had been an issue for less than a decade, "only since enough TNOs were discovered to recognize Plutinos as a sub-class." He argued that "The key lesson from archaeopteryx is to look at Pluto from both sides and not spend decades, or even centuries, disputing the classification" – a distraction from the entire point of classification. More importantly, "If TNOs, and specifically Plutinos, had been known before the discovery of Pluto, our thought processes might have been different, but the history of archaeopteryx suggests that even then the disagreements might have lasted many decades."[21]

A'Hearn also made another perceptive point, that planet definition efforts have been undertaken by two very different groups: on the one hand, planet-formation theorists and extrasolar planet hunters who are concerned about the separation of the largest planets from those failed stars known as brown dwarfs, and on the other hand, cometary, asteroidal, and planetary scientists concerned about the separation of planets from the smaller bodies of the solar system. Indeed, since 1995 planets were being found at an increasing pace around solar type stars. But some of these were borderline objects difficult to distinguish from another class of object, brown dwarfs, precipitating efforts to define planets from the upper mass point of view rather than the lower mass side. Indeed, this led the IAU in 1999 to establish a very active Working Group on Extrasolar Planets, chaired for most of its history by Alan

Boss of the Carnegie Institution of Washington. By 2001, the Working Group had agreed on a "working definition" of a planet as applied to extrasolar planetary systems. As Boss recalled, "We were focused on the upper-mass end: How massive could an object be and still be called a planet? What if a Jupiter-mass object were not in orbit around a star? Would it still be a planet? Our main debate had centered on whether planets should be defined by their formation mechanism or by their ability to undergo nuclear fusion." Given this focus, the Working Group decided that a planet was an object below the limit for thermonuclear fusion of deuterium, calculated as below thirteen Jupiter masses, while objects of 13–75 Jupiter masses would be brown dwarfs, both categories defined without regard to how they had formed. At the lower end of the mass spectrum, the Committee declared "the minimum mass/size required for an extrasolar object to be considered a planet should be the same as that used in our Solar System." Revealingly, they considered that out of their realm, and in the realm of the planetary scientists.[22] But with the planet/brown dwarf mass question, we begin to see that the debate about classification is not confined to Pluto, planets, or minor planets – a supposition that will be amply confirmed in subsequent chapters. Indeed, brown dwarfs would also be implicated in the question of the dividing line between high-mass brown dwarfs and low-mass dwarf stars.

Like the discovery of minor planets at the beginning of the nineteenth century, as more and more Pluto-like objects were discovered in the Kuiper Belt, astronomers continued to debate the question of whether they constituted a new class of object, of which Pluto was the prototype. Other definitions of "planet" would be floated, some accompanied by elaborate arguments and scientific data. In 2006, as IAU deliberations were heating up again for its triennial meeting, Neil Tyson's colleague Steven Soter, for example, submitted a detailed article, "What is a Planet," to the *Astronomical Journal*. It took yet another approach, entirely sensible, based on the origin of the objects in question. Soter concluded that "A planet is an end point of disk accretion around a primary star or substar," the implication being that Trans-Neptunian objects were not such end points.[23] Revealingly, none of this was considered in the now imminent and fateful IAU discussion on Pluto. The precipitating cause of that discussion was not extrasolar planets, but the discovery of a TNO believed to be larger than Pluto.[24]

1.4 Pluto Demoted: Prague 2006

The controversy over the seemingly simple question "what is a planet?" came to a head at the General Assembly of the International Astronomical

Union in Prague in 2006. The General Assembly convenes only every three years, so the question was not undertaken lightly or without preparation. At the request of the Board of the Planetary Systems Sciences Division of the IAU, "in recognition of the fact that discoveries in the Trans-Neptunian region were repeatedly raising the question of 'what is a planet,'" already in March 2004, the IAU Executive Committee established a "Working Group on the Definition of Planet." An international group of nineteen astronomers chaired by Iwan P. Williams from the UK was charged "to investigate the options available and give indications of the level of support and opposition for each if more than one option was emerging."[25] Through the spring of 2005 the group deliberated the definition of a planet via e-mail "ad nauseum," in the words of one panel member, while "every possible argument, constraint, or consideration regarding what should be a planet was presented, seconded, discussed, disregarded, or discarded, with no clear consensus emerging. Achieving a consensus among them was about as hard as trying to herd a group of 19 feral cats into a room with several open doors and windows." Contrary to what one might have expected from logical scientists, "the members tended to have strong opinions on this subject, and they were generally unwilling to change their opinions on the basis of arguments presented by others."[26]

Then in July 2005 came the precipitating event. Astronomers Michael Brown, Chad Trujillo, and David Rabinowitz announced the discovery of 2003 UB313, a Trans-Neptunian they dubbed Xena and later officially named Eris. The team had already found the Trans-Neptunians Quaoar, Orcus, and Sedna, and announced the discovery of Makemake and Haumea on the same day as the announcement of what would now be known as Eris. What is important for the Pluto story is that Eris was apparently slightly larger and more massive than Pluto. Moreover, within a few months Brown and his colleagues would announce that Xena/Eris had a satellite, later named Disnomia. Thus, two of the characteristics (size and satellite) that the Millis group had given as planetary characteristics only six years earlier had been breached.[27] Was Xena/Eris the tenth planet, or was Pluto not a planet? And since many more similar objects were soon known to exist in our solar system beyond Pluto, with more sure to be discovered, what is their status? The IAU wanted a decision from its Working Group.

In August 2005, the IAU Working Group took a vote in which seven members favored planet status for Pluto, seven favored demotion, and seven favored some sort of compromise. Unable to reach consensus after several more tries, in October the group completed its deliberations, transmitted its options to the IAU Executive Committee, and terminated its activities. The options were: (1) A planet is any object in orbit around the Sun with a diameter greater than

2000 kilometers; (2) A planet is any object in orbit around the Sun whose shape is stable due to its own gravity – the so-called hydrostatic equilibrium test; and (3) a planet is any object in orbit around the Sun that is dominant in its immediate neighborhood.

As the General Assembly approached, the IAU Executive Committee, not wanting to choose from the options on its own, established a second "Definition of a Planet Committee," composed of six astronomers and science writer Dava Sobel, well-known for her communications skills with the public, notably in her best-selling book *Longitude*. This time, the goal was clearly to devise a resolution that could be brought to a vote at the General Assembly, which was to begin August 15, 2006. The committee, chaired by Harvard astronomer and historian of science Owen Gingerich, and including the chair of the previous committee (Iwan Williams), met in Paris on June 30 and July 1, and recommended a single definition to the Executive Committee.[28]

Gingerich has given us an inside account of the deliberations of that committee, which examined the results of the previous planet definition committee, but made its own independent assessment and recommendations. Surveying past discussions on the subject, Gingerich saw two fundamentally different ways to define planets, coming from two different astronomical communities, what he called the dynamicists and the structuralists. The former were focused on Pluto's interactions with other bodies, and the latter on the physical nature of Pluto and similar objects. "We needed a physically defensible discriminator," he recalled, "not an arbitrarily chosen dividing line." The committee sided with the structuralists, choosing "roundness," that is to say, "hydrostatic equilibrium," as its criterion for planethood. By this criterion not only Pluto but also several other Trans-Neptunian objects would be planets, though small planets with eccentric orbits, and so distinct from the other planets. The committee suggested they be called "plutons," "the prototype of the newly defined class" of planets.[29] That left Ceres, the first asteroid discovered, definitely round but not Trans-Neptunian, and so in a class of its own, which they called "dwarf planets." The committee fully expected that by its definition Pluto would remain a planet.

The Planet Definition Committee's recommendation was accepted by the Executive Committee and shaped into a draft resolution during the month of July, but still had to be voted on by IAU members at the General Assembly the following August. If this resolution passed there would be twelve planets – the classical eight, the three plutons known at the time (Pluto, Charon, and Eris), and the dwarf planet Ceres. In order not to prejudice the issue in the media before the IAU assembled, the proposal was not released until the day after the General Assembly opened on August 15.

The press officer of the IAU at the time, noting that the issue overshadowed all other science presented at the two-week meeting (at least in the eyes of the media), has given us an inside look at the concerns of the IAU before the vote. An internal working paper before the meeting predicted, "The planet issue has the potential to become a historic event of epic proportions. It may become the hottest astronomy story of the year, or even the decade. It has the potential to change history. Seeing this as a potential historic event, do we fulfill our public duty and inform the world about the process and the decisions openly, or do we keep quiet to protect the slow and thoughtful scientific work process?" The decision was to be open and honest, but in a "crisis communication" mode with specific rules. The IAU worried about lack of communication, political intervention, resentment and resistance within the community, and even the perception of anti-Americanism in the United States. The latter was no small concern, since Pluto was, after all, the only planet discovered in America. The IAU, perceived by some as a European institution since its headquarters was in Paris, worried about individual members withdrawing, the reaction from Lowell Observatory and the Tombaugh family, and resistance in schools to changing textbooks.[30]

Any illusions that the Pluto issue would be settled easily were shattered as the General Assembly itself began. Already at a meeting of the Planetary Systems Science Division on August 18, the dynamicists, feeling ignored by the structuralist approach, added a defining condition that a planet not only be round but also "the dominant object in its local population," effectively eliminating Pluto, Ceres, and other iceballs beyond Neptune. This would leave the solar system with eight planets. The Planetary Systems Division itself reached no consensus (except to reject the term "pluton" for linguistic reasons raised by the Italians), but the planet definition committee reached an agreement for a two-pronged resolution to be voted on: "(1) A planet is a celestial body that (a) has sufficient mass for its self-gravity to overcome rigid-body forces so that it assumes a hydrostatic equilibrium (nearly round) shape, and (b) is in orbit around a star, and is neither a star nor a satellite of a planet, and (2) In our solar system we distinguish between the eight 'classical Planets,' as the dominant objects in their local population, and 'dwarf planets,' which are not." The dynamicists objected even to this compromise because dynamics was given second-tier status, and the drafting of the resolution continued, ending with a definition whereby a dwarf planet was not a planet, in Gingerich's opinion a linguistic absurdity. By this definition, Pluto would not be a planet, in contrast to the expectations of the IAU and its Planet Definition Committee when the General Assembly began.[31]

The fateful vote by the full General Assembly took place on August 24, such voting on resolutions by tradition and necessity always taking place on the last day of the General Assembly after which all the discussion had occurred. It fell to Jocelyn Bell Burnell, a member of the IAU Resolutions Committee, to explain the resolutions, particularly appropriate since almost forty years before she was one of the co-discoverers of the new class of objects known as "pulsars," and so knew something about the pitfalls of declaring an object the first of a new class. She used an actual umbrella to demonstrate how the term "planets" could encompass both the classical planets and the "dwarf planets." But in the end the General Assembly voted that a dwarf planet would not be a planet. Of the approximately nine thousand IAU members, three thousand had attended the General Assembly over its two-week duration. As usual, only about 400 were present for the vote, most having had to return home to other duties. The final resolutions as passed, pointedly titled "Definition of a Planet in the Solar System" to exclude extrasolar planets, read as follows:

> The IAU therefore resolves that planets and other bodies, except satellites, in our Solar System be defined into three distinct categories in the following way:[32]
>
> (1) A planet (note 1) is a celestial body that (a) is in orbit around the Sun, (b) has sufficient mass for its self-gravity to overcome rigid body forces so that it assumes a hydrostatic equilibrium (nearly round) shape, and (c) has cleared the neighbourhood around its orbit.
>
> (2) A "dwarf planet" is a celestial body that (a) is in orbit around the Sun, (b) has sufficient mass for its self-gravity to overcome rigid body forces so that it assumes a hydrostatic equilibrium (nearly round) shape (note 2), (c) has not cleared the neighbourhood around its orbit, and (d) is not a satellite.
>
> (3) All other objects (note 3), except satellites, orbiting the Sun shall be referred to collectively as "Small Solar System Bodies".
>
> Note 1 The eight planets are: Mercury, Venus, Earth, Mars, Jupiter, Saturn, Uranus, and Neptune.
>
> Note 2 An IAU process will be established to assign borderline objects to the dwarf planet or to another category.
>
> Note 3 These currently include most of the Solar System asteroids, most Trans-Neptunian Objects (TNOs), comets, and other small bodies.

> The IAU further resolves:
>
> Pluto is a "dwarf planet" by the above definition and is recognized as the prototype of a new category of Trans-Neptunian Objects (note 1).
>
> Note 1. An IAU process will be established to select a name for this category.

Despite concerns, Pluto was ruled to be a dwarf planet, defined by the IAU as a body that orbits the Sun, is massive enough to be rounded by its own gravity, is not a satellite, and has not cleared its orbit of other materials.[33] Under this definition, the IAU recognized five dwarf planets at the time: Ceres, Pluto, Haumea, Makemake, and Eris. By the 2006 IAU definition, a dwarf planet such as Pluto is not a planet because it has not "cleared its neighborhood" of other materials, nor is it a "small body of the solar system" like asteroids, comets, and trans-Neptunian objects because it is rounded. It is therefore a subplanetary object that falls between the definitions of a planet and a small body of the solar system. Moreover, a dwarf planet is not even necessarily a Trans-Neptunian object because at least one of them (Ceres) is within the orbit of Jupiter and another (Pluto) is inside Neptune for part of its orbit.

The criteria for dwarf planets have been criticized from a number of perspectives. For example, of the five designated dwarf planets, only Ceres and Pluto have actually been observed to be round. Michael Brown has estimated, however, that a rocky object must be about 900 km (the diameter of Ceres) in order to be round, while an icy body would have to be greater than 400 km (the diameter of Mimas, the smallest icy satellite). The roundness criterion, however, highlights the fact that the physical composition of the dwarf planets may vary considerably, from rocky (Ceres) to icy (those in the Kuiper Belt). But the deed was done. By vote of the IAU Pluto was dwarf planet, and, for the first time since 1930, the solar system officially contained only eight *real* planets.

* * *

The IAU decision on Pluto was met with a media storm, huge public reaction, and skepticism among many scientists. In an article, "Pluto: The Backlash Begins," *Nature* magazine reported: "The future of the Solar System – or at least that of some of its nomenclature – may be thrown into turmoil by scientists who are calling for a boycott of a new definition of a planet." Alan Stern objected to the third criterion, added by the dynamicists and the one that resulted in Pluto's demotion, by saying, "We do not classify objects in astronomy by what they are near, we classify them by their properties." Stern pointed out the "clearing" criterion would disqualify Neptune (whose orbit is crossed by Pluto) and Jupiter, which has a raft of Trojan asteroids in its orbit. He was among a dozen scientists who launched a petition, eventually signed by 400 scientists and presented to the IAU president on September 4, stating, "We, as planetary scientists and astronomers, do not agree with the IAU's definition of a planet, nor will we use it. A better definition is needed."[34] Nevertheless, *Encyclopedia Britannica* changed its online articles the same day. NASA said it would abide by the new definition and made the sensible statement that it "will continue pursuing exploration of the most scientifically interesting objects in the solar system, regardless of how they are categorized." Michael Brown's opinion was clear, and is reflected in the title of his book *How I Killed Pluto and Why It Had It Coming*.

The public reaction was beyond expectations, to say the least. Signs were posted, protests were staged, songs were written, and school kids were depressed at the thought of losing Pluto's planethood. "There's such a thing as tradition," the *Washington Post* editorialized, perhaps tongue-in-cheek. "It's just not right to teach generations of kids that there are nine planets, only then to take one away. Sure, it would be inconsistent to keep Pluto as a planet even as astronomers discover other big rocks out there in space. But, as Emerson might have put it, a foolish consistency is the hobgoblin of astronomers." Nearly eight months later, on March 13, 2007, the state of New Mexico, noting that Clyde Tombaugh had lived in the state for much of his life, passed a resolution that "as Pluto passes overhead through New Mexico's excellent night skies, it be declared a planet."[35] Who would have thought such a small object would occasion such a large reaction? Never before was so much attention paid by so many to such a small astronomical object. But the public and scientific reaction demonstrated one clear and surprising fact: astronomy was more a part of popular culture than one would have suspected, and classification mattered. It was, as President Obama might have said, a "teachable moment." The very idea of "demoting" an object clearly raised passionate feelings – not only among the general public, but also among scientists.

The IAU, however, not only stood by its decision but followed up on it, announcing on June 11, 2008 that all dwarf planets with Trans-Neptunian orbits would be called "plutoids."[36] The category includes Pluto, Haumea, Makemake, and Eris. Despite the controversy, the Pluto issue was not brought up at the next General Assembly in Rio de Janeiro in 2009; it is likely Pluto will remain a "dwarf planet" for the foreseeable future.

* * *

Though nothing was clear with foresight, the eighty-year trajectory of the Pluto affair is clear in hindsight. Based on a flawed theory, Clyde Tombaugh discovered an object he and his colleagues believed to be a planet, the outermost such object in the solar system. Very early on, it was suspected the planet was much smaller than theory would have predicted. Almost fifty years later, the discovery of Pluto's first moon, Charon, enabled the first accurate determination of its mass, which turned out to be even smaller than expected. This gave rise to questions about Pluto's planetary status. The discovery of Kuiper Belt Objects beginning fifteen years after that further eroded Pluto's status, especially after objects were found that were larger and more massive than Pluto. But if based on size and mass, where was the line to be drawn between a planet and a distinctly different class of object? Unusually, and after much social as well as scientific input, only a vote of the International Astronomical Union would decide that.

But the questions go well beyond this trajectory. Did Tombaugh discover the first dwarf planet? By almost any definition he did not, since he clearly did not recognize it as such (not to mention that by this reasoning Piazzi discovered the first dwarf planet, Ceres, in 1801!). We have here our first hint that discovery is not a distinct event, but an extended process, beginning with detection and followed by later stages of interpretation and then understanding. Classification is an essential part of the process, and in the case of Pluto a new class was not so much detected as declared. And in that declaration people, politics, and worldviews were involved. The outcome did not simply lie with the nature of things, but with the decisions of scientists. Because the outcome might well have been different, Pluto as a dwarf planet is a clear case of social construction, one that raises the more general question of the open-ended, socially constructed nature of classification.[37]

The Pluto debate assumes transcendent importance because it illuminates, by way of concrete example with real people and real issues minutely documented, many aspects of scientific discovery and the nature of science. With the official IAU definition of a planet in mind, we can see how fraught the classification problem can be, entangled as it is with the history of discovery and the passions of people. Dispassionately put, the status of Pluto and the Kuiper Belt Objects were questions of classification, of moving an object from one class to another with new knowledge in the case of the former, or of classifying newly discovered objects in the case of the latter. It would seem that, just as the status of asteroids had been changed from planets to minor planets in the course of the nineteenth century as more were discovered, so a large number of Trans-Neptunian bodies had to be discovered in order to reveal, or rather to decide and declare, Pluto's true nature. But, it turns out, neither scientists nor the general public are dispassionate. Despite definitions and analogs, it was not clear whether Pluto should be a type of planet or a new class entirely. The IAU opted for the latter, but in a confusing way by declaring Pluto to be in the new class of "dwarf planets," but then declaring that a dwarf planet was not a planet at all!

Planet or not, it is a matter of historical record that it took seventy-six years to recognize, or rather to declare, Pluto as a new class of object. The Pluto debate highlights issues of theory and observation, nomenclature, community negotiation, and the unpredictable vagaries of people and the democratic process. Several communities, including dynamicists, structuralists, and extrasolar planet specialists, had a stake in the outcome of the Pluto affair. Many were not happy with its outcome. In the end it is notable that what precipitated the action of the IAU was the discovery of Trans-Neptunian objects, not the ongoing discovery of extrasolar planets. Even as the Pluto problem was resolved by vote, the classification of planets beyond the solar system remained confused and unresolved, just as one would expect in the early phases of a new field where physical data were sparse.

Finally, the Pluto affair illustrates two distinct types of discovery: purposeful and serendipitous. The discovery of Pluto was not only purposeful, but at the time of its discovery was believed to be guided by theory. Yet its discovery was serendipitous in the sense that the object discovered could not have caused the perturbations in the orbits of Uranus and Neptune. The discovery of Pluto's largest moon Charon was purely serendipitous: James Christy was not even searching for a moon of Pluto when he happened on the apparent "bump" on Pluto. The discovery of Trans-Neptunian objects was purposeful, part of a large-scale search for faint objects in the outer solar system. But its consequences were serendipitous: the downfall of Pluto and a change in the landscape of the solar system that had been ingrained in the minds of the public for generations.

With such a rich and controversial history for one class of object, one can imagine what awaits us among the other classes of astronomical objects discovered over the last 400 years. To them we now turn, with an eye toward the historical problem of distinguishing and defining a new class.

Part II Narratives of Discovery

2

Moons, Rings, and Asteroids: Discovery in the Realm of the Planets

Having dismissed earthly things, I applied myself to the exploration of the heavens.

Galileo, 1610[1]

If ever a discoverer was perfectly prepared to make and exploit his discovery, it was the dexterous humanist Galileo aiming his first telescope at the sky.

Heilbron, 2010[2]

I have announced this star as a comet; but the fact that the star is not accompanied by any nebulosity and that its movement is so slow and rather uniform, has caused me many times to seriously consider that perhaps it might be something better than a comet. I would be very careful, however, about making this conjecture public.

Giuseppe Piazzi, 1801[3]

With the exception of Tycho Brahe's proof in 1577 that comets were celestial phenomena based on their parallax and thus distance, and his inference that the *stella nova* of 1572 was celestial based on its *lack* of parallax, the problem of the discovery and interpretation of new classes of astronomical objects begins substantially with Galileo and the telescopic era 400 years ago. Galileo's telescopic observations revealed what we would today recognize as two new classes of astronomical objects: moons around Jupiter and rings around Saturn. And while Galileo and his contemporaries soon realized the nature of the moons of Jupiter by analogy to our own Moon, the story of Saturn's rings is much more complicated.

As we shall see, both Jupiter's moons and Saturn's rings stand as early examples of what would become commonplace in astronomy: that "seeing" isn't always "knowing," that "detection" does not constitute "discovery," that recognizing a new class of astronomical objects can be a difficult and multifaceted endeavor. In this chapter we begin to dissect the process of discovery in astronomy, in particular as it applies to the discovery of new classes of objects. We shall find it to be a complex and extended series of events consisting most often of at least three components: detection, interpretation, and understanding. This is particularly true of new classes of objects when the observer may have no idea of the true nature of the object. Again and again astronomers ran up against the unexpected in their reconnaissance of the heavens. Their struggle to move beyond mere detection, to enter the difficult realm of interpretation, and to seek physical understanding – often long after the original detection – is a story that has only been told piecemeal, but that deserves systematic treatment because it represents the core of astronomy and the natural history of the heavens.

2.1 Moons and Rings: A Very Strange Wonder

Although the Moon had been observed apparently circling the Earth for millennia, arguably one object does not make a class. Rather, credit for the discovery of a class of such planet-circling objects goes to the Italian natural philosopher Galileo Galilei (Figure 2.1). Galileo's studies at the University of Pisa were in mathematics, and it was as a professor of mathematics that Galileo had been firmly ensconced at the University of Padua since 1592, working primarily on hydrostatics, strength of materials, and the nature of motion. These studies were interrupted in the late summer of 1609 when Galileo learned the general features of a Dutch spyglass that "shows distant things."[4] With his typical if sporadic genius, Galileo applied himself to constructing improved versions of this "spyglass" to the extent that by December 1609, he turned his latest instrument toward the Moon. There he spied its mountains and craters, in opposition to the long-held Aristotelian notion that the Moon was a perfect and unblemished sphere. Then, using his newly made telescope – not the first that he or others had used, but (at 20 power) the best he or anyone had at the time – Galileo observed Jupiter from January 7 to March 2, 1610. Having reported his observations of the Moon, fixed stars, and the Milky Way in his *Siderius Nuncius* published in March of that year, Galileo grandly announced, "It remains for us to reveal and make known what appears to be most important in the present matter: four planets [*planetas*] never seen from the beginning of the world right up to our day, the occasion of their discovery and observation,

Figure 2.1. Galilei Galileo, discoverer extraordinaire, thanks to the invention of the telescope, and to individual and cultural circumstances in Renaissance Italy. This image by Francesco Villamena first appeared in 1613 in Galileo's book on sunspots, a few years after Galileo's first telescopic discoveries, and again in his *Assayer* (1623).

their positions, and the observations made over the past 2 months concerning their behavior and changes."[5] Taking up fully the last half of his treatise, Galileo described his thought processes in exquisite detail.

We are fortunate to have Galileo's description of the night of discovery:[6]

> On the seventh day of January of the present year 1610, at the first hour of the night, when I inspected the celestial constellations through a spyglass, Jupiter presented himself. And since I had prepared for myself a superlative instrument, I saw (which earlier had not happened because of the weakness of the other instruments) that three little stars [*stellulas*] were positioned near him - small but yet very bright. Although I believed them to be among the number of fixed stars, they nevertheless intrigued me because they appeared to be arranged exactly along a straight line and parallel to the ecliptic, and to be brighter than others of equal size.

Galileo then drew the configuration, depicting two "fixed stars" to the East of Jupiter and one to the West. The next night, "guided by I know not what fate," Galileo returned to Jupiter again, and found a very different arrangement: the

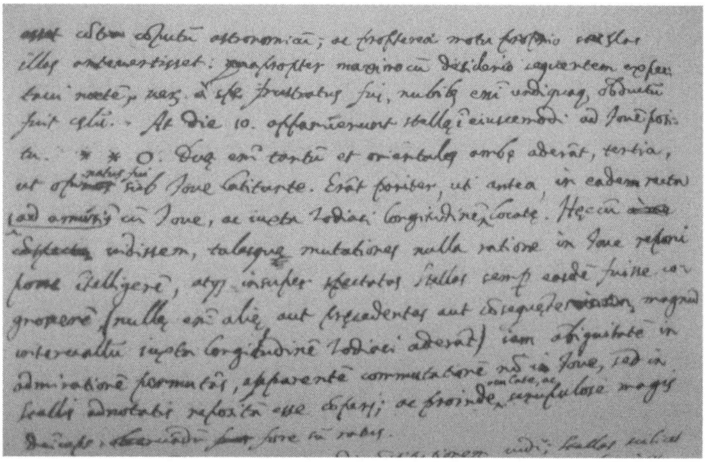

Figure 2.2. Galileo's drawing of Jupiter's satellites on the night of January 10, 1610. It was at this moment that Galileo knew he had discovered something unusual: "I found the change was not in Jupiter, but in said stars." From Galileo's *Opere* (Edizione Nationale, 1892), vol. 3, pp. 35 and 36.

three "fixed stars" [*stellulae*] were now all on the west side of the planet. Galileo was confused:

> Even though at this point I had by no means turned my thought to the mutual motions of these stars, yet I was aroused by the question of how Jupiter could be to the east of all the said fixed stars when the day before he had been to the west of two of them. I was afraid, therefore, that perhaps contrary to the astronomical computations, his motion was direct and that, by his proper motion, he had bypassed those stars [*stellas*].

Galileo eagerly awaited the next night, but it was cloudy. On January 10 only two "stars" [*stellae*] were near Jupiter, both to the east (Figure 2.2). Galileo reasoned that the third was hidden behind Jupiter, due to Jupiter's motion. But then it dawned on him:

> As before, they were in the same straight line with Jupiter and exactly aligned along the zodiac. When I saw this, and since I knew that such changes could in no way be assigned to Jupiter, and since I knew, moreover, that the observed stars were always the same ones (for no others, either preceding or following Jupiter, were present along the zodiac for a great distance), now, moving from doubt to astonishment, I found that the observed change was not in Jupiter but in the said stars [*stellis*].

On January 11, Galileo again saw only two stars to the east of Jupiter, but their relative brightness had changed: "I therefore arrived at the conclusion, entirely beyond doubt, that in the heavens there are three stars wandering around Jupiter like Venus and Mercury around the Sun. This was at length seen clear as day in many subsequent observations, and also that there are not only three, but four wandering stars [*vaga sidera*] making their revolutions about Jupiter."

Thus, only four nights after his first observations, Galileo declared his discovery of new objects in motion around Jupiter. After this realization, Galileo obviously no longer referred to the new objects as fixed stars, having proven by observation that they were not fixed. It is notable that the "moon" or "satellite" nomenclature was not immediate, however; Galileo referred to the system as "Jupiter and his adjacent planets [*planetas*]," and dubbed them the "Medicean stars" [*Medicea sidera*] after his hoped-for patron, Cosimo II de' Medici, whom Galileo had tutored just before Cosimo's accession as Grand Duke of Tuscany in 1609. In the *Sidereus Nuncius* he used, without apparent differentiation, the terms *planetae, sidera, stellae*, and *stellulae* for the new circum-Jovian objects.

As he wrote the *Sidereus Nuncius* in February and early March of 1610, only weeks after his first observations of the new moons, Galileo drew a comparison of the Jovian system of objects and the Earth's Moon, while removing a major argument against the Copernican theory:

> We have moreover an excellent and splendid argument for taking away the scruples of those who, while tolerating with equanimity the revolution of the planets around the Sun in the Copernican system, are so disturbed by the attendance of one Moon around the Earth while the two together complete the annual orb around the Sun that they conclude that this constitution of the universe must be overthrown as impossible. For here we have only one planet revolving around another while both run through a great circle around the Sun: *but our vision offers us four stars wandering around Jupiter like the Moon around the Earth* [italics added] while all together with Jupiter traverse a great circle around the Sun in the space of 12 years.

Already in the *Sidereus Nuncius* the analogy between the Jovian objects and the Moon is explicit. Though he does not specifically say so, Galileo must also have realized that their general physical nature might be the same as the Moon, now known as a result of his own observations during the previous December to be similar to the Earth rather than the perfect unchangeable sphere of Aristotelian thought. Referring to new features besprinkling the lunar surface and previously unobserved, Galileo had written earlier in the *Sidereus Nuncius*, "By oft-repeated observations of them we have been led to the conclusion that we certainly see the surface of the Moon to be not smooth, even, and perfectly spherical, as the great crowd of philosophers have believed about this and other heavenly bodies, but, on the contrary, to be uneven, rough, and crowded with depressions and bulges. And it is like the face of the Earth itself, which is marked here and there with chains of mountains and depths of valleys."[7]

Even though Galileo did not call the Jovian objects moons or satellites, there is no doubt he knew they circled Jupiter and that he had thus discovered what *we* would today call a new class of objects. But more to the point, did Galileo believe *he* had discovered a new class? The distinction is an important one. Terminologically, Galileo was certainly confused. More fundamentally, in these pre-Linnaean days, he most likely did not think in terms of "class," although it is possible he was familiar with Aristotle's classification of animals in his *Historia Animalium*, which used terms such as "genus" and "species." Arguably, Galileo himself did not assign the objects to a new class, at least by the evidence of his mixed terminology. One might argue that the creation of a new class occurred a few months later, when Kepler used the term "*satellitibus*" for the new Jovian objects, and argued that each planet would be served by its own satellites.[8] But nomenclature and conceptualization as a new class of objects are not necessarily coextensive. One might also argue that Galileo certainly had the idea of a new class by 1632, when in his *Dialogo* he has Sagredo ask, "For what reason do you call the four planets surrounding Jupiter moons"? Salviati answers, "They would appear such to someone who looked at them while standing on Jupiter. For they are inherently dark and receive light from the Sun."[9] Even though Galileo, Kepler, and their contemporaries realized something novel had been discovered, the conceptual repertoire of their time likely did not include a concept of class – at least not in the explicit form we will certainly find in William Herschel's time.

In any case, the idea of class must have gradually occurred as more satellites were discovered around other planets. The terrestrial and Jovian moons were the only satellites known in the solar system until Christiaan Huygens discovered Titan around Saturn on March 25, 1655, specifically referring to it as *luna*, its own moon. Only four other moons were discovered in the seventeenth century: Iapetus and Rhea in 1671 and 1672, and Tethys and Dione in 1684, all by Giovanni Domenico Cassini with his large (35-ft) telescope and all around Saturn. Thus, nine satellites were known at the end of the seventeenth century. By the end of the nineteenth century, twenty-one satellites were known to exist other than our own Moon. Today, swelled by spacecraft observations, the total is 167. But after Galileo, never again did astronomers have to make the intellectual leap he struggled with in January 1610: that secondary bodies, which Kepler already dubbed as moons or satellites, can in principle circle a primary planet. That kind of leap is reserved for the discovery of a new class, even if Galileo himself did not use that terminology. Yet, it was not the last time Galileo had to cross that conceptual chasm.

* * *

In a testament to the importance of new technology, Galileo first glimpsed yet another new class of object barely six months later. He first detected what we now know as planetary rings when he turned his telescope toward Saturn on July 30, 1610. Saturn was the most distant of the known planets, and therefore more difficult to observe than Jupiter, even though Galileo now employed a slightly improved 30 power telescope rather than the 20 power he likely used to discover the moons of Jupiter. Galileo was immediately perplexed, noting that "[T]he star of Saturn is not a single star, but a composite of three, which almost touch each other, never change or move relative to each other; and are arranged in a row along the zodiac, the middle one being three times larger than the two lateral ones . . .",[10] whereupon he drew a large circle with a smaller circle on each side. Worried about his priority for the discovery (even though he did not understand it) Galileo famously wrote to his other correspondents a scrambled anagram, which in November he revealed as "Altissimum planetam tergeminum observavi": I have seen the highest planet [Saturn] triple-bodied."

The work of historian Albert van Helden on early observations of Saturn's rings has shown just how difficult the interpretation of a new class of objects can be. Galileo observed Saturn from July 1610 to May 1612, always seeing the same triple-bodied object, which he believed to be its permanent configuration. But when he returned to it in late 1612 he saw the unexpected: "I found him solitary without the assistance of the supporting stars, and in sum, perfectly round and clearly defined as Jupiter." Galileo was flummoxed: "Now what is to be said about such a strange metamorphosis? Perhaps the two smaller stars have been consumed in the manner of sunspots? Perhaps they have vanished or fled suddenly? Perhaps Saturn has devoured his own children? Or else it was an illusion and a fraud with which the glasses have for so long deceived me and so any others who have observed him with me many times?" Galileo accepted none of these possibilities, but rather predicted the "two minor stars" of Saturn would return according to a schedule he gave. As Van Helden speculates, Galileo must have had some model, which he does not specify, as to when and why the objects would return, involving the motion of the lateral bodies about Saturn or a turning of all three on an axis. "Treating the lateral bodies as satellites (although a different type of satellite from those around Jupiter) was only natural, and this conceptualization of the constitution of Saturn, formed

Figure 2.3. Galileo's drawing of what was later explained as Saturn's rings, 1616. Galileo himself had no idea what they were, and their identification as rings was left to Christiaan Huygens 40 years later. *Le Opere* (Edizione Nazionale), vol. 12, p. 276.

during the earliest years of telescopic observations, remained the dominant one for more than thirty years."[11] Saturn's appendages did reappear as Galileo expected, and Van Helden details a whole range of confused explanations over the next forty-five years about how and why they appeared and disappeared as they did, and what they might consist of. Galileo himself continued to observe Saturn (Figure 2.3), and still spoke of the planet "revealing and hiding his collateral globes" in his famous *Dialogo* of 1632, ten years before his death.

It was not until 1655 that the Dutch astronomer Christiaan Huygens gave the correct interpretation of Saturn's collateral bodies as rings, following another famous Saturnian discovery. Huygens's first good telescope, 12 feet in length and magnifying about fifty times, was finished in March 1655, and he almost immediately turned his attention to Saturn. On March 25, he spotted a small star, immediately suspected it was a satellite, and confirmed this over the next few nights of observation. He did not announce it immediately, but first determined its period, then sent out a cryptic anagram in June, and gave its solution that autumn as "Saturno luna sua circumducitur diebus sexdecim horos quatuor" (Saturn is circled by its own moon in sixteen days and four hours). This turned out to be what we today know as Titan, Saturn's largest moon. Several aspects are notable about Huygens's discovery of a Saturnian satellite. First, he did not have to go through the torturous process of reasoning that such a satellite could exist in principle; Galileo had done that forty-five years before with the moons of Jupiter. It was now expected that such a satellite could exist, demonstrating that the discovery of the first object in a class of objects (excluding the Moon itself in this case) is always the most difficult. Second, Huygens was not the first to actually see the Saturnian moon, but the first to interpret it as such. Both Johann Hevelius and Christopher Wren had seen the object near Saturn, but had taken it to be a fixed star. In March 1656, Huygens circulated a printed sheet entitled "De Saturni luna observatio nova" (Observation of a new moon of Saturn) describing his observations in detail. Clearly, as would happen again and again in the history of astronomical discovery, the first person to make the correct interpretation got the credit.[12]

While observing the Saturnian satellite in March 1655, Huygens had also seen the "anses." They were very narrow when he first saw them and disappeared entirely later that year, not returning until October 1656, seven months after Huygens had published his ring hypothesis in the same paper where he announced the satellite, *De Saturni luna*. Therefore, as van Helden concluded, "the problem of Saturn was solved not directly by observation, but rather by reasoning based on certain suggestive observations. In fact, the realization that perhaps a ring was responsible for the changing appearances of Saturn, came to Christiaan Huygens during a period when Saturn's anses were invisible."[13] Like Galileo, Huygens first presented his solution in anagram form, in the March 1656 paper. He gave the solution to the anagram in his *Systema Saturnium*, published in July 1659. "Annulo cingitur, tenui, plano, nusquam cohaerente, ad eclipticam inclinato": He is surrounded by a thin flat ring which does not touch him anywhere and is inclined to the ecliptic.

But how did Huygens come to his ring hypothesis? What were the "suggestive observations"? Again, Van Helden sheds light in his analysis, based on Huygens's own account in *Systema Saturnium*. One clue was that the length of the anses did not change whether they were narrow or wide, but simply became narrower until they disappeared. Another clue was the analog of our own Moon, which led Huygens to believe (perhaps having in mind Cartesian vortices) the anses must have rotational symmetry about Saturn's axis. As van Helden concludes, "At this point, Huygens realized that a ring would satisfy this condition and would also explain Saturn's appearances. Whether this realization came to him as a sudden insight or as the fruit of arduous trial-and-error considerations is not revealed by the author."[14] In any case, as Huygens laid out in great detail in his *Systema Saturnium*, he believed his ring hypothesis not only explained the appearances, but also was physically true (Figure 2.4), even though he had no idea of how it could be true, lacking knowledge of dynamics.

Huygens's ring hypothesis was not immediately accepted, in part because such a ring was unprecedented, and especially because Huygens insisted it had to be a solid ring. In 1658, Christopher Wren wrote, "For Saturn alone stands apart from the pattern of the remaining celestial bodies, and shows so many discrepant phases, that hitherto it has been doubted whether it is a

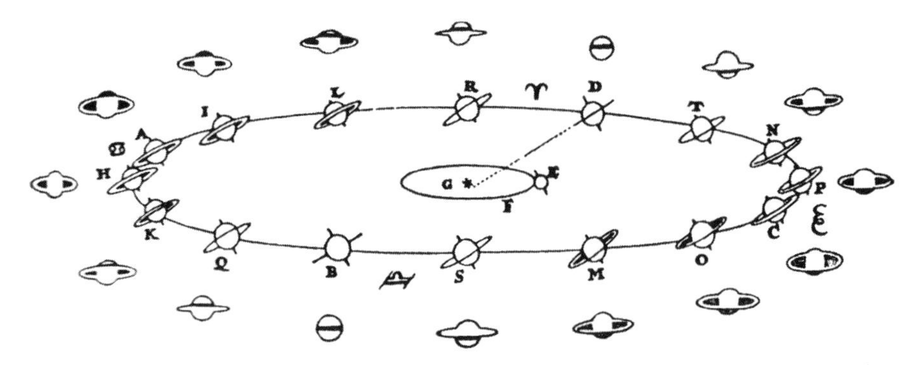

Figure 2.4. Huygens's depiction of how the appearances of Saturn over time can be interpreted as a ring. An understanding of its dynamical nature awaited Maxwell in the nineteenth century. From Huygens, *Oeuvres completes*, pp. xv, 309.

globe connected to two smaller globes or whether it is a spheroid provided with two conspicuous cavities or, if you like, spots, or whether it represents a kind of vessel with handles on both sides, or finally, whether it is some other shape."[15] But by 1670 the theory had gained the upper hand, and after Huygens's predictions for the appearance and disappearance of the ring proved fairly accurate in 1685 and again in 1701, the ring hypothesis was never again seriously challenged.

In Saturn's rings we therefore clearly see a bifurcation in the process of discovery: Galileo's 1610 detection and Huygens's 1655 correct interpretation. Both may be said to be necessary components of discovery. There is, however another step, not consummated until the mid-ninteenth century with James Clerk Maxwell's theoretical understanding that the rings could not be a single solid body.[16] "Understanding" is the third step in the discovery process, and that step itself consists of a continuum of increased understanding. After all, only in the twentieth century did spectroscopy reveal the Saturnian rings to consist mostly of icy particles, followed by spacecraft observations that verified the physical constitution of the rings as almost pure water ice slightly contaminated with silicates. Spacecraft also revealed their dynamical complexity, including braids, waves, spokes, "propellers," and numerous divisions. Many mysteries of the Saturnian rings remain. And the discovery of rings around Uranus (1977), Jupiter (1979), and Neptune (1989) showed that Saturnian rings were not in a class by themselves, but a characteristic of all the giant outer planets of our solar system, and perhaps other solar systems.

Even though understanding is the third component of discovery, this does not imply that the discovery of a new class is an endless process; that would fly in the face of common sense. We discuss how discoveries end in Chapter 6.

2.2 Planets and Minor Planets: A Species Hitherto Unknown

It is a striking fact in the history of astronomy that, aside from the delineation of classes of known planets based on size (see next section) and an explanation for the zodiacal light, it was almost two centuries after Galileo's discovery of moons and rings that the next new classes of astronomical objects were discovered in our planetary system. Although nine planetary satellites were known by the end of the seventeeth century, and a total of thirteen a century later, these were additions to the already well-established class of satellites, led by Galileo's discovery. It was as if biologists had discovered no new species for almost 200 years. To be sure, as we shall see in the next chapter, Galileo himself discovered several new classes in the sidereal realm, and in the late eighteenth century William Herschel began his observational program of the "natural history of the heavens," which constituted the beginnings of sidereal astronomy, revealing a variety of nebular shapes. The latter took many decades to separate into various classes of objects, a subject we take up in the next chapter.

In the solar system, however, things had remained relatively calm. Telescopes and techniques of observation were improved, but it was only in 1781 that Herschel himself, busy with his nebulae, found an unusual object moving in the night sky. Although Herschel did not immediately realize it, this turned out to be the new planet Uranus, the first discovered since the classical planets known to the ancients. Unlike those planets, seen to wander among the fixed stars from time immemorial, Uranus had to be "discovered," as did Neptune in 1846. In the interval between the discoveries of those two planets, five others had been found, and then dozens, and eventually hundreds and thousands. These turned out to be what we now call "minor planets" or "asteroids," but at the time it was not at all obvious that these should be declared a separate class of objects. The discovery of seven "planets" over the sixty-five-year interval from 1781 to 1846, and their eventual separation into two classes of objects, sheds new light on the complexity of discovery.

The Discovery of Uranus

We now know that Uranus had been observed as a starlike object long before Herschel claimed it as a planet in 1781. British Astronomer Royal John Flamsteed observed the object six times in 1690, and the French astronomer Pierre Lemonnier observed it twelve times between 1750 and 1769, including four times on consecutive nights in January 1769.[17] Neither recognized these pre-discovery observations as a planet. On March 13, 1781, Herschel – in his private residence in Bath, England and using a telescope with a 6.5-inch mirror of his own construction – spotted what he described in his observing book as "a curious either Nebulous Star or perhaps a Comet." Herschel had not been searching for such an object, but found it while undertaking observations for stellar parallax. Four days later he found the object had moved, a common characteristic of comets. The article reporting the discovery to the Royal Society, dated April 26, shows Herschel had no inkling he had found a planet; rather, it is titled simply "Account of a Comet." Herschel wrote, "On Tuesday the 13th of March, between ten and eleven in the evening, while I was examining the small stars in the neighborhood of H Geminorum, I perceived one that appeared visibly larger than the rest. Being struck with its uncommon magnitude, I compared it to H Geminorum and the small star in the quartile between Auriga and Gemini, and finding it so much larger than either of them, suspected it to be a comet."[18] Herschel relates how he first spotted the object with a 227-power eyepiece, and found that its diameter increased with higher power eyepieces, not a characteristic of fixed stars. "The sequel has shown that my surmises were well founded," Herschel boasted, "this proving to be the comet we have lately observed." It is notable that although Herschel knew the class of planets existed, none had ever been discovered in recorded history, and he instinctively placed the new object in a class where new members were constantly being discovered – the class of comets.

Based on the orbit and estimated distance of the object, however, some astronomers opined almost immediately that it could be a planet. Herschel had notified the Astronomer Royal, Neville Maskelyne, of his discovery, and on April 23 Maskelyne replied, "I don't know what to call it. It is as likely to be a regular planet moving in an orbit nearly circular to the sun as a Comet moving in a very eccentric ellipsis. I have not yet seen any coma or tail to it." The Russian astronomer Anders Lexell computed a nearly circular orbit far beyond Saturn, leading him to believe it could be a planet. From Berlin Johann Elert Bode described the object as "a moving star that can be deemed a hitherto unknown planet-like object circulating beyond the orbit of Saturn."[19] It was not until 1783, however, that the more cautious Herschel wrote, "By the observation of the most eminent Astronomers in Europe it appears that the new star, which I had the honour of pointing out to them in March 1781, is a Primary Planet of our Solar System."[20] Although Herschel suggested it be named "Georgium Sidus" (George's star) after his patron George III, this seemed too nationalistic to some, and by 1850 even the British had officially accepted the name "Uranus," based on a suggestion of Bode. Thus did a new planet enter the annals of major objects in the solar system, not a new class to be sure, but the first new member of the class of planets in recorded history.

The Discovery of Minor Planets

The lesson of Herschel's discovery of Uranus – that an unknown moving body in the heavens was not necessarily a comet but could be another class of object – was not easily learned, as it turned out in the case of the discovery of minor planets, or asteroids, first detected exactly twenty years after Uranus. This is true even though there was an expectation, based on the relative distances of the planets, that a planet should exist in the space between Mars and Jupiter. This Titius-Bode "law," formulated in 1766 by Johann Daniel Titius and popularized by Bode in the 1772 edition of his popular textbook, found from the mathematical progression in the distances of planetary orbits that a planet should exist in the space between Mars and Jupiter. This suggestion was so compelling that in 1799, Baron von Zach organized a coordinated effort, sometimes called the "Celestial Police," to search for the missing planet. It consisted of twenty-four astronomers throughout Europe, each of whom was given a 15-degree zone of the zodiac to search. "Through such a strictly organized policing of the heavens, divided into 24 sections, we hoped eventually to find a trace of this planet, which had so long escaped our scrutiny, if it did exist and make itself seen," von Zach wrote.[21] It is a matter of historical record that the missing object was not to be found through this coordinated effort, but serendipitously.

On New Year's Day 1801, while working on his great new star catalogue at the observatory in Palermo in Sicily, the Italian astronomer and Theatine monk Giuseppe Piazzi (Figure 2.5) observed an 8th magnitude star in the constellation Taurus. Because Piazzi, aged fifty-four, was using a specialized telescope known as a Ramsden transit circle that moved only along the north-south meridian, he could observe the object for less than two minutes while it crossed the meridian, accuracy being paid for at the cost of observation time. At first he considered it just another of the stars he was observing for positional accuracy in his program to update the 1789 star catalogue of Francis Wollaston. But this object was not in that catalogue, and the very next night he saw the star had moved about 4 arcminutes to the north and west. First he doubted the accuracy of his measurements, but then, "I conceived afterwards a great suspicion that it might be a new star. The evening of the third, my suspicion was converted into certainty, being assured it was not a fixed star." Observing it again on the evening of January 10, "by its motion I easily distinguished my star from the others," he wrote.[22]

As with William Herschel and the discovery of Uranus twenty years earlier, Piazzi believed at first he had found a new comet. On January 24 he transmitted an account of his results to his fellow astronomer, Barnaba Oriani, in Milan, with similar letters to Johann Bode in Berlin and Baron Franz Xavier von Zach.

Figure 2.5. On January 1, 1801, the Theatine monk Giuseppe Piazzi detected an 8th magnitude object. At first he believed it to be a comet, then a planet, despite William Herschel's urging that it was a new class of object that Herschel dubbed "asteroid." The status of this class remained uncertain for a half century.

In the letter to Oriani, Piazzi already expressed doubts about the true nature of the object: "I have announced this star as a comet; but the fact that the star is not accompanied by any nebulosity and that its movement is so slow and rather uniform, has caused me many times to seriously consider that perhaps it might be something better than a comet. I would be very careful, however, about making this conjecture public."[23] The new object was lost in the glare of the Sun after February 12, and attention turned to computing an orbit with the little data at hand, covering a period of 41 days and only 3 degrees of arc on the sky. Piazzi sent his data to Joseph de Lalande in Paris, and by May 31 his colleague Johann Burckhardt had calculated both circular and parabolic orbits. Shortly thereafter, Piazzi himself calculated a nearly circular orbit, but the results based on so little data were uncertain. Overall the predicted positions of the object ranged over 5 degrees of the sky, ten times the diameter of the full Moon!

Piazzi published his results in the summer of 1801, and an exhaustive summary appeared in June in von Zach's *Monatliche Correspondenz.*[24] Piazzi dubbed the object "Cerere Ferdinandea" in honor of Ceres and Ferdinand, respectively the patron goddess of Sicily and the royal patron of the Palermo Observatory.

Even though with so few observations the result was uncertain, and probably taking into account the lack of fuzzy appearance characteristic of comets, Piazzi seemed convinced by now that he had observed a planet. The object, however, had not been recovered since it was lost in February.

In Brunswick, Germany, the famous mathematician Carl Friedrich Gauss saw an opportunity. Having seen a copy of the article in von Zach's *Monatliche Correspondenz* (Monthly Correspondence), he considered Ceres a challenge:

> Nowhere in the annals of astronomy do we meet with so great an opportunity, and a greater one could hardly be imagined, for showing most strikingly, the value of this problem, than in this crisis and urgent necessity, when all hopes of discovering in the heavens this planetary atom, among innumerable small stars after the lapse of nearly a year, rested solely upon a sufficiently approximate knowledge of its orbit to be based upon these very few observations. Could I ever have found a more seasonable opportunity to test the practical value of my conceptions, than now in employing them for the determination of the orbit of the planet Ceres, which during these forty-one days had described a geocentric arc of only three degrees, and after the lapse of a year must be looked for in a region of the heavens very remote from that in which it was last seen?[25]

Only Gauss was up to the challenge. Beginning in October 1801, he developed a new theory that allowed him to use the paucity of data to determine a nearly circular orbit, proving it was no comet. The position of Ceres was predicted and von Zach definitively recovered Ceres on January 1, 1802, exactly one year after its discovery. "Finally, the new primary planet of our solar system has again been discovered and found, like a star fish on the beach," von Zach wrote.[26]

Gauss's work perhaps hastened the identification of Ceres as a planet because his determination placed the distance of the new object in an orbit that accorded with Bode's law. In any case, many (but not all) astronomers quickly adopted the new object as the eighth major planet, as evident already in 1802 from Bode's paper. "On the Eighth Major Planet Discovered Between Mars and Jupiter." On the other hand, in February 1802 Joseph Banks commented on its "little disc of the size of the 1st or 2nd satellite of Jupiter . . . ," implying it did not rank with the other planets. As two recent historians stated,

> Astronomers always seemed to be suspicious of the status of Ceres. It was regarded as being too faint, too small, of too little mass and having an orbit that was too eccentric and of too high an inclination to be worthy of joining the Sun's planetary team.[27]

The odd status of Ceres was compounded when, in March 1802, Heinrich Wilhelm Olbers detected a similar object, now known as Pallas, in the same orbit – an unprecedented situation.[28] As Bode put it writing to Herschel in May of that year, Pallas "is a planet travelling with Ceres, in the same orbit, at the same distance round the sun. Such a thing is unheard of!" Herschel turned his 7-foot, 10-foot, and 20-foot telescopes toward both objects, and determined the sizes of Ceres and Pallas to be 162 and 70 miles respectively, an underestimate by factors of three or more compared to the modern values. In one of the first scientific papers on this new class of objects, dated May 6, Herschel reported his observations and argued they had the characteristics of neither planets nor comets. Therefore, he asserted, "we ought to distinguish them by a new name, denoting a species of celestial bodies hitherto unknown to us . . . From . . . their asteroidical appearance ... I shall take my name, and call them Asteroids . . . These bodies will hold a middle rank, between the two species that were known before; so that planets, asteroids, and comets will in future comprehend all the primary celestial bodies that either remain with, or only occasionally visit, our solar system." Herschel went on to define asteroids as "celestial bodies, which move in orbits either of little or of considerable excentricity [sic] round the sun, the plane of which may be inclined to the ecliptic in any angle whatsoever. Their motion may be direct, or retrograde; and they may or may not have considerable atmospheres, very small comas, disks, or nuclei."[29]

In a letter to Piazzi two weeks later, Herschel proceeded to lobby for his position. He suggested that Piazzi had discovered "a new species of primary heavenly body," a more notable achievement than merely adding another member to the class of planets already known. Herschel's reasoning is worth quoting in full:

> Moreover, if we were to call [Ceres] a planet, it would not fill the intermediate space between Mars and Jupiter with proper dignity required for that station. Whereas, in the rank of Asteroids it stands first, and on account of the novelty of the discovery reflects double honour on the present age as well as on Mr. Piazzi who discovered it. I hope you will see the above classification in its proper light, as so far from undervaluing your eminent discovery it places it, in my opinion, in a more exalted station. To be the first who made us acquainted with a new species of primary heavenly bodies is certainly more meritorious than merely to add what, if it were called planet, must stand in a very inferior situation of smallness.[30]

Hughes and Marsden have traced the course of this negotiation among the astronomical community, with Piazzi himself suggesting that the "*petites planetes*" be called "planetoids." Much of this discussion is a nomenclature rather than a class problem. But it is notable that William Herschel's son John ignored his father's word "asteroids," and counted them as planets in 1833. Others preferred the term "minor planet."[31]

The detection of Ceres and Pallas, followed by Juno in 1804 and Vesta in 1807, was only the beginning of the discovery of the numerous small solar system bodies that we now know between the orbits of Mars and Jupiter. Despite Herschel having dubbed them "asteroids," and despite the fact that he and others felt a new class was called for, during the years that their number was small, the major government almanacs of the world called them planets. This is perhaps because of the conservative nature of government institutions, but following suit, in 1828 most astronomy texts accepted eleven planets - the seven known major planets, plus Ceres, Juno, Pallas, and Vesta. No new objects of this kind were discovered for nearly thirty-nine years after the discovery of Vesta in 1807, and at the end of 1845, with the discovery of Astraea, the number stood at only five. But by the end of 1851, there were fifteen. Only when their numbers grew did astronomers come to a consensus that a new class of solar system objects called minor planets, or asteroids, should be created. The term "minor planet" was first used in the British Nautical Almanac in 1835.[32] Thus, there was a period of nearly fifty years of uncertainty during which many considered these objects to be planets.

The Discovery of Neptune

The controversy over the status of asteroids had not been settled when, in 1846, another planetary discovery shook the astronomical world. Like Uranus, Neptune had also been observed as a star before it was recognized as a planet. In fact, its first observation predates the first observations of what turned out to be Uranus, and was recorded by none other than Galileo Galilei himself, in two drawings on December 28, 1612 and January 27, 1613, made while the object was near Jupiter. The fact that Galileo had observed Neptune was not realized until 1980, when astronomer Charles T. Kowal and historian Stillman Drake found the discovery in Galileo's notebooks. In 2009, Australian physicist David Jamieson claimed that Galileo had not only observed Neptune, but had also detected its motion during the year 1613, an important step toward realizing it was a planet.[33] Galileo, however, did not make that claim; despite his other discoveries, even his Copernican worldview at the time would have made this a difficult leap. His observations of Neptune remain a classic example of pre-discovery.

It was only 234 years after Galileo's first Neptunian observation, on September 13, 1846, that the German astronomer Johann Gottfried Galle, acting on predictions made by Urbain Jean Joseph Le Verrier at the Paris Observatory and transmitted to him at the Berlin Observatory, spotted Neptune and claimed it as a new planet. The circumstances were quite different from the discovery of Uranus or the first minor planets. The difference was an accurate mathematical prediction based on real physics and dynamics. It is true that Bode's law of planetary distances had been enunciated a decade before the discovery of Uranus, and that it had spurred a search for a planet between Mars and Jupiter. But Bode's law was not based on physics or dynamics, and in any case neither Uranus nor Ceres were found as a direct result of Bode's law, both being serendipitous discoveries made during the course of other observations.

In the case of Neptune, irregularities in the orbit of Uranus led already in the early 1800s to the idea of a perturbing body beyond Herschel's planet. By the 1840s, Uranus had made almost an entire orbit since its discovery, and the difference between gravitational theory and observation led the mathematician John Couch Adams in England and Le Verrier in France independently to calculate an orbit for the supposed planet. Le Verrier published his results in a series of papers, read before the French Academy of Sciences and published in its *Comptes Rendus*.[34] On September 18, 1846 he wrote Galle, transmitting his final set of orbital elements and requesting that Galle search for the planet. "Right now I would like to find a persistent observer, who would be willing to devote some time to an examination of a part of the sky in which there may be a planet to discover. I have been led to this conclusion by the theory of Uranus ... you will see, Sir, that I demonstrate that it is impossible to satisfy the observations of Uranus without introducing the action of a new Planet, thus far unknown; and, remarkably, there is only one single position in the ecliptic where this perturbing Planet can be located."[35]

In England, even though Adams had begun his computations first, constructed an orbit by the fall of 1845, and communicated his result in mid-September 1845 to James Challis a mile away at Cambridge University Observatory, neither Challis nor other British astronomers pursued the search until July 29, 1846, spurred by Airy's reading of Le Verrier's first paper. Challis also informed Airy of the calculations, but Airy was skeptical. By contrast, in Berlin, Galle convinced observatory director, Johann Encke, that he should initiate his search on the very day he received Le Verrier's September 23 letter giving his latest and most accurate results. Using the 9-inch Fraunhofer refractor, Galle, assisted by his student Heinrich Louis d'Arrest, made the discovery within the hour and within one degree of Le Verrier's predicted position. The account of the discovery stated that Galle "very soon found a star of about the 8th magnitude, nearly in the place pointed out by M. Le Verrier, and which did not exist in the map. There could be little doubt that this was the new planet, and the observations of the two days following showed that its motion was in the direction of and nearly equal to that of the planet predicted by M. Le Verrier."[36]

Meanwhile, lacking the Berlin Academy's accurate star chart of the region, Challis was forced to observe thousands of stars. He had actually spotted the object twice, on August 4 and 12, but failed to recognize the moving planet. This led to a famous priority dispute, but the British awarded Le Verrier the Copley medal, and historians have in general concluded that Le Verrier and Galle are the true discoverers, the first joint team that showed the power of theory for astronomical observation.

The discovery of Neptune holds several lessons for the discovery process. As one team of historians concluded, "Adams utterly failed to communicate his results forcefully to his colleagues and to the world. A discovery does not consist merely of launching a tentative exploration of an interesting problem and producing some calculations; it also involves realizing that one has made a discovery and conveying it effectively to the scientific world. Discovery thus has a public as well as a private side. Adams accomplished only half of this two-part task." There is also the personal side: "Ironically," the historians wrote, "the very personal qualities that gave Le Verrier the edge in making the discovery - his brashness and abrasiveness, as opposed to Adams's shyness and naiveté, worked against him in the postdiscovery spin-doctoring." And there is the sociological aftermath: "The British scientific establishment closed ranks behind Adams, whereas Le Verrier was unpopular among his colleagues." The latter conclusion is, however, more complex in its fine structure: as Robert Smith has pointed out, for a variety of reasons, "many of the Fellows of the Royal Astronomical Society also came down firmly on the side of Le Verrier against Adams, at least at first." Luck also played a part in the discovery, for it turns out (as it would in the case of the discovery of Pluto) that both Adams and Le Verrier succeeded in getting the predicted longitude because of a "fluke of orbital timing." Had Uranus and Neptune been elsewhere in their orbits, the methods of prediction employed by Adams and Le Verrier would not have resulted in such an accurate prediction.[37]

Although the discovery of Uranus and Neptune shed light on the process of discovery that separated planets from minor planets, they themselves do not constitute the discovery of a new class of objects, only their addition to the family of classical planets already known. At a more refined level, however, the planets themselves were later to be divided into classes. How this was accomplished is taken up in the next section.

2.3 Delineating Planetary Classes: A Semi-Nebulous Condition

With the planets we come to quite a different case than planetary satellites and rings. Prior to the mid-seventeenth century, neither Galileo nor anyone else "discovered" the planets, much less what we would now call the classes of terrestrial and gas giant planets. Planets (the Greek term for wanderers) had been distinguished from the fixed stars by their motions since antiquity. According to this very rough kinematic criterion of classification, the Greeks and Romans considered seven planets to circle the central Earth; in the Ptolemaic ordering of increasing distance, they were the Moon, Mercury, Venus, the Sun, Mars, Jupiter, and Saturn. Lacking knowledge of their physical characteristics, ancient skywatchers had no way of classifying these wanderers other than by their motion or brightness, or potentially by rudimentary calculations of size and distance. Classification by motion proved most useful for the needs of the time, and already in the Ptolemaic system divided the planets into "inferior" (Mercury and Venus) and "superior," depending on the epicyclic properties of their motion.[38] *All of this may be considered what we may call a "pre-discovery" stage with respect to the delineation of physically meaningful classes.*

It was not until Copernicus's heliocentric theory that the Earth became a planet, and the planets potential Earths. For Copernicus, just five classical planets existed, the Sun now holding a special central position, the Moon circling Earth, and the inferior planets now redefined as Mercury and Venus inside the Earth's orbit, and the superior planets Mars, Jupiter, and Saturn outside. We thus learn an important lesson: *classification is time-dependent, particularly in those rare cases when a revolution in worldview takes place*, of the kind Thomas Kuhn has famously described. (This is indeed very rare in astronomy; depending on how one defines "revolution," arguably the only other case is when spiral nebulae were revealed as extragalactic, as we shall see in Section 5.1.) Moreover, *the criteria for classification also depend on the knowledge at hand at any given time, and on the needs of the classifier or the broader community that employs the classification.*

Just how much the Copernican planets were like the Earth was a problem that has occupied astronomers ever since at increasingly refined levels. Rough masses, diameters, densities, and rotation rates could be measured already in the second half of the seventeenth century, and it soon became clear that Mercury, Venus, Earth, and Mars were in a different category from the rest in terms of these properties. Particularly important were Huygens's measurement of apparent planetary diameters, and Cassini's determination of the scale of the solar system.[39] The dichotomy is seen already in the two popular seventeenth-century treatises on the plurality of worlds – Fontenelle's *Entretriens sur la pluralité des mondes* (1686) and Huygens's *Cosmotheoros*, published posthumously in 1698. As Huygens put it after discussing the four inner planets: "If our Earth can claim preeminence of the fore-mentioned planets, for having a Moon to attend upon it, (for its Magnitude can make but a small difference) how much superiour must Jupiter and Saturn be to all four of them, Earth and all? For whether we consider their bulk, in which they far exceed all the others,

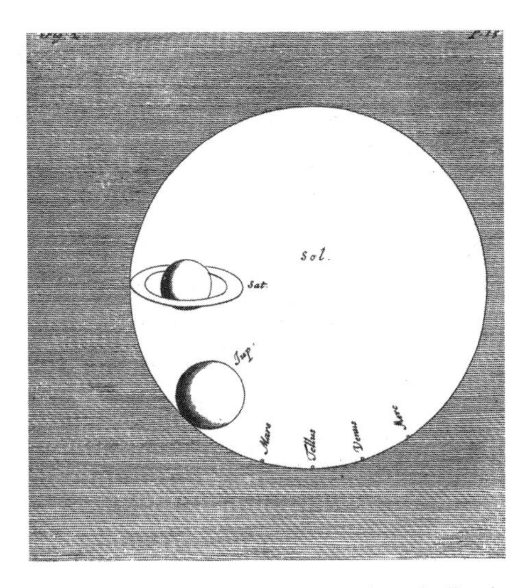

Figure 2.6. Huygens's scale drawing of planetary sizes, indicating a substantial difference between the inner and outer planets based on size alone. This version appeared in his popular posthumous *Cosmotheoros* (1698; *Oeuvres*, vol. 21, p. 802). An earlier version first appeared, together with the known satellites, in the appendix added to his *Systema Saturnium* in 1667 (*Oeuvres*, vol. 15, pp. 374–375).

or the number of Moons that wait upon them, it's very probable that they are the chief, the primary Planets in our System, in comparison with which the other four are nothing, and scarce worth mentioning."[40] Huygens emphasized his point by drawing a scale model of the planets compared to the Sun, similar to one he had already added to the appendix of his *Systema Saturnium* in 1667 (Figure 2.6). Clearly the terrestrial planets were separated from Jupiter and Saturn in terms of size by the second half of the seventeenth century. But there was as yet no idea of the true nature of the two giant planets, as evidenced by the fact that Huygens believed them to be habitable.

The essential realization that some planets were more Earth-like than others in terms of composition rather than size was made only gradually between the seventeenth and nineteenth centuries with advances in telescopic observation. As Agnes Clerke wrote in 1885, "Crossing the zone of asteroids on our journey outward from the sun we meet with a group of bodies widely different from the 'inferior' or terrestrial planets. Their gigantic size, low specific gravity [density relative to water], and rapid rotation, obviously from the first threw the 'superior' planets into a class apart; and modern research has added qualities still more significant of a dissimilar physical constitution."[41] As more became known about the physical nature of the planets, the basis for classification switched from distance, motion, and size to gross compositional characteristics.

Although the separation of the terrestrial and outer planets based on motion and gross physical characteristics was thus well known by 1700, the separation of a class specifically known to be gas giants came much later.

For terrestrial planets the essential parameter was their rockiness, or, to put it in more modern terms, the presence of silicates. It was gradually realized that the planets differ widely in other ways. Mercury is alternately scorched and frozen; Venus has temperatures of 900°F, a thick carbon dioxide atmosphere, and sulfuric acid rain; Mars has very little atmosphere at all. This demonstrates how much the physical characteristics of an object may vary and still be considered to belong to the same class. But there are limits to the boundaries of class, and it is here that the gradual delineation of a new class of gas giants becomes illuminating.

Though we now take them for granted, the existence of giant planets composed largely of gas was one of the great surprises in the history of astronomy. We hold that a gas giant planet is one whose main constituents are hydrogen and helium, and whose gaseous/liquid component is larger than the core, which may be rocky or metallic. In the deeper interior above the core, the majority of the hydrogen and helium exists in liquid or metallic form due to the effects of high pressure. Jupiter and Saturn are the two gas giants in our solar system by this definition, which not only distinguishes gas giants from terrestrial planets, but also (for some astronomers) from the ice giants. It wasn't until the advent of spacecraft observations that astronomers realized at a more refined level of analysis that Uranus and Neptune may be different enough from Jupiter and Saturn in structure and composition to constitute a possible new class, or perhaps subclass, of "ice giants."

The seemingly simple question of when the outer planets like Jupiter and Saturn came to be considered a class of "gas giants" separate from the terrestrial planets is more complicated than delineation by size. The title "gas giant" implies a much more sophisticated understanding than the simple title "giant planet" because size is much easier to determine than structure and composition. To be sure, Jupiter's density was known with considerable accuracy even in the eighteenth century.[42] At mid-century, Bessel calculated it had 388 times the mass of the Earth but only 1.35 times the density of water (the figures today are 318 Earths and 1.33 times the density). Bessel's estimates represented only one-quarter of the Earth's density, enough to hint that it was quite different than a terrestrial planet, and Saturn's density was found to be only 0.69, meaning it would actually float on water. Indeed, in his treatise on the plurality of worlds, William Whewell conjectured that Jupiter may be composed largely of water, or water ice, and that Saturn must consist largely of "vapour."[43]

But the problem was that all calculated densities were average densities, and thus were no guarantee of the structure of the planets. The differential rotation of Jupiter's atmospheric features could be measured, a remarkably rapid ten hours. But these might be only atmospheric features covering a solid surface, masking what, by analogy with the terrestrial planets, would be its underlying rocky composition. Lacking knowledge, astronomers sometimes assumed to a first-order approximation that Jupiter's composition was similar to the Sun, since the Sun and planets arose out of the same protosolar nebula. Indeed, Jupiter was most often compared to the Sun rather than the other planets in terms of composition. But the Sun itself was not believed to be totally gaseous either at this time.

Only after the mid-nineteenth century did research on the planet turn from measurement of physical parameters such as its position, diameter, and rotation rate, to more qualitative descriptions of its form and substance, descriptions that would eventually lead to a knowledge of its true nature.[44] Only then could astronomers apply the new arsenal of spectroscopy, thermodynamics, and kinetics to the mystery of Jupiter's structure and composition. Surprisingly, the same held true of the Sun, which until 1859 was believed to be a solid object with a luminous atmosphere. By the 1860s, laboratory evidence and physical theory indicated that even the Sun's interior could be gaseous.[45] By the 1870s, some astronomers and popularizers such as R. A. Proctor were concluding that Jupiter was far from Earth-like, though they still believed there was a "surface" somewhere below the clouds. The question was how far below. By 1876, astronomers were speculating the surface could be thousands of miles below the visible cloud features. As the British astronomer George H. Darwin (second son of Charles Darwin) wrote, "that planet must be very much denser in the centre than at the surface. Is it not possible that Jupiter may still be in a semi-nebulous condition, and may consist of a dense central part with no well-defined bounding surface? Does not this view accord with the remarkable cloudy appearance of the disk, and the remarkable belts?"[46]

The uncertainty of this speculation even 50 years later is evident in a review by none other than the young (and later famous) physicist Harold Jeffreys. "The densities of the outer planets are comparable with that of water. It has been held that their low densities afford evidence that large fractions of their volumes are gaseous, but the argument is open to suspicion." The four outer planets have roughly equal densities, Jeffreys argued, but very different masses, and "the low density affords as good an argument for the hypothesis that these planets are solid as against it. It compels us, however, to suppose that these planets are all composed of matter very different from the chief constituents of the earth."[47] That "very different matter" was not necessarily gas, Jeffreys argued, it might just as well be ice.

Exactly what those constituents were therefore remained a matter of conjecture. Building on the work of Jeffreys, in 1932 Rupert Wildt identified ammonia and methane in the spectrum of Jupiter, and he and others went on to note that its low density was consistent with a dominant composition of hydrogen similar to the Sun. Even then, Jeffreys suggested in 1934 that Jupiter had a rocky core, and Wildt calculated that Jupiter's inner core comprised 50 percent of the diameter of the planet, with an atmosphere perhaps 10,000 km deep.[48] By the 1940s, observations of the masses, radii, and rotation periods of the giant planets, together with new knowledge of the high-pressure behavior of matter, constrained the composition of Jupiter and Saturn to be predominantly hydrogen. Hydrogen, however, is difficult to detect at planetary temperatures, and it was not actually detected until 1960, by Carl Kiess and his colleagues. The presence of helium awaited the Pioneer 10 encounter in 1973.[49]

Even in 1940, astronomers talked in terms of the atmospheres of the giant planets, arguing only that "The giant planets Jupiter, Saturn, Uranus and Neptune possess atmospheres of considerable extent and density." It is revealing that the term "gas giant" appears to have first been used by science fiction writer James Blish in his story "Solar Plexus," as anthologized in 1952.[50] It was not present in the original 1941 version of the story. Thus, in Blish's mind, undoubtedly a reflection of the new scientific realities, the outer planets as gas giants became a common concept only at mid-twentieth century, even if their exact composition was still unknown.

The story of planetary classes in the solar system is therefore one of a progression, and many intermediate classifications before gas giants became a widely recognized class. In the Ptolemaic system there were the inferior and superior planets based on motions. In the Copernican system the Earth became a planet and the planets, Earths. As soon as even rough sizes were calculated, already by 1700 it became evident there was a division between the terrestrial planets and the outer planets. But only around 1950 was it widely accepted that these were "gas giants," even if some of their constituents still needed to be determined. This long but insightful story reinforces the view that *classification is time-dependent, even when basic physical properties rather than revolutions in world-view are involved.* To put it another way, *the fundamental meaning of a particular class may change over time, an attribute we shall see in the next chapter with the nebulae.*

The delineation of gas giants as a class is not quite the end of the story of planetary classes. By mid-twentieth century, based on the gross similarity of their features, the planets Uranus and Neptune had taken their place as gas giants, presumed to have similar structures as the gas giants. Their true physical nature as a possible separate class remained unknown until the Voyager 2 spacecraft observations provided data that constrained models of their interiors. A hydrogen-rich atmosphere is believed to extend from the observed cloud tops to 85 percent of Neptune's radius, 80 percent in the case of Uranus. Today the average density of Uranus is known to be 1.27, and Neptune 1.638, not so different from Jupiter. Models show the density and pressures of the deep interior mimic laboratory experiments of an artificial icy mixture known as "synthetic Uranus," and so it is likely icy. The total amount of hydrogen and helium in Uranus and Neptune is about two Earth masses, compared to 300 Earth masses for Jupiter.

Although other models exist, in 1999 astronomer Mark Marley declared, "Given the relatively small amounts of gas compared to ices in Uranus and Neptune, these planets are aptly termed 'ice giants,' whereas Jupiter and Saturn are indeed 'gas giants'." By 2006, Heidi Hammel declared, "Planetary scientists now appreciate that the planets Uranus and Neptune differ from Jupiter and Saturn in more than just size: their interiors differ in composition and phase; their zonal winds differ in magnitude and direction; and their magnetic fields differ in structure. Hence we refer to these mid-sized planets as 'Ice Giants' to distinguish them from their larger 'Gas Giant' cousins."[51] It is notable that in both cases the declaration of a new class was made on the basis of physical characteristics, including composition. Average density was not a good enough discriminator. While some considered ice giants to be a subclass of gas giants, full class status appears to be winning out by common usage among astronomers. The situation will undoubtedly be illuminated as the sample size of planets is increased with the discovery of the properties of exoplanets.

The delineation of the planets of our solar system into classes introduces yet another new element into discovery. The terrestrial, gas giant, and ice giant classes were not "detected" in the same way Galileo detected planetary moons and Piazzi detected what turned out to be minor planets. Rather, following a long pre-discovery phase with respect to the physically meaningful classes during which the planets were seen as mere moving points of light, *the classes were inferred based on new knowledge, first of size in the case of terrestrial and giant planets, then of composition in the case of gas and ice giants.* As we shall see in Chapter 4, the same holds for classes of stars. In both cases there is still ample room for interpretation and understanding following the inference of the new classes.

2.4 Trans-Neptunian Objects: The Hypothesized Population

Beyond the orbit of Neptune are thousands of objects that have come to be known collectively, and quite logically, as Trans-Neptunian Objects, or TNOs for short. They comprise, as Alan Stern has suggested, the "third zone" of the solar system, beyond the zones of the terrestrial and giant planets.[52]

Despite the difficulties of their detection, a long history of speculation exists about such objects. Theories of planetary accretion, as well as comets and asteroids known to exist in the inner solar system, indicated that similar bodies should be found in the outer solar system. The technology of the time, however, made them very difficult to detect, as the painstaking discovery of the first Trans-Neptunian object, Pluto, proved in 1930.[53]

Such difficulties did not prevent informed speculation. As mentioned in Chapter 1, both the American astronomer Gerard Kuiper and the Irish economist, military engineer, and late-blooming astronomer Kenneth Edgeworth noted in the 1940s and 1950s that there was no reason to believe that the solar system ended with Neptune or Pluto. In 1943 and 1949, Edgeworth proposed that the cloud which condensed into the solar system must have extended to greater distances than Pluto, that the condensations in the outer regions did not condense into planets, but "retained their individuality," leaving the outer regions of the solar system "occupied by a very large number of comparatively small bodies." He also suggested that comets might originate in this reservoir beyond Pluto, different from the Oort Cloud proposed by the Dutch astronomer Jan Oort in 1950. In 1951, Kuiper argued that such objects would be scattered to the Oort cloud and be perturbed back into the inner solar system as comets.[54] Harvard astronomer Fred Whipple and others developed these ideas in the 1960s, and by 1980 a distinct comet belt was again advanced as the source of short-period comets, as was the Oort Cloud for long-period comets.[55] The short-period belt was first named the "Kuiper Belt" in 1988, and that term, or the alternative "Edgeworth-Kuiper Belt" (especially among the British) gained rapid usage thereafter.[56] The discovery of any such objects, however, remained elusive; just how elusive can be appreciated by the fact that a given asteroid would be 10,000 times fainter if moved from 3 to 30 astronomical units.

The chances of a successful detection improved immensely with the arrival of charge-coupled device (CCD) technology in the 1990s, enabling detection of extremely faint objects that we now recognized as Kuiper Belt Objects (KBOs), along with a few possible inner Oort Cloud members. As we saw in Chapter 1 in connection with the serious onset of the Pluto debate, the first unambiguous KBO, dubbed 1992 QB1, was discovered on August 30, 1992 after a five-year search by astronomer David Jewitt and his student, Jane Luu. It was detected as a very faint 22nd magnitude object with slow retrograde motion, using the University of Hawaii's 2.2 meter telescope on Mauna Kea (Figure 2.7). "Here we report the discovery of a new object, 1992 QB1, moving beyond the orbit of Neptune," they wrote. "We suggest that this may represent the first detection of a member of the Kuiper Belt, the hypothesized population of objects beyond Neptune and a possible source of the short-period comets."[57] They

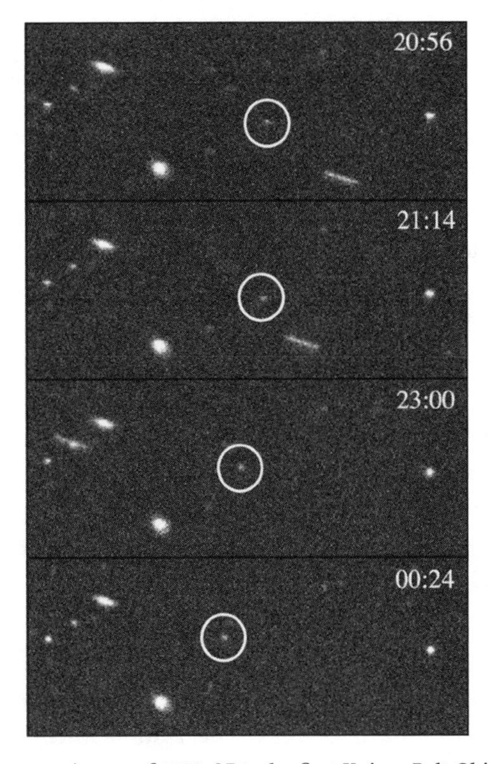

Figure 2.7. Discovery image of 1992 QB1, the first Kuiper Belt Object, aside from Pluto itself, detected by David Jewitt and Jane Luu. The extremely faint 22nd magnitude object (circled) was detected with the University of Hawaii's 2.2 meter telescope on Mauna Kea, equipped with a charge-coupled device (CCD) detector. In these four panels covering about 3.5 hours, it was moving at three seconds of arc per hour. By comparison, Pluto at the time of its discovery was 15th magnitude. The longest streaked object to the lower right of the upper two panels is an asteroid in the field. Courtesy David Jewitt.

estimated its distance at 41 astronomical units, and its diameter at about 250 km, roughly one-eighth the size of Pluto. Six months later they found a similar object, named 1993 FW, and six months after that announced two more. Another team almost simultaneously announced another two. Within a span of twelve months, the Kuiper Belt had gone from theory to reality. By 1995, the Hubble Space Telescope saw its first evidence of the KBO population, and in 2009 it discovered the smallest KBO known at the time, 3,200 feet in diameter, 4.2 billion miles away. Since then more than a thousand have been discovered, undoubtedly the tip of the Kuiper Belt iceberg.[58]

On October 9, 1996, during a wide area search for TNOs, Jewitt, Luu and their colleagues also discovered the first object in the so-called scattered disk of the Kuiper Belt, dubbed 1996 TL66.[59] As with 1992 QB1, they found the object using a CCD detector on the 2.2 meter telescope on Mauna Kea, as part of the same survey. Data from the Spitzer Space Telescope indicates its diameter is about 575 km, compared to 2300 km for Pluto. 1996 TL66 has a perihelion distance of 35 AU and a diameter of about 600 km. The same survey found three more scattered KBOs in 1999. Scattered-disk objects are distinguished from more stable KBO orbits in that they have orbital eccentricities as high as 0.8 (where zero is perfectly circular), inclinations as high as 40 degrees (where zero is the plane of the ecliptic), and perihelia extending as close as 30 AUs and aphelia beyond 100 AUs. Thus, while the closest approach of scattered-disk objects to the Sun may overlap those objects with more stable orbits in the Kuiper Belt disk, at the most distant points of their orbits, they are two to three times the distance of Pluto. Because of their high eccentricities, their semi-major axes may exceed 50 AU. Such pioneers in the field as Jewitt himself have called them "scattered Kuiper Belt Objects"; whether they should be considered a separate class of objects is open to definition and interpretation, but based on their likely physical composition, they seem to be KBOs. With the advance of telescope detector technology, such objects have been discovered in increasing numbers ever since.

No Trans-Neptunian objects have been unambiguously detected in the much more distant Oort Cloud, which remains a largely theoretical but highly likely location of millions more TNOs. However, some astronomers, including its discoverer Michael Brown, have argued that the object known as Sedna (after the Inuit goddess of the sea) is an inner Oort Cloud object; the IAU's Minor Planet Center still classifies it as a scattered-disk object, a subset of the Kuiper Belt.[60] Discovered on March 15, 2004, Sedna is three times farther than Pluto, about 90 astronomical units, the most distant known object in the solar system. It takes 10,500 years to orbit the Sun, compared to 248 years for Pluto. Its likely diameter is between 1200 and 1600 kilometers. Some astronomers prefer to call it simply a "detached object" in the sense that it is detached from Neptune's influence. Other detached objects have also been found closer than Sedna, but none have been found farther, most likely due to the difficulties of detection. Many more, however, are sure to be found in the future.

This menagerie of objects raises all sorts of questions about classification; indeed, as we saw in Chapter 1, it precipitated the Pluto "crisis." The largest Trans-Neptunian Objects now known are those officially designated dwarf planets, including Pluto, Eris, Makemake, and Haumea (Ceres is also classified as a dwarf planet - and a minor planet!). They are rivaled in size by a few other objects in the Kuiper Belt, including Orcus, Quaoar, Varuna, and Ixion. Haumea, discovered by Michael Brown of Caltech in 2004, and by J. L. Ortiz in 2005 in Spain (yet another priority controversy), is now known to have two moons. Eris, at first thought to be the largest known dwarf planet but in reality about the same size as Pluto (2300 km in diameter), was discovered in January 2005 by Brown. Makemake was also discovered by Brown, on March 31, 2005. About fifty Trans-Neptunian Objects are candidates for dwarf planet status, but it is not yet known if they are in hydrostatic equilibrium, one of the criteria for dwarf planet status.[61] Aside from Pluto, the largest KBOs not (yet) declared dwarf planets are about 850 miles in diameter, half that of Pluto. The population of Trans-Neptunian objects in the Kuiper Belt with diameters larger than 100 km is likely to exceed 70,000, with tens of millions more down to 1 km size. Today the International Astronomical Union officially designates Trans-Neptunian Objects as part of the "Small Bodies of the Solar System," along with minor planets and comets.[62] *The discovery of small solar system objects beyond Pluto thus illustrates an important characteristic of modern astronomical classification: an object may belong to more than one class.* Pluto, for example, is simultaneously a dwarf planet, a Trans-Neptunian Object (most of the time!), and a Kuiper Belt Object. As we shall see, this characteristic is not common beyond the solar system.

Moreover, sophisticated attempts have been made to develop a taxonomy of Trans-Neptunians, largely on the basis of color, probably a reflection of their physical characteristics.[63] All categories of Trans-Neptunian Objects observed so far appear to have roughly the same low densities and composition of frozen ices such as water and methane. However, the so-called scattered-disk objects tend to have a more white or grey appearance, compared to a reddish color for the classical Kuiper Belt Objects; this may be due to methane having frozen over the entire surface as opposed to Pluto, for example, where the methane may have frozen only at higher elevations, resulting in a lower surface brightness. Whether this feature is enough to distinguish scattered-disk objects as a new class of objects remains an open question. Dynamical properties would constitute a stronger argument for class status separate from KBOs, but even on those grounds astronomers have often declared scattered-disk objects to be part of the Kuiper Belt. Like so many classes of objects, it is a matter of definition, but one that becomes clearer if physical composition is the defining criterion for class status, a criterion that in this case is reinforced by dynamical considerations.

Such classifications are important beyond the individual objects they designate. Like the other small bodies of the solar system, and as originally suspected by Edgeworth and Kuiper, TNOs may shed light on the nature of the original protoplanetary disk of our solar system and others. Kuiper Belt–type disks have been resolved around about a dozen other stars, where they are also known as debris disks. As Jewitt has pointed out, "The main importance of the Kuiper Belt is that it provides a very local example of a highly evolved circumstellar disk. Processes that are probably general to the circumstellar disks of solar-mass stars can be discerned already in the size and orbital element distributions of the Kuiper Belt."[64] Classifications in our solar system may therefore also transfer to planetary systems in the universe at large. But that will be only a beginning, for if extrasolar planet detections have demonstrated anything, it is that planetary systems may be very different from ours. Planets may exist in extraordinary variety, and even the debris disks already discovered are being classified based on how far they have advanced from protoplanetary to post-planetary.

* * *

We have discussed enough objects in the realm of the planets to provide a taste of some of the issues astronomers have grappled with through history and continue to address today. What constitutes the discovery of a new class of objects? How does one define "class"? What criteria should be used for classification? And how do classes and classification systems become accepted? We have only begun to see how the discovery of new classes of objects is affected by theory and technology, and by individual and sociological considerations. Before tackling those questions in more detail in subsequent chapters, however, we increase our sample size in the natural history of discovery by turning to the discovery of new classes in the realm of the stars.

3

In Herschel's Gardens: Nebulous Discoveries in the Realm of the Stars

> But what a field of novelty is here opened to our conceptions! A shining fluid, of a brightness sufficient to reach us from the remote regions of a star of the 8th, 9th, 10th, 11th or 12th magnitude, and of an extent so considerable as to take up 3, 4, 5, or 6 minutes in diameter! Perhaps it has been too hastily surmised that all milky nebulosity, of which there is so much in the heavens, is owing to starlight only.
>
> William Herschel, 1791[1]

> Here in dividing the different parts of which the sidereal heavens are composed into proper classes, I shall have to examine the nature of the various celestial objects that have been hitherto discovered, in order to arrange them in a manner most comfortable to their construction.
>
> William Herschel, 1802[2]

If discovery in the realm of the planets was difficult, in the realm of the stars the problems were only multiplied. While Pluto, representing the outer edge of the solar system as it was known in 1930, was about forty times the distance of the Earth from the Sun (40 astronomical units, or 4 billion miles), the nearest star was 25 trillion miles, a factor of nearly 7000 times more distant. Put another way, rather than light minutes or light hours for the planets, the nearest stellar distances were measured in light-years, and the more distant ones in hundreds or thousands of light-years. Whereas a light beam from Pluto would take a few *hours* to travel to Earth, the nearest starlight (other than our own Sun!) would take 4.2 *years*. In terms of distance, astronomers studying the stars clearly entered a new realm in discovery and interpretation when it came to techniques of observation and methods of inference.

Astronomers at the time, however, knew little about the distances they were dealing with. The fixed stars of time immemorial notwithstanding, so-called sidereal astronomy did not become an important part of astronomy until the eighteenth century, when observers such as the French astronomers Nicholas-Louis de la Caille and Charles Messier focused on objects in the realm of the stars, and especially when William Herschel began using his newly constructed large telescopes in England. The Copernican theory implied vast stellar distances, but those distances remained unknown until 1838 with the first determinations of stellar parallax, and even then only the distances to the nearest stars could be measured.[3] A few years earlier Auguste Comte famously highlighted the composition of the stars as the very prototype of a problem that could never be solved. But the development of spectroscopy in the 1850s and 1860s enabled just that, a stunning feat in the annals of human thought that unlocked the secrets of the sidereal realm – but only slowly and grudgingly.

It is a matter of historical record that it was the so-called nebulae that first began to yield their secrets to sidereal astronomers in the century before stellar spectroscopy was developed; because some of them were extended objects, they were more easily studied than the stars themselves. As we shall see in the next chapter, the stars turned out to be much more complex than their pinpoints of light indicated, for they existed in a great variety of types, requiring knowledge of their distances and their spectra before stellar classes could be unraveled. Meanwhile Herschel began to observe not the stars themselves but new classes of objects that appeared to be *between or among* the stars, and in doing so inaugurated the natural history of the heavens.

The diversity in the realm of the stars was thus first revealed by studying those fuzzy objects known as nebulae. Placing ourselves in the shoes of our ancestors, we must not yet confer any meaning on this word, which is Latin for "cloud" and appeared to be just that, though obviously of a different variety than atmospheric clouds. A few nebulae had been noticed in the pre-telescopic era, the naked-eye Magellanic Clouds of the Southern Hemisphere being a famous example. Even Ptolemy and Copernicus had listed a half dozen "nebulous stars" in their catalogues,[4] and such objects proliferated in the wake of the telescope. Galileo himself never admitted the existence of real nebulosity, reporting already in his *Sidereus Nuncius* of 1610 that the Milky Way itself "is nothing else than a congeries of innumerable stars distributed in clusters. To whatever region of it you direct your spyglass, an immense number of stars immediately offer themselves to view." Moreover, "what is even more remarkable – the stars that have been called 'nebulous' [*nebulosae*] by every single astronomer up to this day are swarms of small stars placed exceedingly closely together." He drew the "nebula called Orion's Head" and "the nebula called

Praesepe," showing how they could be resolved into stars. Galileo may thus be credited with the discovery of star clusters, even if he could have no idea that they were gravitationally bound.[5]

Inexplicably, however, Galileo did not observe what we today call the Orion Nebula, located well below Orion's "Head." As U.S. Naval Observatory astronomer E. S. Holden remarked in 1878 in his exhaustive and beautifully illustrated monograph on the Orion Nebula, it is curious that for an object just visible to the naked eye under good conditions, the Orion Nebula was not reported in the pre-telescopic era. This gave rise to the idea that the Orion Nebula may have brightened since Galileo's time, but the more likely explanation is that Galileo had resolved the nebulous stars in Orion's Head, and had no desire to present a counterexample.[6] In any case it did not take long for others to see it. While the French astronomer Nicolas Claude Fabri de Peiresc is often credited with observing the Orion Nebula in 1610, the claimed observation is ambiguous; the Jesuit astronomer Johann Baptist Cysat is more likely to have been the first to see it in 1618 from Ingolstadt, Austria, where he made the observations of the comet of 1618–1619 for which he is best known. Neither knew its true nature. Christiaan Huygens studied the Orion Nebula in more detail later in the seventeenth century and produced the first recognizable drawing in his book on Saturn in 1659. He perceived twelve stars instead of three, of which seven shone "as if through a cloud," so that the space around them appeared much brighter than the rest of the sky. Numerous observations were made of it prior to the advent of William Herschel's great reflectors.[7]

Surprisingly, only in the eighteenth and nineteenth centuries did astronomers systematically scan the heavens for new objects, including nebulae. At first this was largely a by-product of the attempt to discover new comets; Charles Messier famously compiled his late -eighteenth-century catalog of fuzzy objects because they confused his search for comets. Only a few years later, along with his double star work, nebulae dominated William Herschel's systematic natural history of the heavens. We now believe those objects seen by Messier and Herschel comprise no fewer than seven distinct classes of objects, as defined by their physical composition, including truly gaseous nebulae today classed as planetary nebulae and hot hydrogen clouds; dusty clouds classed as dark nebulae and reflection nebulae; and open and globular star clusters actually composed of stars. (Two more classes of nebulae, cool atomic clouds and molecular clouds, were never seen by Messier or Herschel but were added later.) Last but not least, some of the nebulae that Messier and Herschel did see turned out to be an entirely new realm of objects – congeries of stars known as galaxies – as far removed from our own Galaxy as the sidereal realm was from the planets. Separating all these apparent nebular phenomena into classes was an endeavor that would test the patience and ingenuity of astronomers for two centuries, constituting one of the great stories in the history of astronomy.

3.1 A Shining Fluid: Separating Star Clusters from Nebulae

The most obvious interpretation of the nebulae was that they were star clusters, as Galileo had shown already in 1610 in the case of the Praesepe and Orion clusters, ignoring the Orion Nebula itself for whatever reason. The existence of such clusters had already been hinted at by the naked-eye observation of a few apparent concentrations of stars in the sky, what we today call "open clusters." The observation of the brighter open clusters, in the sense of first being seen though not understood, occurred already in prehistoric times. Foremost among these was the Pleiades, a compact bright group visible in the Northern Hemisphere preceding the spectacular rise of Orion and Sirius. In classical literature Homer mentions the Pleiades in the *Odyssey*, where it is said of Odysseus: "The master mariner steered his craft, sleep never closing his eyes, forever scanning the stars, the Pleiades and the Plowman [the constellation Boötes] late to set and the Great Bear that mankind also calls the Wagon." In the Bible, the Book of Job reads, "Canst thou bind the sweet influences of Pleiades, or loose the bands of Orion?" The Greek poet Hesiod and the Roman poet Ovid also mention the Pleiades.[8] Ancient mariners considered the heliacal rising of the Pleiades a sign that the sailing season had begun.

We would not, however, claim that anyone in antiquity "discovered" open clusters of stars; rather, this is best characterized as a pre-discovery phase. Galileo certainly telescopically detected them, arguably the first step in the discovery of their true nature. Only in the mid-eighteenth century were the stars comprising at least some of the bright clusters even believed to be physically associated, an interpretation that required some understanding of gravitational attraction as proposed by Isaac Newton a century before. The eighteenth-century clergyman John Michell was one to champion this view, and also to argue that all nebulae would be resolved into stars if only observed with a large enough telescope.[9]

But that some eighteenth-century telescopic observers still viewed clusters of stars as potentially distinct from the fuzzier nebulae is apparent in La Caille's *Sur les etoiles nebuleuses du Ciel Austral*, published as a *Memoire* of the Academie Royale des Sciences in 1755. Observing from Cape Town, La Caille catalogued forty-three Southern Hemisphere nebulae and star clusters, and divided them into three classes: nebulae without stars, nebulous stars in clusters, and stars accompanied by nebulosity. While this shows that La Caille believed nebulosity could be distinguished from star clusters, these classes bear no resemblance to

Figure 3.1. William Herschel, discoverer of the "luxuriant gardens" in the realm of the stars. With his self-constructed telescopes Herschel detected more new classes of astronomical objects than anyone since Galileo 170 years earlier. Extant portraits of Herschel are discussed in A. J. Turner, "Portraits of William Herschel," *Vistas in Astronomy* 32 no. 65 (1988): 74. Portrait by J. Russell, courtesy AIP Emilio Segrè Visual Archives, E. Scott Barr Collection.

the modern classification of nebulae. More important, La Caille did not advance the argument that fundamentally different classes of nebulae exist beyond mere appearance.[10] Similarly, the distinction between stars and nebulae is evident in the title of Messier's 1781 *Catalogue des Nébuleuses & des amas d'Étoiles* [Catalogue of Nebulae and Star Clusters]. Moreover, already in that same catalogue open clusters were distinguished by shape, not composition, from what we today call globular clusters. But the distinction of the two objects based on shape again did not mean they were fundamentally different classes of objects, and in any case Messier did not seem to be interested in classifying them except to distinguish them from comets. We know today that 33 of the 103 objects in the Messier catalogue are open star clusters, 29 are globular clusters, and the rest are true gaseous nebulae or galaxies. But with Messier as well as La Caille, the natural history of the heavens had not yet quite begun.[11]

It would be William Herschel (Figure 3.1) who would be credited with resolving what he first called "round nebulae" into stars, distinguishing them as a class of star cluster different from open clusters such as the Pleiades, and lending credence to the idea that perhaps all nebulae could be resolved into stars. But it would also be Herschel who would later show that not all nebulae could be so resolved, thus demonstrating with some probability the existence of a new class of truly nebulous objects. These achievements go far beyond mere curiosity and annotation of appearances in observing books, and represent the beginnings of systematic classification in astronomy, even if Herschel was limited by the techniques of his time to interpreting morphology, sometimes leading to erroneous conclusions. All of this makes the first reconnaissance of the nebulae so interesting for the historian.

Because Herschel is so pivotal to the story of the nebulae, it is important to understand how he came to the problem of the nebulae, and to illuminate the circumstances of his discoveries. Born in Hanover, Germany in 1738, William was the son of a Hanoverian Guard bandmaster, and thus was interested in music from an early age. Already at age eighteen he was posted briefly to England as an oboist in his father's band, and in 1757 he immigrated there, living in London and Leeds and eventually settling in 1766 in one of its small but famous cities, Bath. At first he made a living as a musician, teaching, composing, directing, and playing. But he had also grown interested in astronomy, and began experimenting with telescope construction. This became his true passion, and working together with his sister Caroline, his astronomical career took off. In 1779, he began a series of preliminary surveys of the sky, repeating them year after year at increasingly fainter magnitudes using larger and larger telescopes. As described in the last chapter, it was during one of these sweeps in 1781 that he discovered the planet Uranus, an achievement that brought him so much fame he became astronomer to King George III.[12]

At first Herschel's sweeps were mainly to search for double stars, but in December 1781, William Watson, Jr., a friend in Bath with connections to Astronomer Royal Neville Maskylene and the Royal Society of London, presented him with a copy of Messier's second Catalogue. Seeing a chance to use his superior telescopes to study these objects, in October 1783 Herschel began his twenty years of systematic sweeps for nebulae, using his new 20-foot reflector with an 18.7-inch speculum mirror, first from Datchet, briefly from Old Windsor, and then (after March 1786) from Slough, all near Windsor Castle. Although Herschel never fathomed the true nature of all the nebulae, together with the southern sweeps of the sky by his son John Herschel, this "astronomical collecting" between 1783 and 1860 led to the great catalogues of deep sky objects, and finally to their correct interpretation. Taken together, the rich diversity of nebulae detected during this period brought to a head for their discoverers the problem of what constitutes a new "class" of object. The same diversity raised the question of classification systems.

These were questions William Herschel was constantly asking himself as he made his observations between 1783 and 1818, four years before his death. Herschel regularly reported his observations in the *Philosophical Transactions of the Royal Society*, including five papers "On the Construction of the Heavens," published in 1784, 1785, 1789, 1811, and 1814. Taken together, these works (along with his catalogues on nebulae and star clusters and their introductions) comprise the corpus of Herschel's thoughts on the sidereal realm.[13] Their importance to the story of the nebulae is evident in the fact that when Herschel began his observing programs, only about one hundred nebulous objects were known, mostly as represented in Messier's catalogue and some already resolved into star clusters, while during his career Herschel discovered and cataloged 2500 more.[14] Despite many other achievements during his career, making sense of this variety was Herschel's greatest challenge.

The progression of Herschel's thought on the classification of nebular and stellar phenomena is clear. In his first "Catalogue of One Thousand New Nebulae and Clusters of Stars," published in 1786, Herschel divided these phenomena into eight classes. His motivation for doing so is worth quoting in full:

> In the distribution of the nebulae and clusters into classes, I have partly considered the convenience of other observers: thus, in the first class, the degree of brightness of the nebulae has been the leading feature, as most likely to point out those which their several instruments may give them expectation to reach. The first class, therefore, contains the brightest of them; the second, those that shine but with a feeble light; and in the third are placed all the very faint ones. Besides this general division, I have added a fourth and a fifth class, which contain nebulae that, on different accounts, seemed to deserve a more particular description than I had allotted to the three former divisions.

Thus, Herschel had "the convenience of other observers" in mind as he labeled these classes I ("bright nebulae"), II ("faint nebulae"), III ("very faint nebulae"), IV ("Planetary nebulae"), and V ("very large nebulae"). The remaining three classes were "clusters of stars sorted by their apparent compression": VI ("very compressed and rich clusters of stars"), VII ("pretty much compressed clusters of large or small stars"), and VIII ("coarsely scattered clusters of stars").[15]

These eight classes, distinguished on observational grounds, are not to be confused with the four "forms" Herschel distinguished from a more "theoretical view," one that included the possible action of gravity and represented a step on the way to a more natural classification system. And it is in this context that Herschel distinguished another class of star cluster distinct from the already well-known open star clusters. In his 1785 paper "On the Construction of the Heavens," Herschel had described under the heading "formation of nebulae" that the first of four "Forms" of nebulae, by the action of gravity, "form themselves into a cluster of stars of almost a globular figure."[16] But Herschel was more interested in composition than morphology, or "Forms," in distinguishing real classes, and it was in his "Remarks on the Construction of the Heavens" in 1789 that Herschel really made his case for globulars as spherical clusters of stars distinct from open clusters such as the Pleiades. The phenomenon to be interpreted, as observed through the telescope, Herschel wrote, "is that of a number of lucid spots, of equal lustre, scattered over a circular space, in such a manner as to appear gradually more compressed towards the middle; and which compression, in the clusters to which I allude, is generally carried so far, as, by imperceptible degrees, to end in a luminous center, of a resolvable blaze of light."[17]

To us today, this is an excellent description of what we know as a globular cluster. But Herschel had to argue every step of the way that these phenomena represented a real cluster, not an apparent one. The amateur astronomer John Michell, he pointed out, had computed odds of nearly 500,000 to 1 that the six stars in the Pleiades could simply appear to be clustered. That the much more numerous stars of a globular would appear clustered by accident, Herschel asserted, is "so highly improbable that it ought to be entirely rejected." After a number of other steps Herschel concluded, "And thus, from the above-mentioned appearances, we come to know that there are globular clusters of stars nearly equal in size, which are scattered evenly at equal distances from the middle, but with an encreasing accumulation towards the center."[18] It was the first use of the term "globular cluster"; more importantly it was the first correct interpretation of the physical constitution of what we now call globular clusters. A new class had been established, based not on morphology but on physical nature. The progression, from eight purely observational classes to classes based not only on morphology but also on physical nature, is notable.

In this 1789 paper, Herschel also remarked that the different stages of compression of the nebulae might be an indication of their age, the more dispersed globulars not yet having condensed like the more compact "planetary nebulae" that he had observed since 1782, objects which "may be looked upon as very aged, and drawing on towards a period of change, or dissolution." This led Herschel to his famous biological analogy of the heavens as a luxuriant garden full of a variety of species in different stages of growth. Namely, "This method of viewing the heavens seems to throw them into a new kind of light. They now are seen to resemble a luxuriant garden, which contains the greatest variety of productions, in different flourishing beds; and one advantage we may at least reap from it is, that we can, as it were, extend the range of our experience to an immense duration. For, to continue the simile I have borrowed from the vegetable kingdom, is it not almost the same thing, whether we live successively to witness the germination, blooming, foliage, fecundity, fading, withering, and corruption of a plant, or whether a vast number of specimens, selected from every stage through which the plant passes in the course of its existence, be brought at once to our view?" Herschel was to repeat this biological analogy more than once in his writings, further testimony to a mind rooted in the natural history tradition.[19]

* * *

We know from the work of historians Michael Hoskin, Robert Smith, and Simon Schaffer that distinguishing nebulae as a fundamentally different class from star clusters was a considerably more difficult task than distinguishing open clusters from globular clusters.[20] This is true not just as a matter of "pure observation," a construct that does not exist in the real world, but also from the mental interpretations and inferential gyrations that inevitably go with it, especially when observing something never before seen. One would have thought that Herschel's working hypothesis would be that all nebulae could be resolved into stars, given enough telescope power, for this not only "saved the appearances" but also drew on the already well-established class of stars. Indeed, as Hoskin remarks, this hypothesis saved the appearances too easily, for any unresolved nebula could be explained away as requiring a bigger telescope to achieve resolution, and because Herschel could sometimes convince himself that he saw stars where we know today none could have been seen. For a time Herschel did indeed claim to have resolved almost everything in Messier's catalogue, including M 31, the Andromeda Nebula, which we today know to be a galaxy, but which Herschel could not possibly have resolved into stars even with his great telescopes.[21]

In fact the historical record is more complicated: prior to 1784, Herschel believed in true nebulosity due to apparent changes in the Orion Nebula, which he had observed sporadically beginning in March 1774. Herschel reasoned that anything observed changing in this way over such short time scales could not be a distant cluster of stars. But then, from 1784 to 1790, Herschel seems to have equated all nebulae with star clusters, somehow blocking from his mind the observed changes in Orion. Despite his earlier suspicions of the true nebulosity of the changing Orion Nebula, Herschel was skeptical of the existence of actual nebulosity in the objects that astronomers had suspected as such in the past. "Cloud or nebulous stars have been mentioned by several astronomers; but this name ought not be applied to the objects which they have pointed out as such; for, on examination, they proved to be either mere clusters of stars, plainly to be distinguished with my large instruments, or such nebulous appearances as might be reasonably supposed to be occasioned by a multitude of stars at a vast distance," he wrote in 1791.[22]

This makes all the more remarkable the fact that in the same paper of 1791, Herschel argued that true nebulosity did exist, not in the case of those objects previously claimed, but in the case of one peculiar object. That observation was of what he termed "planetary nebulae," an unfortunate nomenclature deriving from their round planetary appearance through the telescope. Herschel had believed they were star clusters in the late stages of evolution – and therefore not a new class of object but a form of one already known. But in November 1790, he observed what we today regard as a planetary nebula with a visible central star, and he classified both together as a "nebulous star": "A most singular phaenomenon! A star of about the 8th magnitude, with a faint luminous atmosphere, of a circular form, and of about 3' [arcminutes] in diameter. The star is perfectly in the center, and the atmosphere is so diluted, faint, and equal throughout, that there can be no surmise of its consisting of stars; nor can there be a doubt of the evident connection between the atmosphere and the stars." Herschel was forced to conclude he was observing "a star which is involved in a shining fluid, of a nature totally unknown to us." In short, seeing a star surrounded by nebulosity convinced him of the existence of true nebulosity, and he furthermore believed it demonstrated the star was condensing out of the nebula under the force of gravity.[23] Thus did Herschel first declare a new class of *nonstellar* objects among the nebulae, an achievement for which he has been credited through the ages, as when astronomers Simon Newcomb and Edward Holden stated in their astronomy textbook of 1881 that William Herschel made "the first exact statement of the idea that, beside stars and star-clusters, we have in the universe a totally distinct series of objects, probably much more simple in their constitution."[24]

In this connection, Schaffer has made the important point that Herschel's work on nebulae can only be understood in the light of his methodology of natural history, in particular in light of a taxonomy of nebulae of his own making. In this "dramatic break with their traditional conceptions of the practice of astronomy," he argued, Herschel was influenced by the biological natural history tradition of Bath, in particular the Bath Philosophical Society where he presented his first papers. Hoskin has made the further point that Herschel's brother, Dietrich, was an amateur entomologist who introduced William to butterfly collecting, and more broadly to natural history. This methodology and his taxonomy, Schaffer argues, led him to make claims that went beyond his observations. Referring to Herschel's seemingly abrupt changes of mind on the nature of nebulae, Schaffer argues, "The clue to unravel these difficulties lies in the integration of his observational programmes within the natural historical project which defined his astronomy. Herschel surveyed the field of the heavens in order to isolate species from each individual specimen, and connect those species into a series. An isolated observation of a single specimen only had meaning insofar as it had been so integrated. That was the implication of his attitude to natural types. Far from 'a usual veneration for the unvarnished observational facts,' Herschel's attitude was that of the natural historian to the interpretation of his specimens."[25]

In using the concept and terminology of "natural type," Schaffer introduces the structuralist philosophy of the French philosopher Michel Foucault. A description of that work is beyond the scope of the present narrative, but Shaffer's point is well taken: Herschel interpreted his observations within a broader framework, one of his own making when it came to astronomy, but owing much to other natural historians in the realms of biology and geology who preceded him with this methodology. The great number, variety, and form of nebulae, it turned out, gave great scope to the natural history of the heavens, as well as to the idea of natural celestial types analogous to biological species. These natural types had to be arranged in an orderly series based on physical law, and this is exactly what Herschel did, invoking the concept of gravity acting on these types.[26]

The same methodology was employed for all the other nebulous objects. Herschel indeed observed a great variety of shapes, illustrated in his 1811 paper "Astronomical Observations Relating to the Construction of the Heavens" (Figure 3.2). As we have seen, already in his three catalogues of nebulae and star clusters of 1786, 1789, and 1802, Herschel divided these nebulae into eight classes, based on appearance, as follows: bright nebulae, faint nebulae, very faint nebulae, planetary nebulae, very large nebulae, very condensed and rich clusters of stars, compressed clusters of small and large stars, and coarsely scattered clusters of stars. Herschel realized the limits of these apparent classes into which objects were placed as the observations were made, calling them "little more than an arrangement of the objects for the convenience of the observer compared to the disposition of the books in a library, where the different sizes of the volumes is often more considered than their contents."[27] These eight classes constitute what we would today call an artificial classification system.

Herschel's interest in the physical nature and relationships of these classes was unrelenting, and when delineating a more proper set of classes in the introduction to his third catalogue of 1802, he wrote, "Here in dividing the different parts of which the sidereal heavens are composed into proper classes, I shall

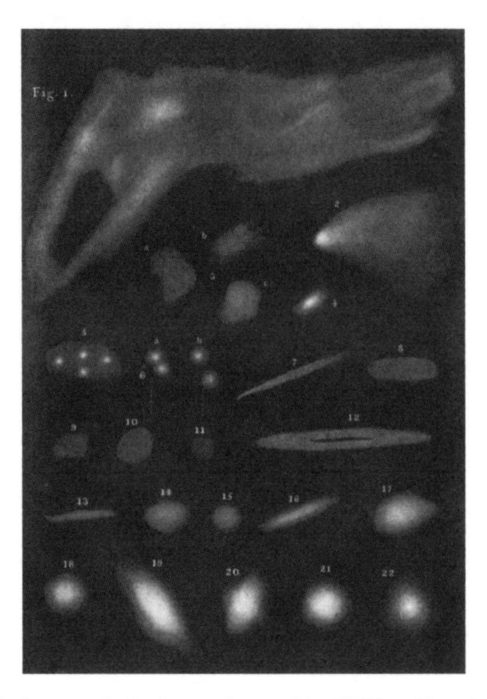

Figure 3.2. Nebulae morphologies as observed by William Herschel, from his 1811 paper "Astronomical Observations Relating to the Construction of the Heavens," *Scientific Papers*, vol. 2 (London, Royal Society, 1912), p. 480. This plate is one of two depicting forty-two morphologies; the other is on p. 496 of the paper. Herschel early on divided nebulae into eight classes based on appearance, but later defined twelve classes based on what he saw as the physical nature of the objects rather than their phenomenology. They bear little resemblance to modern classes, but those labeled 5 and 6 are what Herschel described as multiple nebulae and double nebulae (now multiple and double galaxies), while 18, 21, and 22 are globular clusters of stars.

have to examine the nature of the various celestial objects that have been hitherto discovered, in order to arrange them in a manner most comfortable to their construction."[28] Herschel did so by delineating what he considered a natural progression represented by twelve classes, "proceeding from the simplest to the more complex arrangements": insulated [isolated single] stars, double stars, multiple stars, stars clustering, groups of stars, and star clusters, followed by nebulae not resolved into stars, stellar nebulae, milky nebulosity, nebulous stars, planetary nebulae, and planetary nebulae with centers. Although constituting what we would today call a natural classification system, this does not mean Herschel reached his goal of physically meaningful classes. Indeed, we now know that most of the nebulae in Herschel's massive paper of 1811 were galaxies.[29] Aside from speculation, already indulged in by Thomas Wright of Durham and Immanuel Kant, there was no way Herschel or anyone at his time could prove the true nature of the great variety of nebulae. As we shall see below, this required more tools than Herschel had at his disposal, notably spectroscopy. And as we shall see in Chapter 5, such proof for galaxies would not come until the 1920s with the work of Edwin Hubble, who still referred to them as comprising "the realm of the nebulae."

In terms of nebulae, then, in the end Herschel's achievement (aside from resolving globular clusters into stars and his discovery of some 2500 nebulous objects!) is that he first separated star clusters from true gaseous nebulae, definitively in his mind in the case of planetary nebulae, and by inference in the case of diffuse nebulae such as that in Orion. That was where things stood in 1790, and at Herschel's death in 1822. It was beyond the technology of his time to go any further, in determining either their nature or composition. A quarter century after that, William Parsons, third Earl of Rosse (Figure 5.4) finally outdid Herschel and built a large 6-foot reflector (Figure 9.4) at Parsonstown, Ireland. Rosse believed by end of his life he had resolved the Orion Nebula into stars, misled by the observation of many faint stars in its central core. The implications were profound, for many had seen it as a test of the nebular hypothesis for the origin of solar systems. J. P. Nichol, professor of astronomy at Glasgow University, regretted that "Every shred of that evidence which induced us to accept as a reality, accumulations in the heavens of matter *not stellar*, is for ever and hopelessly destroyed."[30] In this case, however, technology, perhaps combined with wish-fulfillment, had led astronomers astray.

* * *

Lord Rosse's observations of Orion notwithstanding, forty-two years after Herschel's death his claim for a class of truly gaseous nebulae was confirmed in spectacular manner when on August 29, 1864, the British amateur astronomer William Huggins used a spectroscope to observe the bright planetary nebula in Draco today known as the "Cat's Eye."[31] The origins of spectroscopy are beyond the scope of our narrative, but suffice it to say that Huggins made an observation that remains one of the most powerful tools in astronomy today: while a star exhibits a continuous rainbow spectrum from red through violet (crossed with some dark absorption lines as William Wollaston and Joseph Fraunhofer had shown already for the Sun in the early nineteenth century), a cloud of warm gas will emit light only at specific wavelengths characteristic of its composition, giving rise to bright lines at those wavelengths against a black background, a phenomenon that came to be known as an "emission spectrum." Huggins had already shown that the spectra of the Sun and fixed stars were similar in the sense of displaying such a continuous spectrum crossed by dark lines. His problem now was "to ascertain whether this similarity of plan observable among

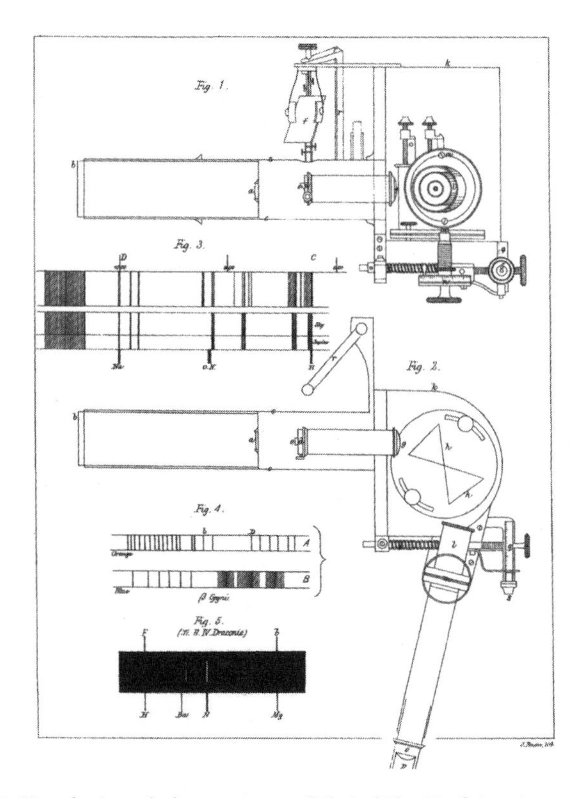

Figure 3.3. Huggins's emission spectrum (labeled Fig. 5) of the planetary nebula in Draco, August 29, 1864. Although eventually seen as the first definitive proof of the existence of gaseous nebulae, even this proof was extended over many years. Today the object is known as the Cat's Eye nebula. Huggins's result could only have been obtained with his new spectroscopic equipment, shown here schematically in his Figures 1 and 2. Figure 3 is a comparison of the solar spectrum with Jupiter, and his Fig. 4 is a comparison of the spectra of double stars Albireo A and B. From Huggins, *PTRSL*, 154, plate X.

the stars, and uniting them with our sun into one great group, extended to the distinct and remarkable class of bodies known as the nebulae."[32]

In order to do so, Huggins turned his telescope to "some of the most enigmatical of these wondrous objects," which Herschel had placed in the class of "planetary nebulae." The spectrum of the object in Draco that Huggins observed with some difficulty on August 29 consisted of three bright lines without the continuous spectrum characteristic of stars (Figure 3.3). The spectra of seven out of eight other planetary nebulae Huggins observed showed the same emission spectrum. Huggins concluded that instead of a solid or liquid body transmitting light through an atmosphere that absorbed certain wavelengths (such as the Sun), "we must probably regard these objects, or at least their photo-surfaces, as enormous masses of luminous gas or vapour."[33] Huggins's use of the word "probably" is important in this early stage of astronomical spectroscopy, for he went on to admit that "it is indeed *possible* that suns endowed with these peculiar conditions of luminosity may exist, and that these bodies are clusters of such suns." Huggins did not think so, but he hedged his bets.

Huggins was more triumphant when he recalled these observations more than three decades later: "The riddle of the nebulae was solved. The answer, which had come to us in the light itself, read: Not an aggregation of stars, but a luminous gas. Stars after the order of our own sun, and of the brighter stars, would give a different spectrum; the light of this nebula had clearly been emitted by a luminous gas."[34] As Barbara Becker has emphasized in her biography of Huggins, this latter-day retelling of his discovery likely shrouds many steps of interpretation, alas, now largely lost to history. In some ways his observations of emission lines raised more questions than they answered at the time. But Herschel would have been amazed that an entirely new technology beyond his imagining would so definitively resolved the riddle of the nebulae - even if not as quickly as Huggins would have us believe.[35]

The same observations that proved the gaseous nature of the planetary nebulae also hinted that the composition of diffuse nebulae was gaseous. In the four years after his first discovery Huggins examined the spectra of about seventy nebulae, including diffuse nebulae such as found in Orion, and found from their emission spectra that one-third of them were gaseous nebulae as opposed to star clusters. Thus, almost a century after Herschel's pioneering work on planetary nebulae, and 250 years after Peiresc or Cysat first detected the Orion Nebula, what we today call diffuse nebulae were definitively separated from star clusters as a distinctive class of objects - as Herschel had suspected early in his career based on apparent observed changes. Huggins's subsequent observations of these nebulae showed emission lines of hydrogen, helium, carbon, nitrogen, and oxygen. The predominance of their hydrogen composition was not yet evident; the strongest emission line in planetary nebulae was later shown to be due to oxygen, but the strongest line does not necessarily translate to the most abundant element. The realization that diffuse nebulae such as Orion were ionized hydrogen regions would take another fifty years, not least because the atomic physics had yet to be worked out. Meanwhile, diffuse nebulae would be further distinguished into many classes of objects as time went on, a story to which we now turn.

3.2 Dust in the Gardens: Dark Nebulae, Reflection Nebulae, and the Composition of Gaseous Nebulae

With the spectroscopic proof of nebulosity in the sidereal realm, one might have thought the story of the nebulae was over. Far from it. But what else could nebulae be if not a system of unresolved stars or a gas, either diffuse

Figure 3.4. Isaac Roberts's image of the Pleiades, taken in 1886 and described in *MNRAS* 47 (1887): 89–91. Slipher's spectroscopic observations in 1912 indicated this was a new class of object: a reflection nebula, later shown to consist mostly of dust rather than gas. Institute of Astronomy Library, Cambridge University.

or surrounding a central star as in the "planetary" nebulae? It would be fifty years before astronomers found out, much to their surprise. Quite aside from the question of the composition of nebular gas, it turns out that the universe had also conspired to create diffuse clouds composed of dust.

The story of how dusty nebulae were first separated from gaseous nebulae begins in the mid-nineteenth century. In 1859, using a four-inch refractor, comet and asteroid hunter Wilhelm Tempel discovered a nebula around Merope, one of the stars near the Pleiades. With the improvement of photography, in the 1880s Paul and Prosper Henry in Paris, and Isaac Roberts in England, first revealed the full extent of the nebula surrounding the entire Pleiades star cluster (Figure 3.4). Using analogy as an argument, many astronomers believed "all the clustered Pleiades constitute, as it were, a second Orion trapezium in the midst of a huge formation of which Tempel's nebula is but a fragment."[36] In other words, it was another gaseous nebula, but so faint that this could not be proven with the spectroscope.

It was only in 1912, when Vesto M. Slipher succeeded in showing that the spectrum of the nebula around Merope was the same as that of the Pleiades star cluster itself, and thus shone by reflected light, that the existence of a dusty class of nebulae was established. In doing so Slipher – familiar to us from

Chapter 1 as the director of the Lowell Observatory and Clyde Tombaugh's boss at the time of the discovery of Pluto in 1930 – made use not only of his state-of-the-art spectrograph attached to a 24-inch refractor but also of another powerful tool of spectroscopy: if light from a hot source passes through a cooler gas, atoms in the cloud absorb light at wavelengths determined by the cloud's composition. As stellar spectroscopists well knew by then, this is exactly what happens in the case of a star, where light from the hot interior passes through the star's cooler outer layers, giving the "absorption spectrum" that Wollaston and Fraunhofer had observed in the solar spectrum early in the nineteenth century, even though they did not yet know the cause. In the case of the nebula surrounding the Pleiades, Slipher found absorption lines of hydrogen and helium superimposed on "a true copy of that of the brighter stars of the Pleiades [containing] no traces of any of the bright lines found in the spectra of gaseous nebulae." In other words, the nebula was reflecting the light of nearby stars, producing a continuous spectrum crossed by the absorption lines of its constituents. Slipher speculated that the nebula could therefore not be gas, but "disintegrated matter similar to what we know in the solar system, in the rings of Saturn, etc.," and which "shines by reflected star light." Four years later he showed the same effect for the nebula known as Rho Ophiuchi.[37] These were what soon came to be called "reflection nebulae," consisting of dust, first separated from the gaseous nebulae shining by their own light, whether diffuse or planetary.

An important question, however, is whether Slipher himself believed this to be a new class of object at the time. He opened his 1912 article with the following words: "The nebula in the Pleiades would doubtless naturally be placed in the class of gaseous nebulae, where it was assigned by Miss Clerke, since in its prominent characteristics it resembles more the Great Nebula in Orion – the typical gaseous nebula – and the filamentous nebulae in Cygnus, NGC 6960 and 6992, than it does the more numerous class of spiral nebulae; but in consequence of its faintness, its spectrum hitherto seems not to have been investigated. In the course of similar observations of other nebulae, I have secured with a spectrograph attached to the 24-inch refractor spectrograms of this nebula which reveal a spectrum of more than common interest."[38] With the observation of only one object, Slipher himself was not quite ready to declare the nebula surrounding the Pleiades a new class in so many words. Rather, he looked for something in the heavens with which to compare it, and concluded the article by wondering if it was the same kind of object as the spiral nebulae such as Andromeda, which "might consist of a central star enveloped and beclouded by fragmentary and disintegrated matter which shines by light supplied by the central sun." In this, of course, he was wildly mistaken, and only when new members of the class were found was new class status conferred. The term "reflection nebula" did not come into common use until the 1930s; looking back from 1937 Otto Struve declared that with Slipher's observations of 1912, "a new type of nebulae had been discovered, which consisted neither of separate stars nor of a luminous gas."[39] But it is clear that both Slipher and Struve were following precisely Herschel's program of seeking new classes of objects as part of the natural history of the heavens – and furthermore attempting to demonstrate not only their morphology but also their physical composition.

Nor was theirs the last word on dusty nebulae. As it turned out, another manifestation of dust in the universe was the dark regions seen already by naked-eye and telescopic observers, notably as great dark bands across the whitish Milky Way. During his sweeps of the sky, however, Herschel recorded more local manifestations of this phenomenon, and his reaction to these phenomena illustrates yet another instance of the difficulties of separating classes. According to his sister and assistant Caroline, coming upon one such spot in Sagittarius (now known as the Ink Spot) she heard him "after a long, awful silence exclaim 'hier ist wahrhaftig ein Loch im Himmel' [here is a hole in the heavens!]. In exhorting John Herschel to observe the object soon after his arrival at the Cape of Good Hope, Caroline records that because of its "uncommon appearance" William returned many times to the object, and believed it might be "something more than a total absence of stars." In his 1785 "On the Construction of the Heavens," however, Herschel noted, "one of the richest and most compressed clusters of small stars I remember to have seen, is situated just on the western border of it, and would almost authorize a suspicion that the stars, of which it is composed, were collected from that place, and had left a vacancy." Herschel also noted a similar phenomenon in another location.[40] Here is yet another twist in the detection phase of discovery, for Herschel did not believe he had detected an object, but a non-object.

The American astronomer E. E. Barnard was the first systematically to gather photographic evidence of such regions in the 1890s and early 1900s, and was the first to catalogue such objects in 1907. Like others before him, as late as 1906 Barnard believed most of these features of the Milky Way could be explained as vacant areas where no stars existed; while some considered them dark masses nearer to us than the Milky Way, "in most of these peculiar features the easiest and apparently the correct solution is that they are real vacancies among the stars," Barnard wrote.[41] But there were some areas that did not quite fit this explanation, and Barnard opined, "these last are quite a different class of objects and should be studied differently." Over the next few years Barnard toyed with the idea of dying nebulae that no longer emitted light, just as there were dying stars that no longer emitted light. As William

Figure 3.5. Object or Non-Object? After believing for many years the standard interpretation that dark areas were "vacancies among the stars," in 1913 Barnard began to interpret the evidence as dark nebulae composed of "masses of obscuring matter." Herschel had already observed such "holes in the heavens" in 1785, and as a case of pre-discovery, dark bands across the whitish Milky Way had been observed from antiquity. Was Herschel or Barnard the discoverer of what we now know as dark nebulae? Shown here is the dark nebula Barnard 86 in Sagittarius, independently found by Barnard in 1883, today known as "the Ink Spot" (B86) in Barnard's catalogue of such areas. From George Ellery Hale, *The Depths of the Universe* (New York: Charles Scribners' Sons, 1924), p. 47.

Sheehan has pointed out, most other astronomers lacked interest in these dark areas, with a few notable exceptions such as J. C. Kapteyn, who suggested in 1909 that "an enormous mass of meteoritic matter" might obscure starlight. Though Kapteyn himself wavered, Barnard came more and more to the view of dark opaque nebula. According to Sheehan, the decisive turning point came in 1913. On a particularly clear and moonless night, Barnard observed "a group of tiny cumulous clouds scattered over the rich star-clouds of Sagittarius. The phenomenon was impressive and full of suggestion. One could not resist the impression that many of the small spots in the Milky Way are due to a cause similar to that of the small black clouds mentioned above – that is, to more or less opaque masses between us and the Milky Way"[42] (Figure 3.5).

Convinced that he had "conclusive evidence that masses of obscuring matter exist in space," Barnard went on to write in his expanded 1919 catalogue that "what the nature of this matter may be is quite another thing." Barnard himself was not a spectroscopist, and so relied on the recent work of Slipher: "Slipher has shown spectroscopically that the great nebula about Rho Ophiuchi is probably not gaseous; that is, it does not have the regular spectrum of a gaseous nebula. The word 'nebula,' nevertheless, remains unchanged by this fact, so that we are free to speak of these objects as nebulae. For our purposes it is immaterial whether they are gaseous or non-gaseous, as we are dealing only with the question of obscuration."[43] In this case, therefore, it might seem that Barnard did not care so much about classification of objects as their effects. Nevertheless, the dusty nature of the dark nebulae was clinched by the work of Henry Norris Russell. In citing the work of Barnard and Russell, by 1924 George Ellery Hale wrote, "we are thus called upon to recognize the existence of a new class of astronomical objects, which cover extensive and widely distributed areas of the heavens."[44]

Just how widespread soon became clear, for in Barnard's grander scheme he too had embraced the natural history of the heavens; by 1927 he had catalogued 349 dark objects. And the trend toward finding more members of the new class continued; by 1962 B. T. Lynds used the Palomar Sky Survey to catalogue more than 1800 dark nebulae, ranging from large objects to smaller ones by then known as Bok globules. In other words, in the elaboration of natural history, types of dark clouds within the classes were now being distinguished.

The realization that some nebulae consisted primarily of dust also helped solve the puzzle of the true constitution of gaseous nebulae like Orion, and illuminated the physics at play producing their emission spectra. This realization came in 1922, when astronomer Edwin Hubble - soon to become better known for his studies of extragalactic nebulae rather than those inside the Galaxy - announced that nebulous regions with emission line spectra were found near hot O and B1 spectral type stars, while reflection nebulae with continuous spectra were found near cooler stars.[45] Hubble credited Henry Norris Russell with advancing the theory that nebulae with emission spectra, both diffuse and planetary, "are excited to luminosity by radiations from involved or neighboring stars." Russell had indeed advanced this theory in the *Proceedings of the National Academy of Sciences* in 1922, stating that "the luminosity of gaseous nebulae is probably due to excitation of the individual atoms by radiations of some sort (ethereal or corpuscular) emanating from neighboring stars of very high temperature. In the Orion nebula the stars of the Trapezium (theta Orionis) appear to be the source of excitation."[46] Russell added that this did not mean that the luminous gas constituted most of the matter of the nebula, but simply that it was the most excited by the stellar radiation. Hubble tested this theory in his landmark 1922 paper and found it to fit the observations. In 1926, Donald

Menzel and Herman Zanstra independently explained the primary mechanism by which emission lines in a gaseous nebula are produced as ionization by absorption of stellar ultraviolet radiation followed by recombination.[47]

Hubble's work revealed another subtlety when he showed not only that emission nebulae occurred around hot O-type stars, while reflection nebulae were found around cooler stars, but also that the B1 spectral type harbored an equal number of emission and reflection nebulae.[48] This demonstrated that reflection nebulae are a mixture of gas and dust: the emission nebula around the hotter stars overwhelms the reflection nebulae that one would otherwise see from scattering in the dust, but in the cooler stars the reflection nebula from the dust component dominates. If there is no nearby star, the dust blocks the distant starlight and the cloud is dark. Gaseous and dusty nebulae may therefore coexist as a kind of hybrid class in which the two embrace each other, a common occurrence that takes place even in the famous Orion Nebula.

As yet, however, the nature of the ionized gas in emission nebulae was unknown; even though spectra showed hydrogen, helium, carbon, nitrogen, and oxygen lines, there was no indication of which gas predominated. The revelation that hydrogen was the key began in 1934, when Eddington argued that in most regions of space hydrogen exists in a non-ionized state, in other words, as electrically neutral hydrogen consisting of a single proton surrounded by a single electron, the simplest element in the universe. Four years later Otto Struve and Christian Elvey observed extended areas of the Galaxy exhibiting so-called Balmer emission lines, a unique fingerprint characteristic of hydrogen, and so argued that those particular clouds, distinct from Eddington's neutral hydrogen, were mainly ionized hydrogen. These clouds exhibiting emission lines came to be known as H II regions, to distinguish them from what came to be called neutral H I regions (though the chemist's notation of simply 'H' makes more sense).[49] This work was immediately followed up by Danish astronomer Bengt Strömgren, who studied H II regions in more detail and published his results in 1939. He originated the concept now called the "Strömgren sphere," the sphere of hydrogen that can be kept ionized around a star of given luminosity. He greatly refined the concept in his classic paper of 1948.[50]

While today the terms "emission nebula" and "H II region' are both used, the latter is more specific and was assured of continued usage after astronomer Stewart Sharpless, having published "A Catalogue of Emission Nebulae" in 1953, published his second catalogue of such nebulae in 1959 under the title "A Catalogue of H II Regions." There he explained that "the term 'H II region' (Strömgen, 1948) is used here instead of the term 'emission nebula.' An H II region is an entity defined not only in terms of the ionized gas but also in terms of the hot stars which are responsible for the ionization."[51] Both H I and H II regions proved important for broader problems in astrophysics, for they helped trace the spiral arms of the Galaxy, first with the H II regions in the optical spectrum, and several years later by radio observations of the cooler H I regions.[52] And dust, both in nebular form and diffused throughout space, would prove important for the properties of interstellar extinction and interstellar reddening, which affect the determination of stellar and galactic distances.

3.3 Two New Classes of Gaseous Nebulae - Cool Hydrogen Clouds and Molecular Clouds

The cool hydrogen (H I) regions already mentioned in connection with the work of Eddington, Struve, and Elvey in the 1930s are commonly seen to constitute yet another class of nebula, for a cool hydrogen region is very different from a hot ionized hydrogen region in its physical properties. In his classic 1948 paper, Strömgren had noted that Struve and Elvey's work published in 1938 had shown that "From the observed strengths of interstellar emission lines, we know that hydrogen is by far the most abundant element in interstellar space." And in that same paper he had further distinguished ionized H II regions from non-ionized H I regions.[53] But the initial evidence for cool gas in interstellar space came long before Eddington, Struve, Elvey, or Strömgren.

The first evidence of interstellar gas not in visible nebular form from emission or reflection came in 1904, when the German spectroscopist Johannes Hartmann deduced "stationary absorption lines" of ionized calcium (Ca II) in the spectrum of the binary star delta Orionis. Hartmann, also an instrumentalist better known today for the "Hartmann test" for telescope optics, was using the 30-inch refractor at the observatory in Potsdam, under the famous spectroscopist Hermann Vogel. Unlike the shifting spectral lines exhibited by the stars due to the Doppler shift caused by their binary orbit motion, the absorption lines of ionized calcium (also known as the K line) did not shift back and forth through the binary orbit. What might seem a subtle effect to some was for Hartmann a startling phenomenon demanding explanation. After eliminating other possibilities, Hartmann wrote, "We are thus led to the assumption that at some point in space in the line of sight between the Sun and delta Orionis there is a cloud which produces that absorption . . . [it is] very probable from the nature of the observed line, that the cloud consists of calcium vapor."[54] V. M. Slipher and Walter S. Adams supported this claim and soon found more cases, but Hartmann's interpretation was not immediately accepted, and the nature of the calcium lines remained a puzzle for many years. Some astronomers thought the gas was located near the star rather than in interstellar space. But in 1924 and 1925, J. S. Plaskett and Struve argued that Hartmann's observations and others showed that vast clouds of calcium, perhaps thrown out by stellar prominences, exist throughout space. In 1926, Eddington showed conclusively from a theoretical standpoint that the lines came from interstellar gas.[55]

The discovery that cool gas clouds were primarily interstellar hydrogen, and that calcium was only a tracer, came slowly. Following the discovery by Cecilia Payne, William McCrea and others in the late 1920s that stars were composed mainly of hydrogen, it was logical to consider hydrogen the most abundant element in the universe. By 1932, Strömgren showed that hydrogen probably dominated stellar cores also. Once it was realized that the space between the stars was not empty, it took no great stretch of the imagination to deduce the existence of interstellar neutral hydrogen in those regions not too close to a star that would ionize the hydrogen - those H II regions we have already discussed. Eddington made the suggestion already in 1934 that "in a normal region of interstellar space the hydrogen will be entirely un-ionized, and indeed in molecular form," but it was the work of Theodore Dunham and Struve in the late 1930s that clearly identified the clouds as primarily neutral hydrogen.

Dunham's work showed further that the number of free electrons in space must be much larger than expected, and Struve deduced that "the most promising source is interstellar H, the existence of which - reasonable on general grounds - has recently been suggested by spectrographic observations at the McDonald Observatory," namely twenty-two regions of the Milky Way that show emission lines of hydrogen. Struve calculated that interstellar gas clouds would have a million times more hydrogen than calcium atoms. Because hydrogen provided free electrons to help maintain the ionization visible as the Ca II absorption lines observed by Hartmann and others, but was not hot enough to produce observable Balmer absorption lines characteristic of H II regions, for decades the gas was believed to be clouds of calcium rather than hydrogen.[56]

Just five years after Struve's publication indicating the abundance of neutral hydrogen in interstellar space, in 1944 the twenty-five-year-old Dutch astronomer Hendrik van de Hulst predicted that neutral hydrogen should produce radiation at a frequency of 1420 MHz, corresponding to a wavelength of 21 centimeters, due to two closely spaced "hyperfine" energy levels in the hydrogen atom. This frequency was in the radio region of the spectrum, and radio astronomy was still very much in its experimental stages. Nevertheless, on March 25, 1951, the American astronomer Harold Ewen and the American physicist Edward Purcell, both at Harvard, detected this radiation from space emanating from the interstellar medium. They used a detector of their own construction, built largely with their own money. (Today their detector is displayed at the National Radio Astronomy Observatory in Green Bank, West Virginia.) Over the next three months two other groups confirmed the 21-cm radiation with their own detections, the Dutch in May and the Australians in June. The American and Dutch results were published in the same September issue of *Nature*, as well as a brief note on the Australian results.[57]

Because neutral hydrogen is found predominantly in the spiral arms of a galaxy, it has played an important part in the history of astronomy: it was observations of the 21-cm line that allowed the arms of the Milky Way to be mapped with radio telescopes in the 1950s. Early observations during the 1950s concentrated on emission from the galactic plane because of its brightness. Since that time, 21-cm emission studies have been a key probe of the structure and dynamics of the Galaxy, and the subject has undergone a renaissance since the large-scale surveys of the last decade. Such surveys now trace a dynamic interstellar medium with structures on all scales, demonstrating the recycling of matter between stars and the interstellar medium.[58]

In 1990, J. M. Dickey and F. J. Lockman used radio observations to show that H I gas was also located in the galactic halo, later estimating that as much as half the mass of the neutral halo may be in the form of hydrogen clouds.[59] The 21-cm technique has since been applied to many other galaxies as well. The neutral hydrogen 21-cm line is also famous because of its association with SETI, the Search for Extraterrestrial Intelligence. Because it is a prominent line emitted by the most abundant element in the universe, the physicists Giuseppe Cocconi and Philip Morrison put it forward in their famous paper in 1959 as a likely frequency (later dubbed "magic frequencies") on which extraterrestrials might communicate.[60]

* * *

The nebulae held yet one more surprise – that contrary to all expectation because of harsh interstellar environments, clouds composed not of neutral or ionized gas but of molecules, could exist in space. Despite the lack of expectations, the possible existence of molecules in space was deduced already in the 1930s when Mt. Wilson astronomers P. W. Merrill, Theodore Dunham, Jr., and Walter S. Adams discovered spectral lines that could not be identified with atomic transitions.[61] In 1937, the Belgian spectroscopist Polydore Swings and Leon Rosenfeld (a younger associate of Niels Bohr) calculated that diatomic compounds such as OH, CH, NH, O_2, CO, and CN should occur in interstellar space, and in 1940 Andrew McKellar, an astronomer working at the Dominion Astrophysical Observatory in Canada, identified three astrophysical spectral lines with CH, CN, and NaH molecular spectra as produced in the laboratory. "If these identifications are proved true," McKellar wrote, "they are of considerable interest and importance in that they constitute the first definite evidence of the existence of molecules in interstellar space. Furthermore, they demonstrate the presence of carbon and nitrogen in interstellar space and provide direct observational basis for the view, held by astronomers for many years, that there must be an abundance of hydrogen in the vast spaces between the stars."[62]

After this discovery, however, almost a quarter century of inactivity passed. As astronomer James Kaler later wrote, "Most astronomers were fairly well convinced that interstellar molecules were not very important - difficult to make in the low densities and temperatures of space, and easily destroyed by high-energy stellar radiation."[63] Nevertheless, theorists Charles Townes and Joseph Shklovskii predicted that some lines of interstellar molecules might appear in the radio spectrum, and in 1963 absorption lines from the hydroxyl radical (OH) were identified in the supernova remnant Cassiopeia A.[64] In 1968, Townes and his colleagues found ammonia (NH_3) emissions toward the center of the Galaxy, followed by water one year later and the first organic molecule, formaldehyde (H_2CO). The discovery of carbon monoxide (CO) emission in 1970 proved especially valuable, because it traces molecular hydrogen, otherwise unobservable at temperatures less than 100 degrees Kelvin.[65] Using CO as a tracer, in 1975 N. Z. Scoville and P. M. Solomon, following the first survey of CO emission in the galactic plane, reported that a large fraction of interstellar hydrogen is in molecular form. Ever more complex molecules like ethyl alcohol were discovered, and by 1985, sixty-eight interstellar molecules had been reported.[66]

Perhaps most astonishing of all was the reported detection in 2003 of interstellar glycine (NH_2CH_2COOH), the simplest amino acid, one of the building blocks of life. The detection was reported in three sources, including the hot molecular cloud in the galactic center known as Sagittarius B2 (Sgr B2), and was based on the observation of twenty-seven lines in nineteen different spectral bands. However, in an indication of the difficulty of identifying such complex molecules, in 2005 Lewis Snyder, one of the pioneers in the field of interstellar molecule detection, concluded that "key lines necessary for an interstellar glycine detection have not yet been found."[67] At this level of complexity the ensemble of lines is modeled, as opposed to single-line detection for less complex molecules. Nevertheless, to date about 160 molecular species have been identified in interstellar molecular clouds with up to thirteen atoms, and the number and type of (generally similar) molecular species in comets is rapidly catching up.[68]

Today we know that most interstellar molecules are not found in isolation, but rather in molecular clouds, which form when interstellar gas clouds become dense enough, perhaps a thousand times the density of an atomic cloud. Hydrogen and more complex molecules are indeed fragile, and this density helps shield them from ultraviolet radiation, which tends to tear them apart in the vicinity of a star, where H II regions are formed. We now know molecular clouds have diameters ranging from less than 1 light-year to about 300 light-years and contain enough gas to form 10 to ten million stars like our Sun. At least two types of this class are distinguished based on size: small molecular clouds less than a few hundred times the mass of the Sun, and giant molecular clouds that exceed the mass of millions of suns.

In 1947, the Dutch-American astronomer Bart Bok first drew attention to the smaller type of cloud now called Bok globules; as we have seen, several dozen of them were already evident in photographic atlases such as E. E. Barnard's catalog of dark nebulae, but Bok singled them out for special attention as possible places of starbirth. In 1950, South African astronomer A. D. Thackeray found such globules in the open cluster IC 2944, also known as the Running Chicken Nebula, and they have been an object of study as possible star formation sites ever since.[69] Little is known about their nature, but they are only a few thousand astronomical units across and often associated with H II regions. Because star formation is taking place in them and they are rich in molecules, they are usually classified as a type of molecular cloud rather than as a dark nebula. While the smaller Bok globules may give birth to double or multiple stars, giant molecular clouds over their long lifetimes may form stars by the hundreds, thousands, or millions. The Hubble Space Telescope revealed such a stellar nursery in the three gaseous pillars of the Eagle Nebula.

We know today that molecular clouds are dominated by molecular hydrogen, followed by carbon monoxide, which is much easier to observe. Using CO as a tracer, in 1977 Solomon, D. B. Sanders, and Scoville estimated about 3000 Giant Molecular Clouds in the Galaxy, with dimensions of 10–80 parsecs (30–250 light-years).[70] The most abundant and famous molecular clouds in our Galaxy are the Orion Molecular Cloud and the Sagittarius B2 Cloud near the center of our Galaxy. Like H I and H II regions, molecular clouds are found primarily in spiral arms, but molecular hydrogen is concentrated much more to the center of the galaxy. Molecular clouds have also been found in other galaxies.[71] A typical spiral galaxy contains about 1000 to 2000 Giant Molecular Clouds and many more smaller ones. The study of molecules in interstellar clouds, as well as in comets and other astrophysical environments, has given rise to the field of molecular astrophysics, sometimes called astrochemistry.

* * *

Only by a circuitous path, then, did we arrive at our knowledge that the interstellar medium consists of about 99 percent gas and 1 percent dust, and that the gas component consists of three main classes of objects: cool atomic clouds composed mainly of neutral hydrogen (H I); hot ionized clouds also composed mainly of hydrogen, known as H II regions and observed as emission nebulae;

and cold molecular clouds composed largely of hydrogen (H_2) with a sprinkling of other molecules. The molecular clouds form perhaps 25 percent of the interstellar medium, and at about 20 degrees K constitute its coldest component. Planetary nebula constitute a fourth class of gaseous nebulae, circumstellar and spectacular, but relatively minor in terms of total nebular mass. The dusty nebulae known as reflection nebulae and dark nebulae, depending on whether they reflect or obscure, might be considered one class because they likely have similar composition. But astronomers persist in separating them into two classes, both for physical reasons and because the separation is apparently useful for their analysis.

Despite the preponderance of hydrogen in all gaseous nebulae, we should not lose sight of the fact that about 25 percent of interstellar gas in general is composed of helium. Nor should we forget that interstellar gas is extremely rarefied, with a density of about one atom per cubic centimeter. Moreover, about 95 percent of interstellar hydrogen is H I, since the transition to ionized H II requires a nearby star. As Strömgren argued with his concept of Strömgren spheres, H I and H II regions are therefore often found adjacent to each other near stars, and the extent of the neutral hydrogen converting to ionized hydrogen depends on the luminosity and temperature of the star.

The discovery of so many distinct classes of nebulae should not obscure the fact that they are often associated with each other, depending on the circumstances of their environment. Thus, diffuse nebulae – the term commonly used in modern astronomy to refer to any nebula with irregular outlines (as opposed to planetary nebula) – are commonly mixtures of gas and dust clouds that become visible when they appear as emission, reflection, or absorption (dark) nebulae, depending on their environment. This is certainly true of the Orion Nebula and the associated Orion Molecular Cloud out of which stars are forming. A molecular cloud also harbors interstellar dust on which the gas molecules are believed to form. All these associated nebulae have in common the characteristic that they are stellar or pre-stellar nebulae, as opposed to post-stellar nebulae like supernovae remnants and planetary nebulae.

It is also notable, if not surprising in retrospect, that the nebulae were discovered by a variety of techniques, depending on their nature. Visual observations first detected what turned out to be H II regions and planetary nebulae, though only spectroscopy confirmed their true nebular nature. Because a "cold" H I region has a temperature of around 100 K, compared to 10,000 K for an H II region, it is detected not by optical emission lines as in H II regions, but by radio observations of the so-called 21-cm emission lines, or 21-cm absorption lines if a hotter object is in the background. And, because molecular spectral lines are also in the microwave region of the spectrum, radio astronomy is the chief technique for discovering molecules in space. In the Space Age, nebulae are often observed at multiple wavelengths, each of which reveals something about the nature of the objects.

Finally, nebulae have turned out to have an importance that eluded even Herschel. Both H I and H II regions, for example, proved important because they trace the spiral arms of the Galaxy, and provide a window on numerous other phenomena associated with the interstellar medium. Molecular clouds are the perfect site for star formation, because gravitational attraction within the cloud overcomes the small outward pressure due to low temperature, providing stellar nurseries that have now been imaged in detail. Planetary nebulae represent a step on the way to the evolutionary endpoint of a low-mass star like the Sun when the dying star, having expanded to a red giant, ejects its outer envelope. The star itself eventually becomes a white dwarf, while the expanding ejected envelope is what we see as a planetary nebula, some 500 to 1000 times the size of the solar system at the orbit of Pluto. In short, today nebulae have taken their place in the story of cosmic evolution, an essential part of cosmic ecology in the birth, life, and death of stars.

All of this was hard-won knowledge in the realm of the stars. The separation of the stars themselves into classes such as "red giant" and "white dwarf" occurred by a very different route than the more direct observation of the nebulae that constitute the interstellar medium. We now turn to that story, reserving for Chapter 5 the incredible saga of the separation of some of the "nebulae" into the entirely new realm of the galaxies.

4

Dwarfs, Giants, and Planets (Again!): The Discovery of the Stars Themselves

> It is hardly exaggerated to say that the spectral classification now adopted is of similar value as a botany, which divide the flowers according to their size and color. To neglect the c-properties in classifying stellar spectra is nearly the same thing as if the zoologist, who has detected the deciding differences between a whale and a fish, would continue in classifying them together.
>
> Ejnar Hertzsprung, 1908[1]

> There seem, therefore to be two series of stars, one very bright and of almost the same brightness, whatever the spectrum, the other diminishing rapidly in brightness with increasing redness ... These series were first noticed by Dr. Hertzsprung, of Potsdam, and called by him "giant" and "dwarf" stars.
>
> Henry Norris Russell, 1913[2]

> The message of the Companion of Sirius, when decoded, ran: "I am composed of material 3,000 times denser than anything you have come across; a ton of my material would be a little nugget you could put in a matchbox." What reply can one make to such a message? The reply which most of us made in 1914 was – "Shut up. Don't talk nonsense."
>
> Arthur S. Eddington, 1927[3]

As historian David DeVorkin has written, by the early twentieth century, American astrophysicists were becoming world leaders in astronomical natural history thanks to the unparalleled power of astrophysics to reveal the nature of celestial bodies. "Akin to the naturalist, the typical American professional astronomer was collector and classifier. Instead of museum shelves and cases, astronomers stored their systematic observations in plate vaults and letterpress log books, and displayed them in catalogues sponsored by universities and observatories."[4] In particular, as more and more stellar spectra were gathered, they hinted at numerous variations in the nature of the stars. Once stellar physics was understood later in the twentieth century the reasons became clear: stellar structure depended on mass, temperature, and luminosity, and the range of all of these physical variables was enormous. Moreover, stars existed in a variety of different ages, and (it turned out) in a variety of stages. Stars were born, lived, and died; once the concept of stellar evolution was accepted, the problem was determining which stars were in which stages and how the physics worked under varying conditions, among the greatest puzzles in the history of science. In short, William Herschel's gardens, first explored with his work on the nebulae, were luxuriant beyond his wildest dreams when it came to the stars themselves.

Over the last half of the nineteenth century and the first half of the twentieth century this astrophysical collecting led to the discovery of new classes of stars. At first these were arranged according to various features in their spectral lines, but then in a more physically meaningful way according to their size and luminosity, resulting in the first stages of discovery of dwarfs and giants between 1905 and 1913, then supergiants (1917), subgiants (1930), and subdwarfs (1939). Moreover, over many decades the analysis of spectra led to the interpretation and understanding of what turned out to be the evolutionary states of these classes of stars, as well as additional classes of evolutionary endpoints or penultimate endpoints: supernovae, white dwarfs, and neutron stars. Black holes, arguably not objects but "singularities," were such bizarre endpoints of massive star evolution that for many decades after they were theoretically proposed, most astronomers refused to believe they existed. The observational confirmation of *stellar* black holes was a long time coming; because such objects exist in their most massive form at the centers of galaxies, we discuss them in the next chapter on discovery in the realm of the galaxies.

Meanwhile, each of the classes of stars has its own unique story of discovery and classification. While necessarily broaching stellar classification systems here, we reserve full discussion of that subject for Chapter 8, concentrating here on the process of the discovery of stellar classes. We conclude the chapter by analyzing quite a different phenomenon, the discovery of stellar systems.

4.1 Giants and Dwarfs: Making Sense of the Natural History of the Stars

It is obvious to anyone peering at the night sky that some stars *appear* different than others, some being brighter and some fainter. Each has its fascination, whether spectacular Sirius following in Orion's wake, or the fainter stars of the Pleiades cluster, sometimes used as a test of eyesight, observed from antiquity preceding Orion's rising. But clearly these brightness differences might be due only to differences in distance, which were largely unknown until the first measurements of stellar parallax in 1838, and mostly unknown for the remainder of the century. The key to determining whether real differences existed among the stars was the determination of these distances, but even more, the rise of astronomical spectroscopy, truly a landmark in the history of astronomy.

While William Wollaston and Joseph Fraunhofer had first detected the dark lines of the continuous solar spectrum in the early nineteenth century, it was only with the work of G. R. Kirchhoff and R. W. E. Bunsen in Heidelberg that the meaning of the lines even began to be understood. In November 1859 Bunsen wrote to a colleague, "At the moment I am occupied by an investigation with Kirchhoff which does not allow us to sleep. Kirchhoff has made a totally unexpected discovery, inasmuch as he has found out the cause for the dark lines in the solar spectrum and can produce these lines artificially intensified both in the solar spectrum and in the continuous spectrum of a flame, their position being identical with that of Fraunhofer's lines. Hence the path is opened for the determination of the chemical composition of the Sun and the fixed stars with the same certainty that we can detect chloride of strontium, etc., by our ordinary reagents."[5]

Over the next two years Kirchhoff discovered the basic principles of spectral analysis: when passed through a prism, light from a hot opaque solid, liquid, or compressed gas will produce a continuous rainbow spectrum with no lines; a hot transparent gas will produce bright lines characteristic of the gas; a cooler gas placed in front of the hot opaque object will absorb light at exactly the same wavelengths as it emits if observed only as a hot gas. The Fraunhofer lines in the solar spectrum constituted the third case, as the cooler, less dense atmosphere of the Sun itself absorbed those wavelengths characteristic of its gases, extracting those wavelengths from the light produced deeper inside the Sun. And as we saw in the last chapter, only a few years after the work of Kirchhoff and Bunsen, William Huggins proved that some nebulae were truly gaseous because they produced bright "emission" lines – the second case.

If the Sun could produce an absorption spectrum, so could the other stars. A stellar spectrum could be observed for any star that was bright enough; such spectra were "collected" at an increasing pace as telescopes improved and spectroscopic and photographic techniques advanced in the late nineteenth and early twentieth centuries, and starlight from fainter and fainter stars could be analyzed. That history has been written in detail elsewhere; as we shall see in more detail in Chapter 8, numerous competing classification systems were constructed for stellar spectra, but it was the stellar spectra classification system at Harvard that triumphed. There, beginning in the mid-1880s, Harvard College Observatory director E. C. Pickering supervised a program to mass produce stellar spectra using wide-field astrographs equipped with a prism placed in front of the objective lens, a method that yielded hundreds of spectra at once, by contrast with a spectrograph on the eyepiece end. The result was thousands of photographic plates, each with hundreds of spectra. Their analysis was undertaken by a corps of women under the direction of Pickering, led in its early years by Williamina P. Fleming, a teacher who had worked briefly as Pickering's housekeeper.[6] They classified each spectrum according to an alphabetical scheme ranging from A through Q, relying principally on the so-called Balmer lines of hydrogen and the H and K lines of calcium, class A exhibiting the strongest hydrogen lines. In 1890, they published their first catalogue of spectra for more than 10,000 stars, double the number produced before that time.

Meanwhile, Antonia C. Maury, another of Pickering's staff, would play a very different role. The granddaughter of John W. Draper (the first person to photograph the Moon in 1840) and the niece of Henry Draper (who obtained the first photograph of the Orion Nebula in 1880 and whose widow funded the Draper spectroscopy program at Harvard), she was a student of Maria Mitchell at Vassar, graduating in 1897 and thereby gaining some measure of independence. She had found, through a more detailed study of a group of high-quality spectra, that brighter blue stars showed some interesting anomalies in terms of their spectral line widths: the hydrogen lines were narrower and the metallic lines stronger than in the much more frequent "normal" blue stars. These abnormal stars she referred to as spectral subdivision "c" in 1897. While Pickering did not pay much attention to this finding, it would prove crucial within the decade for the subject of physically meaningful stellar classes.[7]

On another front, yet another Pickering staffer, Annie J. Cannon, who had attended Wellesley College from 1880 to 1884 and was virtually deaf for most of her life, consolidated Fleming and Pickering's original A through Q classification system, arranging it in a continuous sequence based on color. This yielded the system we now know as spectral types O, B, A, F, G, K, M. During a long career at Harvard, Cannon classified the spectra of almost 400,000 stars.[8] Her rearranged sequence was also based on evolutionary considerations having to do with the connection of the O and B stars with nebulae. Although suspected, as yet there was no proof that the color sequence was actually a temperature sequence – that awaited the work of Henry Norris Russell and especially the physical explanations of the Indian physicist Meghnad Saha, who in a series of papers in 1920 and 1921 devised an equation to calculate the degree of ionization of an element in a gas at a given temperature and density in terms of the atomic structure of that element. By 1924, the Henry Draper Catalogue emanating from Harvard College Observatory had classified 225,000 stars to 8.5 magnitude, much fainter than the 6th magnitude limit of the naked eye under the best circumstances. It provided the best database of the natural history of the stars.[9] But the classification of spectra was only at the phenomenological level, hinting at something physically meaningful underlying the spectral phenomena.

In order to determine the true nature of a star one needed to know not only its spectrum, but also its absolute magnitude, from which its true brightness, or "luminosity," could be determined. This required a knowledge of stellar distances, which had to be estimated through stellar proper motions (on the assumption that smaller proper motions implied greater distance) or determined through trigonometric parallaxes, which existed for only a few hundred stars at the turn of the twentieth century.

The story of the discovery of two classes of stars – we would now say the story of the separation of dwarfs from giants – is largely the story of two men: the Danish chemist and astronomer Ejnar Hertzsprung and the American astronomer Henry Norris Russell (Figures 4.1a and 4.1b). To be sure they relied heavily on spectroscopic data from Harvard and distance data obtained from others, in addition to their own work. And as usual, there were hints of what was to come before their own work. For example, based purely on proper motions the Irish amateur astronomer William Monck proposed a luminosity effect in a paper published in 1895, where he stated "I suspect . . . that two distinct classes of stars are at present ranked as Capellan, one being dull and near us, and the other bright and remote like the Sirians." Monck gave Alpha Centauri and Procyon as examples of the first class, and Canopus as the second class; we now know the first is indeed a dwarf and the latter a giant (he was wrong about Procyon, now classed as a "subgiant"). Monck's paper had no effect; more than a century later astronomer Allan Sandage conjectured this was because it had only one intelligible paragraph! The Dutch astronomer J. C. Kapteyn found a similar relationship. There were also conjectures in the popular astronomy literature about "h" red stars like Arcturus.[10] But these were based on sparse data; what was needed was a more systematic treatment, above all based on empirical data.

Enter Ejnar Hertzsprung. Building on Antonia Maury's observation that stars of the same spectral type could have different spectral line widths, in 1905 and 1907 he wrote two papers in an obscure photographic journal that in effect revealed the existence of giant and dwarf stars of the same spectral temperature.[11] It turned out that Maury's narrow-lined stars (designated "c")

Figure 4.1A and 4.1B. Henry Norris Russell (left) and Ejnar Hertzsprung, co-discoverers of giant and dwarf stars, both seen here in later life. Their discovery during their younger days, first known as the "Russell diagram" (Figure 4.2) and later as the "Hertzsprung-Russell diagram" (Figure 4.3), took place by inference rather than detection. The Russell photo shows him measuring spectra at Mount Wilson, circa late 1920s. Photograph by Margaret Harwood (who is herself seen in Figure 8.3). Credits: (Left) University of Chicago Yerkes Observatory, courtesy AIP Emilio Segrè Visual Archives, Tenn Collection. (Right) Dr. Dorrit Hoffleit, Yale University, courtesy AIP Emilio Segrè Visual Archives, Tenn Collection.

had smaller proper motions and thus were more distant, suggesting that they were much more luminous than the nearer ones. Maury's distinction had heretofore been ignored, but Hertzsprung wrote Edward C. Pickering, Maury's boss at Harvard, like Herschel using a biological analogy: "It is hardly exaggerated to say that the spectral classification now adopted is of similar value as a botany, which divide the flowers according to their size and color. To neglect the c-properties in classifying stellar spectra is nearly the same thing as if the zoologist, who has detected the deciding differences between a whale and a fish, would continue in classifying them together."[12] Hertzsprung's papers contained tabulated data, but (crucially as it turned out for their acceptance) no diagrams until 1910 and 1911. Pickering remained unconvinced of the importance of this work, believing that the c-properties found by Maury and used by Hertzsprung to guide his work were spurious (and protecting the Draper classification). The German astronomer Karl Schwarzschild, a correspondent and mentor of Hertzsprung, called attention to Hertzsprung's "Giganten" stars, but to little avail at first.[13]

By 1910 Henry Norris Russell was independently convinced that giant stars existed, based on direct and inferred parallaxes. Russell learned of Hertzsprung's work from Karl Schwarzschild later in the year, and was in correspondence with Hertzsprung himself by the end of that year.[14] The conviction that giant stars really existed was strengthened in 1913 when Russell and Shapley made direct distance determinations of giants belonging to eclipsing binaries, indicating radii 300 times the Sun's and 20 solar masses, and thus densities a million times less than the Sun's.[15] In June 1913, Russell addressed the Royal Astronomical Society in London and laid out the current evidence for two classes of stars, the giants and dwarfs. Plotting a graph of absolute magnitude versus spectral type for all stars with "fairly reliable parallaxes" – which he had been doing in his lecture notes since 1907 – Russell pointed out the peculiar results: the lower left corner representing "faint white stars" was vacant "except for one star whose spectrum is very doubtful." All the faint stars, located to the lower right of the diagram, were very red, at the K and M spectral types. On the other hand, in the upper part of the diagram were bright red stars such as Arcturus, Aldebaran, and Antares.

In short, Russell saw a pattern in his diagram: "There seem, therefore to be two series of stars, one very bright and of almost the same brightness, whatever the spectrum, the other diminishing rapidly in brightness with increasing redness." Russell was generous in acknowledging Hertzsprung: "These series were first noticed by Dr. Hertzsprung, of Potsdam, and called by him 'giant' and 'dwarf' stars. All I have done in this diagram is to use more extensive observational material."[16] Russell concluded his lecture with the hope that he would get his results into print the following year, and indeed, following a more detailed lecture on the subject at the Atlanta meeting of the American Astronomical Society in December, the first published diagram appeared in both *Nature* and *Popular Astronomy* in 1914 (Figure 4.2).[17] The diagram, first known as the Russell Diagram but after 1933 (at the urging of the Danish astronomer Bengt Strömgren) as the Hertzsprung-Russell diagram, became a fundamental tool in stellar astronomy.[18] Not only did it first graphically depict the differences between giants and dwarfs, as more data came in it would reveal other stellar classes and plot their evolutionary relationships.

Russell, however, was already interested in the evolutionary meaning of his two series: indeed, there had been speculation of an evolutionary sequence of the Harvard spectral types, and Russell introduced his RAS talk as "some studies bearing upon stellar evolution, beginning with the relation between the spectral types of the stars and their real brightnesses." Russell's interest therefore went beyond the "collecting" aspects of natural history to physical meaning and evolution; for him, collecting spectra was only a means toward an end. Russell's idea of stellar evolution, however – in which the "main sequence" of stars was a cooling branch in which even the giant stars eventually cooled – did not survive the mid-1920s. In his landmark book, *The Internal Constitution of the Stars* (1926), the irascible Arthur Eddington spoke of "the Giant and Dwarf

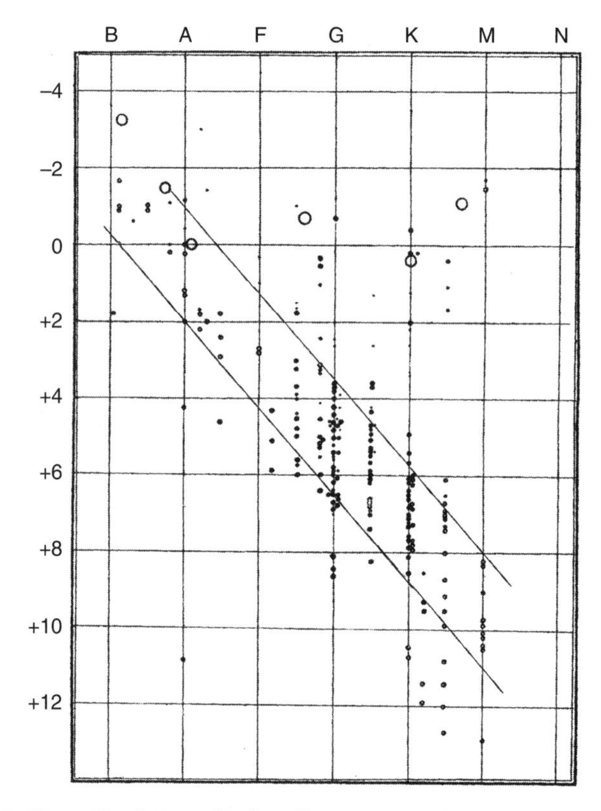

Figure 4.2. Henry Norris Russell's first diagram separating the classes of dwarf and giant stars. Harvard spectral types are shown at the top, and absolute magnitudes along the left side. Russell believed the absolute magnitudes represented luminosities 7500 times that of the Sun at the top, and 1/5000th of the Sun's luminosity at the bottom. Today we know the "Main Sequence" between the two diagonal lines represents dwarf stars, while the much rarer stars at upper right represent giant stars. Note the single star at bottom left: still believed to be an anomaly at the time this was published, it is now known to be the harbinger of a new class of stars – white dwarfs. The plot, today known as a "Hertzsprung-Russell Diagram," was constructed in the spring of 1913, and published simultaneously in *Nature* and *Popular Astronomy* in 1914.

Theory of E. Hertzsprung and H. N. Russell," and while admitting it had gained widespread acceptance, found "it is difficult to accept the giant and dwarf theory in its entirety."[19] He was referring to the evolutionary implications; indeed, a true knowledge of stellar evolution required a more robust understanding of stellar physics, and that would not come until the late 1930s.

Further work on determining the nature of stars, in other words, true stellar luminosities, depended on the determination of more stellar distances. This enterprise received an enormous boost when Walter S. Adams and Arnold

Kohlschütter at the Mt. Wilson Observatory found a method for determining stellar distances using spectral lines, first published in 1914. Unaware of the work of Monck, Hertzsprung, or Russell, they concluded that "Certain lines are strong in the spectra of small proper motion stars. The use of the relative intensities of these lines results for absolute magnitudes in satisfactory agreement with those derived from parallaxes and proper motions."[20] Their work using this method, now known as spectroscopic parallax, increased the number of known stellar distances a hundred-fold.

It did not take astronomers long to realize that not all giants were created equal. By 1917, William H. Pickering (Edward's brother) noted that a certain star in Orion must be a "giant among giants," and that "perhaps we may properly describe these stars as consisting of many giants and a few supergiants."[21] Thus, over time was a new, if extremely rare, class of stars delineated. It would not be the last new class. Adams and Kohlschütter's monumental volume *Spectroscopic Absolute Magnitudes and Distances of 4179 Stars*, published in 1935, not only gave clear delineations of the giant and dwarf classes, but also indications of a new class called subgiants. The discovery of these subgiants – as well as yet another class dubbed "subdwarfs," each proved important for unraveling the true nature of stellar evolution.[22]

4.2 Subgiants and Subdwarfs: The Road to Evolutionary Understanding

As we have seen, since the beginning of star classification stellar evolution had been an important, even driving factor. But these early ideas of stellar evolution, represented by Russell and many other astronomers, mainly had to do with gravitational contraction and simple thermodynamic ideas about stars heating and cooling as ideal gases. This model indicated the giant stars were an early stage of stellar evolution, leading to dwarf stars like our Sun. These simple ideas were upended in 1924 after Eddington established a mass-luminosity relation for stars, and it would take another thirty years before astronomers accepted that red giants represent just the opposite: old, highly evolved stars.[23]

Part of this unraveling of the true nature of the giant stars was the discovery – or perhaps more accurately the designation – of the new classes of subgiants and subdwarfs. The separation of subgiants as a class distinct from giants and main sequence stars was difficult and accomplished only several decades after the Hertzsprung-Russell diagram was first devised, when the absolute magnitudes of stars were more accurately determined. Although there were inklings of a separate subgiant sequence as early as 1917, Gustaf Stromberg first used the term "subgiants" in 1930.[24] The catalog of the entire Mt. Wilson program of spectroscopic parallaxes published in 1935 included an HR diagram that clearly showed a separate sequence between the main sequence and giant stars, but it was populated by only ninety stars out of the 4179 plotted. The authors cautiously wrote that "the existence of a group of stars of types G and K somewhat fainter than normal giants has been indicated by the statistical studies of Stromberg . . . Although these stars may not be entirely separated from the giants in absolute magnitude, there is some spectroscopic evidence to support the suggestion."[25] The existence of subgiants was strengthened in 1936 when reliable trigonometric parallaxes (and thus distances and luminosities) of some subgiants became available.

W. W. Morgan and his colleagues at Yerkes Observatory also saw these differences independently. With the further development of spectroscopic techniques, luminosity effects could be detected ever more accurately by the width and intensity of stellar spectra lines, giving rise to a more formal two-dimensional temperature and luminosity system, known as the MKK (Morgan-Keenan-Kellman) system. This system incorporated the Harvard spectral classification, now known to be a temperature sequence, and included not only dwarfs, giants, and supergiants, but also subgiants and a new class known as bright giants. The first edition of what became the canonical MKK catalog of luminosity spectral types, published in 1943, included five stars as the defining examples of subgiants: Beta Aquilae, Eta Cephei, Gamma Cephei, delta Eridani, and Mu Herculis.[26]

The place of subgiants in stellar evolution was not resolved until the early 1950s when astronomers, including Allan Sandage, used the old stars found in the globular clusters M3 and M92 and the younger stars found in the open cluster M67 to infer the relation between subgiants and main sequence stars.[27] Walter Baade had found in the 1940s that these two types of star clusters were the homes of two different "populations" of stars – older "Population II" stars in globulars, and young "Population I" stars in open clusters. We now know this is true because the higher mass stars in globulars evolve faster and burn out, while open clusters are much more recently formed and therefore contain young stars. Sandage later characterized the discovery of the relation of subgiants to main sequence stars as "a serendipitous discovery" that "arose from the solution of the independent problem posed in 1948 by [Walter] Baade in his outline of stellar population programs for the 200-inch Palomar telescope." In particular, by making observations to fainter magnitudes in globular clusters where the star distances were approximately the same and the luminosities therefore well known, astronomers found that "the main sequence was attached to a stubby subgiant sequence, which in turn merged continuously into the giant branch . . . It seemed clear that the subgiant and giant-branch stars had once been main-sequence stars; they had evolved those two branches as they aged." Sandage characterized this discovery as a landmark in the study of stellar evolution: "This moment germinated, I believe, the first ideas of how the various sequences in the HR diagram were connected 'continuously.' It was then, and remains now, the ineffable moment in observational astronomy for me."[28]

A theoretical understanding of what was going on was suspected by Strömgren already in the 1930s. Advances in stellar physics showed that stellar energy could be explained by nuclear fusion reactions that converted hydrogen to helium. In 1938, physicists Hans Bethe and Carl von Weizsäcker independently showed that upper main sequence stars accomplished this conversion through a process involving carbon, nitrogen, and oxygen, the CNO cycle. In the same year, following up on a suggestion by German physicist Carl von Weizsäcker, physicists Hans Bethe and Charles Critchfield showed the proton-proton chain, which fuses four protons into a helium nucleus, explained energy production in the lower main sequence stars like the Sun.[29] Interestingly, the discovery of these two separate fusion processes did not result in the formal declaration of two new classes, one upper main sequence and one lower main sequence, since both had hydrogen-burning cores producing helium. Moreover, although the CNO and proton-proton process dominated the upper or lower part of the sequence, respectively, there was some of each process going on in most stars, especially in what are today called "intermediate main sequence" stars located between the upper and lower main sequence. And in terms of MKK classification, the spectral lines generated at the star's surface were paramount, not internal fusion processes.

The important point for the subgiant story is that theory indicated that once hydrogen burning ceased, no stable configuration existed for a star. Martin Schwarzschild soon showed that when a star with a core about 10 percent of its total mass converts all its hydrogen to helium, the core contracts and releases gravitational energy so that hydrogen begins burning in a shell around the core. The star expands in radius and leaves the main sequence. The existence of subgiants was thus explained, and stars were shown to expand as they age rather than contract, as Russell and most other astronomers had thought. "By mid-1952," Sandage wrote, "it was clear that the evolution of a star off the main sequence and into the subgiant sequence is secure, both from the Mount Wilson/Palomar observational side and now from the theoretical perspective as well."[30] A similar result in 1955 from the open cluster M67 explained the subgiant sequence in population I stars. Subgiants thus played a major role uncovering the process of stellar evolution as we understand it today.[31] The modern H-R diagram is thus a key to stellar evolution (Figure 4.3).

* * *

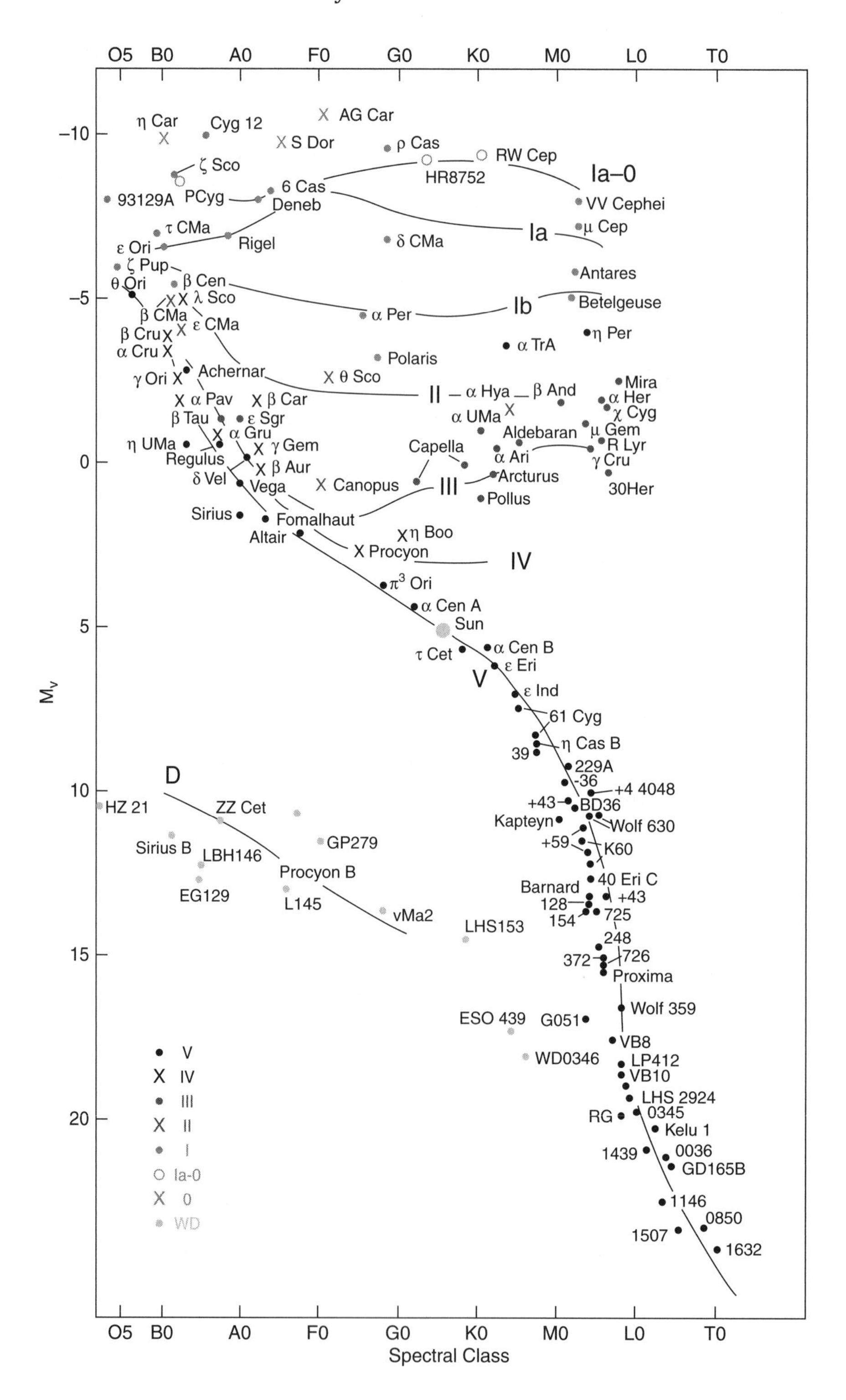

O5 B0 A0 F0 G0 K0 M0 L0 T0
η Car
Cyg 12
AG Car
S Dor
ρ Cas
RW Cep
HR8752
Ia–0
ζ Sco
93129A
PCyg
6 Cas
Deneb
VV Cephei
τ CMa
μ Cep
ε Ori
Rigel
δ CMa
Ia
ζ Pup
Antares
θ Ori
β Cen
λ Sco
Ib
Betelgeuse
β CMa
α Per
ε CMa
η Per
β Cru
α TrA
α Cru
Polaris
γ Ori
Achernar
θ Sco
Mira
II
α Hya
β And
α Her
α Pav
β Car
χ Cyg
β Tau
ε Sgr
α UMa
μ Gem
α Gru
Aldebaran
η UMa
γ Gem
Capella
R Lyr
Regulus
α Ari
γ Cru
β Aur
δ Vel
Arcturus
Vega
Canopus
III
30Her
Pollus
Sirius
Fomalhaut
Altair
η Boo
Procyon
IV
π3 Ori
α Cen A
Sun
α Cen B
τ Cet
ε Eri
V
ε Ind
61 Cyg
η Cas B
39
229A
D
-36
+4 4048
+43
BD36
HZ 21
ZZ Cet
Kapteyn
Wolf 630
GP279
+59
K60
Sirius B
LBH146
Procyon B
40 Eri C
Barnard
EG129
128
+43
L145
154
725
vMa2
LHS153
248
372
726
Proxima
Wolf 359
ESO 439
G051
VB8
WD0346
LP412
VB10
LHS 2924
RG
0345
Kelu 1
1439
0036
GD165B
1146
0850
1507
1632
V
IV
III
II
I
Ia-0
0
WD
−10
−5
0
5
10
15
20
M_V
Spectral Class

The first hint of yet another class of stars came in 1913, when astronomer Walter S. Adams reported that two high-velocity stars, known as Lalande 5761 and 28607, had peculiar A-type spectra with weak hydrogen lines. "It is a singular coincidence that two stars, both of peculiar but similar spectra, should be characterized by such velocities," he wrote, estimating those velocities as receding at 144 and 170 kilometers per second, very close to the current measured value.[32] By 1934, Adams and Alfred Joy had found three more similar stars. Because their luminosities were fainter than the main sequence, but brighter than the white dwarfs, Adams and his team called them "intermediate white dwarfs." Because of their high velocities, these dwarfs were most often detected in proper motion surveys.

In 1939, Gerard Kuiper coined the term "subdwarf" to designate this group of high proper motion stars, which has nothing to do with white dwarfs that we know today. Kuiper used the first month of operation of the 82-inch telescope at the MacDonald Observatory in Texas to determine accurate spectral types for large proper motion stars. In delineating a new class of objects, he commented that these stars "extend almost along the whole main sequence. Since these stars merge into the main sequence and are much more similar to main-sequence stars than to white dwarfs (probably also in the interior), the name 'subdwarfs' is suggested for this class of stars, in analogy with 'subgiants.' This name will prevent the confusion of these stars with the white dwarfs proper which are very much fainter."[33] Due to sparse data about their true nature, subdwarfs were not designated luminosity class VI in the original MKK/Yerkes system in 1943. Astronomer Nancy Roman, who contributed to the MKK system, first designated them luminosity class VI in 1955. In the early 1960s, Keenan and Morgan suggested to the International Astronomical Union that class VI be

---

Figure 4.3. The modern Hertzsprung-Russell diagram, plotting absolute visual magnitude and Harvard spectral types for the brightest and closest stars, supplemented with others from James Kaler's *Extreme Stars*. As more data became available over the decades after the original HR diagram, in addition to the giants and main sequence dwarfs seen in Figure 4.2, subgiants, supergiants, and subdwarfs were also separated. The MKK system defined them as luminosity classes I (supergiants) through V (dwarfs). In 1971, Keenan argued a class 0 "hypergiants" should be added, and they are increasingly used as a distinct class (here labeled 'X' at the very top left), though without official sanction. The white dwarfs are a separate sequence labeled "D." The cool red L and T spectral types at right were added in 1999 with further discoveries, some in the infrared. Russell's original diagram (Figure 4.2) covered only luminosities from –4 to +12, the central part of this diagram. With permission from James B. Kaler, *Encyclopedia of the Stars* (Cambridge: Cambridge University Press, 2006), figure 6.12.

adopted, but Keenan also demurred by suggesting the matter be "left open for discussion." But the designation never caught on and was not adopted by the IAU. Subdwarfs today are more often designated "sd."[34]

Today subdwarfs, or more accurately "cool subdwarfs," are considered to be an odd type of star that run about one magnitude below the main sequence in the Hertzsprung-Russell diagram. They are usually classed as main sequence stars, since they are hydrogen burning, although near the end of this phase of energy generation. It is now known that the relative faintness of subdwarfs, amounting to one or two magnitudes less than normal main sequence stars, is due to their composition. In particular it is due to their low "metallicity," meaning they have a lower content of what astronomers refer to as "metals," elements other than hydrogen and helium. This causes them to be hotter and bluer than most main sequence stars of the same luminosity. This effect for subdwarfs was first demonstrated in a famous and controversial paper in 1951 by Lawrence Aller and Joseph W. Chamberlain. They showed that the two high-velocity stars first detected by Adams in 1913 were actually F spectral type, not A spectral type. This in turn had major implications for the stars' composition, indicating that their iron abundances, for example, were only 1/100th that of the Sun, what would today be defined as a metallicity of -2 (2 orders of magnitude less than the Sun). "This was the beginning – the breaking down of this prejudice that all stars had the same composition," Aller noted.[35]

Aller and Chamberlain's paper did indeed open the way to numerous studies of metal-poor stars. We now believe that metallicity is in general a difference between the two stellar populations that Walter Baade detected in the 1940s, and which (as we have seen) also played a role in delineating the subgiants via the very different HR diagrams of old globular and the relatively younger open clusters, including the intermediate age cluster M67, which showed both populations. In this case, subdwarfs are found not only to be old, but also to have high velocities and highly elliptical orbits because they are not confined to the galactic disk, instead orbiting in the bulge and halo of the Galaxy. Their lower metallicity indicates they are earlier generation stars that have not been greatly enriched by heavier elements. Subdwarfs thus played an important role in our current understanding that stars in different parts of the Galaxy have different ages, that the Galaxy is becoming more metal-rich with age as heavier elements are generated inside stars, and that we are, in fact, starstuff.

Because subdwarfs were never officially incorporated into the canonical MKK system (or its successor, the MK system), they explicitly highlight the questions "What is a class?" "Who decides if there should be a new class?" and "How do new classes become accepted?" For that matter, the creation of a class of 'bright giants" in the MKK system, intermediate between giants and supergiants, raises the same questions, as does the creation of classes in other realms. We reserve such analysis for Chapter 8.

4.3 White Dwarfs and Neutron Stars: Evolutionary Endpoints

The existence of white dwarfs first became known as an observational anomaly. Henry Norris Russell has given us the story of the first recognition of white dwarfs in 1910 involving himself, E. C. Pickering, and Williamina Fleming at Harvard College Observatory. Russell wrote:

> The first person who knew of the existence of white dwarfs was Mrs. Fleming; the next two, an hour or two later, Professor E. C. Pickering and I. With characteristic generosity, Pickering had volunteered to have the spectra of the stars which I had observed for parallax looked up on the Harvard plates. All those of faint absolute magnitude turned out to be class G or later. Moved with curiosity I asked him about the companion of 40 Eridani. Characteristically, again, he telephoned to Mrs. Fleming who reported within an hour or so, that it was of Class A. I saw enough of the physical implications of this to be puzzled, and expressed some concern. Pickering smiled and said, "It is just such discrepancies which lead to the increase of our knowledge." Never was the soundness of his judgment better illustrated.[36]

The companion to the star 40 Eridani (first seen, in another example of a "pre-discovery" observation, by William Herschel in 1783 and now known as 40 Eridani B) did not follow the pattern of other known stars - it was a faint, relatively hot star in contrast to the many faint cool stars known. It appeared well to the bottom left of the main sequence stars on what is now known as the Hertzsprung-Russell diagram - the object Russell had referred to in his 1913 address at the RAS as representing "faint white stars," an area of the diagram vacant "except for one star whose spectrum is very doubtful (Figure 4.2)." Russell classified it as a dwarf star, and Hertzsprung called it a "dark white star."

The second object that turned out to be a white dwarf was Sirius B. In 1844, Wilhelm Friedrich Bessel suggested that the perturbed motion of the bright star Sirius was caused by an unseen companion. In 1862, telescope maker Alvan G. Clark first observed this companion to Sirius A while testing one of his telescopes; it was ten magnitudes fainter than its primary star, a factor of 10,000 in luminosity. It was so faint that some astronomers suggested it was actually a large planet.[37] In 1914, Walter S. Adams obtained the first spectrum at Mt. Wilson Observatory, showing that it was spectral class A, similar to 40 Eridani B.[38]

The third of the classical white dwarfs is van Maanen 2, located in the constellation Pisces, discovered in 1917 by Dutch astronomer Adriaan van Maanen, working at Mt. Wilson on high proper motion stars. Now known to be at a distance of only fourteen light-years, it was the first white dwarf discovered that is not part of a multiple star system.[39] The fact that all three white dwarfs were relatively close to the Sun hinted that they were common objects throughout the Galaxy.

Astronomer Willem J. Luyten coined the term "white dwarf" in 1922 as part of his work at Lick Observatory studying the high proper motion of nearby stars. In a series of papers that year, Luyten first spoke of "faint white stars" of high proper motion, and then "white dwarfs," extending the contemporary terminology of "yellow dwarfs" and "red dwarfs." The term was popularized by the theoretician Arthur S. Eddington in 1924.[40] By 1950, more than one hundred were known; fifty years later, more than 2000, and now more than 10,000 white dwarfs have been catalogued, thanks both to large surveys and especially the Sloan Digital Sky Survey. Because they are the evolutionary endpoint of low-mass stars, many more undoubtedly exist but are too faint to be seen.

The true nature of white dwarfs emerged only slowly. Eddington captured their incredible nature when he wrote, "The message of the Companion of Sirius, when decoded, ran: 'I am composed of material 3,000 times denser than anything you have come across; a ton of my material would be a little nugget you could put in a matchbox.' What reply can one make to such a message? The reply which most of us made in 1914 was – "Shut up. Don't talk nonsense.'"[41]

How to explain this in terms of physical theory? The astronomer J. B. Holberg has detailed the long road to the discovery of white dwarfs, including the interplay between theory and observation, and the understanding of their remarkable nature.[42] In 1926, the British mathematician Ralph Fowler applied the new quantum mechanics to white dwarfs and concluded that its electrons and nuclei were packed into the smallest volume possible, resisted only by the degeneracy pressure produced by electrons, as required by Fermi-Dirac statistics.[43] Five years later, reasoning that white dwarf electrons would be moving near the speed of light, Chandrasekhar applied relativistic degeneracy to show that the compaction of white dwarfs could not continue forever. He calculated the upper mass of a white dwarf, now known as the Chandrasekhar limit, to be slightly less than one solar mass; it is now known to be 1.4 solar masses – beyond that it would collapse.[44]

Not everyone agreed with Chandrasekhar's conclusion, least of all the great British theorist Arthur Eddington. In his book *Empire of the Stars*, Arthur Miller details the heated controversy between Eddington and Chandrasekhar regarding white dwarfs and the fate of more massive stars: Chandra embraced relativistic degeneracy as the basis of his work, while Eddington believed it was a mathematical result with no astrophysical meaning. As Matthew Stanley has shown, for Eddington the pursuit of physical understanding in astrophysics was primary, while Chandra put his faith in rigid deduction. In the end it was Chandra who first glimpsed the truth about white dwarfs and more massive stars as endpoints of stellar evolution when he wrote in 1935, "for a star of small mass the natural white dwarf stage is an initial step towards complete extinction. A star of large mass [greater than the upper limit for white dwarfs] cannot pass into the white dwarf stage, and one is left speculating on other possibilities."[45] For Eddington this seemed a physical absurdity; he did not yet grasp how strange the universe could be.

* * *

Unlike white dwarfs, neutron stars did not first appear as an observational anomaly, but as a theoretical possibility. Three decades before their discovery, they were theorized by Walter Baade and Fritz Zwicky in 1934 as objects formed from supernovae.[46] Jocelyn Bell (later Burnell) and Anthony Hewish first discovered pulsars, which turned out to be the theorized neutron stars, in 1967 at the Mullard Radio Astronomy Observatory in Cambridge, England. It was a serendipitous discovery in the sense that Hewish was actually looking for quasars using the technique of interplanetary scintillation.[47] This twinkling effect occurs as a result of diffraction of radio waves when they pass through the solar wind in interplanetary space, and it is most pronounced for radio emission from compact objects such as quasars. In search of quasars, which had only been discovered a few years before, in the summer of 1967 Hewish and his students set up an array of 2048 dipole radio antennae covering 4.5 acres near Cambridge, the size of fifty-seven tennis courts. The array was so large because scintillation due to plasmas is most pronounced at long wavelengths; Hewish used a wavelength of 3.7 meters, and commenced the survey in July.

Graduate student Jocelyn Bell had sole responsibility for operating the telescope and analyzing the data, with Hewish supervising. The data were recorded on long strips of paper that were visually inspected. "Six or eight weeks after starting the survey I became aware that on occasions there was a bit of 'scruff' on the records, which did not look exactly like a scintillating source, and yet did not look exactly like man-made interference either," Bell wrote. "Furthermore I realized that this scruff had been seen before on the same part of the records – from the same patch of sky (right ascension 1919)." Bell showed the charts to Hewish, and by the end of September, Hewish suspected they had located a flare star similar to the M-dwarfs under investigation by Bernard Lovell at Jodrell Bank. By November more observations had been made, and Bell recalled, "As the chart flowed under the pen I could

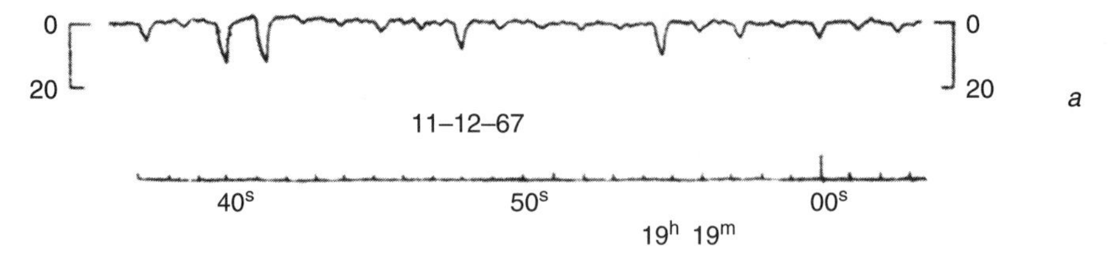

Figure 4.4. Evidence for a pulsating radio source, a class of objects now known as pulsars. This record of when the source was unusually strong was obtained on December 11, 1967, and was published in the discovery paper. It covers only about 30 seconds of data, and the vertical scale is 10^{-26} watts/meter^2/hertz. On such small effects great discoveries are made. The object was soon theorized to be the previously predicted "neutron star." Reprinted by permission from Macmillan Publishers Ltd: Anthony Hewish, S. Jocelyn Bell, John D. H. Pilkington et al. "Observation of a Rapidly Pulsating Radio Source," *Nature* 217 (February 24, 1968): 709–13.

see that the signal was a series of pulses, and my suspicion that they were equally spaced was confirmed as soon as I got the chart off the recorder" (Figure 4.4).[48]

The next step was the interpretation of the data. Hewish at first thought such regular pulses must be man-made. Radar reflected from the Moon, satellites in peculiar orbits, and local effects were eliminated when another telescope confirmed the results, and it was established that the source was outside the solar system, but inside the Galaxy. The "Little Green Men" hypothesis was raised. As Hewish recalled, "the short duration of each pulse suggested that the radiator could not be larger than a small planet. We had to face the possibility that the signals were, indeed, generated on a planet circling some distant star, and that they were artificial. I knew that timing measurements, if continued for a few weeks, would reveal any orbital motion of the source as a Doppler shift, and I felt compelled to maintain a curtain of silence until this result was known with some certainty. Without doubt, those weeks in December 1967 were the most exciting in my life."[49] The hypothesis was rejected when Doppler shifts in the signal showed only the orbital motion of the Earth, not of a planet with extraterrestrials. The discovery of similar signals coming from Cassiopeia A also mitigated this possibility, since two civilizations would not likely be signaling at the same frequency. In February 1968, the data were reported in *Nature*, where the authors speculated that the signals could be caused by radial pulsations of white dwarfs or neutron stars.[50]

Theorist Thomas Gold at Cornell quickly developed the model that explained pulsars as the rotating neutron stars that had been predicted by Baade and

Zwicky; other papers in *Nature* immediately following Gold's discussed vibrating white dwarfs or a satellite orbiting a neutron star as the explanation. Gold had predicted the pulse period should increase with time, and as Hewish put it, this prediction "soon received dramatic confirmation with the discovery of the pulsar in the Crab Nebula. Further impressive support for the neutron star hypothesis was the detection of pulsed light from the star, which had previously been identified as the remnant of the original explosion. This, according to theories of stellar evolution, is precisely where a young neutron star should be created. Gold also showed that the loss of rotational energy, calculated from the increase of period for a neutron star model, was exactly that needed to power the observed synchrotron light from the nebula." By 1969, astronomers at the Steward Observatory in Tucson, Arizona made the first optical identification of a pulsar – the central star of the Crab Nebula.[51] Within a year after the first discovery, twenty-seven more radio pulsars were found.

4.4 Brown Dwarfs and Exoplanets

Whereas astronomers over time had declared the existence of two planetary classes in our solar system (terrestrial planets and gas giants) with a third (ice giants) bidding fair to join the family, they had not yet demonstrated the existence of more than one member in the class of "planetary systems." Aside from one pulsar planetary system announced in 1992, a system so bizarre as to defy membership in the class, the solar system was, as far as anyone could tell, a unique phenomenon. With the resurrection of the nebular hypothesis, by the second half of the century most astronomers did not really believe this, however, holding rather that relatively small planets could simply not be detected around their much brighter and more massive parent stars, at least with the technology of the time. The belief that there were more members of the class, closely tied to the ever-popular question of extraterrestrial life, became one of astronomy's Holy Grails.

Two premature claims gave astronomers hope. As early as 1943, independent observational claims were made for the existence of two planetary systems around nearby stars. Peter van de Kamp, a student of Hertzsprung who had come to the United States in 1923 and fifteen years later ended up at Sproul Observatory at Swarthmore College, was undertaking an astrometric program of measuring parallaxes of stars. This required biannual observations of each star, and lent itself to the detection of low-mass companions by looking for any perturbations in the proper motions of stars. In 1943, Kaj Strand, another student of Hertzsprung working under van de Kamp, announced he had discovered perturbations indicating a planetary companion to the star 61 Cygni, famous as one of the first stars to have its parallax measured in 1838. Using photographic observations from the Potsdam, Lick, and Sproul observatories covering the years 1914–1918 and 1935–1942, Strand announced in no uncertain terms, "The only solution which will satisfy the observed motions gives the remarkably small mass of 1/60 that of the sun or 16 times that of Jupiter. With a mass considerably smaller than the smallest known stellar mass (Kruger 60B = 0.14 [solar masses]), the dark companion must have an intrinsic luminosity so extremely low that we may consider it a planet rather than a star. Thus planetary motion has been found outside the solar system."[52] Almost simultaneously, Dirk Reuyl and Erik Holmberg, based primarily on observations made at the Leander McCormick Observatory in Charlottesville, Virginia, announced they had discovered a planetary companion around the star known as 70 Ophiuchi. Though they spoke of only a "third body" and not a "planet," the deduced mass for the third body was between 0.008 and 0.012 solar masses (compared to 0.016 for Strand's claimed planet). That this was in the planetary mass range escaped no one.[53]

The reaction to these discoveries was considerable. Immediately on publication of the results, Henry Norris Russell sat down and wrote an excited account that appeared in the June issue of *Scientific American*. The following month, in an article entitled "Anthropocentrism's Demise," Russell put the results in a broader context for the same magazine, arguing that the new evidence required a reversal of the previous opinion that planetary systems were rare. And during the same period, he wrote another article examining from a theoretical viewpoint the physical characteristics of stellar companions of small mass, in which he concluded regarding a body such as claimed by the new observations that "it is well within the bounds of accepted usage to call the new body a planet."[54]

The increased interest in the subject of low-mass companions generated by these discoveries is evident in a modest symposium on "Dwarf Stars and Planet-Like Companions" held in late 1943 under the auspices of the American Astronomical Society. Six participants, including Strand, Russell, and van de Kamp, presented papers. Representing the two photographic astrometric approaches to the subject, van de Kamp reviewed work on unseen companions of single stars and Strand did the same for unseen companions of double stars, while Russell elaborated on his previous theoretical discussion of the physical characteristics of low-mass companions. An important offshoot of the latter was the question of the difference between stars and planets; in other words, when did a low-mass star become a planet candidate? Russell concluded that an object less than 1/20th the size of our Sun would have a surface temperature of about 700 K and would be invisible even under the best circumstances. Van de Kamp the following year adopted Russell's value of 1/20 (0.05) of the Sun's mass "as a conventional borderline between visible stars and the *per se* invisible bodies which we shall designate by the term *planet*. It is the *amount of mass*, therefore, which determines whether a body should be classified as a 'star' or as a 'planet'; size is of secondary significance."[55] But confirmation and further progress in the search for planetary systems would be slow and difficult.

Twenty years later the subject had not advanced much except for the occasional discovery of lower and lower mass stars, and van de Kamp's (spurious) claim in 1963 of a planet around Barnard's star. Again the question was raised at what point a star's mass would be so low that it would not sustain nuclear fusion. In 1963, astronomer Shiv Kumar, at the University of Virginia's McCormick Observatory with its astrometric tradition on this subject, described stars that might exist with less than 0.08 solar mass, and termed them "black dwarfs."[56] But because "black dwarf" was already being used by some astronomers to describe hypothetical cooled white dwarfs, astronomer Jill Tarter coined the term brown dwarf in her 1975 dissertation, to denote substellar objects that could not sustain nuclear fusion. Though others proposed names like "planetar" and "substar," the term "brown dwarf" stuck. "It was obvious that we needed a color to describe these dwarfs that was between red and black. I proposed brown and Joe [Silk] objected that brown was not a color," in terms of the primary colors of the spectrum, Tarter recalled.[57]

Because of the difficulty of the observations, the discovery of actual brown dwarfs was a long time coming. In 1988, Eric Becklin and Ben Zuckerman discovered a low-mass star known as GD 165 B, cooler than the well-known M dwarfs, the first of what turned out to be a class of low-mass stars now known as L dwarfs, still undergoing hydrogen fusion. Their technique involved imaging the extremely dim object, and their ambiguity regarding the nature of the object was evident: "We have discovered an infrared object located about 120 AU from the white dwarf GD165," they wrote. "With the exception of the possible brown dwarf companion to Giclas 29–38 which we reported last year, the companion to GD165 is the coolest (2,100 K) dwarf star ever reported and, according to some theoretical models, it should be a sub-stellar brown dwarf with a mass between 0.06 and 0.08 solar masses. These results, together with newly discovered low-mass stellar companions to white dwarfs, change the investigation of very low-mass stars from the study of a few chance objects to that of a statistical distribution. In particular, it appears that very low-mass stars and perhaps even brown dwarfs could be quite common in our Galaxy."[58] Several hundred such objects, now known as "L dwarfs," have subsequently been found by a variety of methods.

In 1989, Harvard astronomer David Latham and his colleagues reported a possible brown dwarf using a very different technique, the change in the line-of-sight, or "radial velocity," of a star with an unseen companion. The size of the

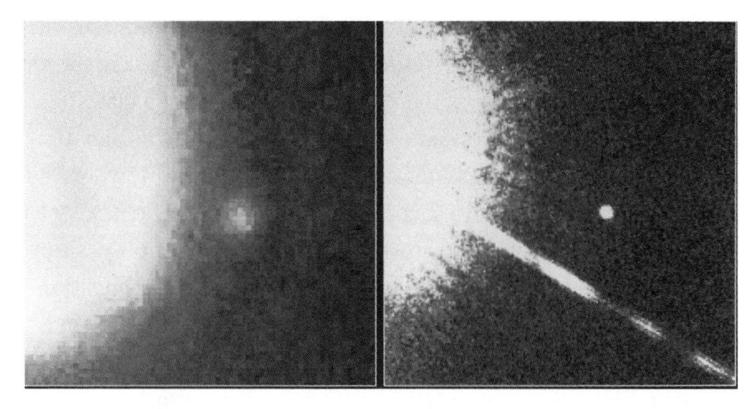

Figure 4.5a and 4.5b. Evidence for brown dwarfs. The first unambiguous photographic discovery of a brown dwarf (Gliese 229B), observed (left) on October 27, 1994 with adaptive optics using the 60-inch reflector at Palomar, and confirmed (right) with the Hubble Space Telescope on November 17, 1995. The image on the left is labeled "discovery image" in the NASA press release dated twelve days later. A spectrum obtained with the 200-inch Hale telescope at Palomar also provided essential information: an abundance of methane, found in giant planets, but not ordinary stars. Left, courtesy T. Nakajima (Caltech), S. Durrance (Johns Hopkins). Right, courtesy S. Kulkarni (Caltech), D. Golimowski (Johns Hopkins) and NASA.

tug indicated a companion to the star HD 114762, which they calculated could have a mass as small as .001 of the Sun, some eleven Jupiter masses. "Thus the unseen companion of HD 114762 is a good candidate to be a brown dwarf or even a giant planet," they concluded, allowing that there was less than a 1 percent chance that this companion could be massive enough to burn hydrogen stably. The uncertainty was due to the unknown orbital inclination of the object with respect to its star as viewed from Earth, the "M sin (i) factor," where *M* is the mass and *i* is the inclination. Because of this factor, Latham and his colleagues cautioned that the object was most likely not an extrasolar planet, but a brown dwarf.[59] Even now, great uncertainty surrounds the nature of this object. It may be a brown dwarf or an L dwarf star, but some astronomers still believe it may be the first extrasolar giant planet discovered. In any case, neither Becklin and Zuckerman's object nor Latham's were unambiguously the long-sought substellar brown dwarfs. Like extrasolar planets, brown dwarfs were still in danger of remaining hypothetical objects as the 1990s began.

On October 27, 1994, using an adaptive optics coronagraph on the 60-inch telescope at Mt. Palomar, Caltech astronomer T. Nakajima and his colleagues imaged an even lower temperature object, Gliese 229B, orbiting the bright nearby M dwarf Gliese 229 (Figure 4.5a). "Here we report the discovery of a probable companion to the nearby star Gl 229, with no more than one tenth the luminosity of the least luminous hydrogen-burning star," they wrote. "We conclude that the companion, Gl 229B, is a brown dwarf with a temperature of less than 1,200 K, and a mass 20–50 times that of Jupiter." S. Durrance and D. Golimowski confirmed the Gliese 229B discovery with a now-famous Hubble Space Telescope image on November 17, 1995 (Figure 4.5b).[60] This turned out to be the first brown dwarf discovered and imaged, the prototype of the class of objects now known as "T dwarfs." The rapid acceptance of this object as the first unambiguous brown dwarf was due to the presence of methane in its atmosphere, found during follow-up observations with the Palomar 200-inch telescope.

In the meantime, astronomers in the Canary Islands led by Rafael Rebolo discovered an object known as Teide 1 in the Pleiades, a likely host for brown dwarfs because of its young age. Their claim was unambiguous: "Here we report the discovery of a brown dwarf near the centre of the Pleiades. The luminosity and temperature of this object are so low that its mass must be less than 0.08 solar masses, the accepted lower limit on the mass of a true star." Teide 1 is sometimes also referred to as the first verified brown dwarf.[61] Rebolo pioneered in the lithium test for brown dwarfs; just as the presence of methane is an indication of brown dwarf status because methane cannot survive in a star undergoing fusion, so stars also rapidly deplete lithium. The presence of lithium is therefore also a test for brown dwarfs, though older and more massive brown dwarfs may have burned their lithium.

We now believe that brown dwarfs are objects intermediate in mass between planets and stars, too large to have formed as planets, too small to sustain hydrogen fusion. Although they have been called "a poor excuse for a star," they are embraced by stellar astronomers and have even found a place in the standard stellar classification system. They range in mass from thirteen to eighty times the mass of Jupiter, about 8 percent of a solar mass, but most are about the size of Jupiter. They are completely boiling, convective objects. Brown dwarfs are so difficult to detect because of their very low luminosity, which during the first hundred million years or so derives from gravitational contraction, after which they become even fainter. Their temperatures of 1000 degrees Kelvin and less dictates that they radiate primarily in the infrared region of the spectrum, and are especially amenable to detection by infrared telescopes. Brown dwarfs can undergo deuterium and lithium fusion during their first 10 million years.

Brown dwarfs appear at the far bottom right of the H-R diagram, the latest extension of the MK system beyond the M and L dwarfs to what are now called "T dwarfs." (In a demonstration of the difficulties of classification, there is, however, some crossing over of brown dwarfs into the L and even M dwarfs.) In the discovery process it is often difficult to distinguish high-mass brown T dwarfs from low-mass L stars, and low-mass brown dwarfs from large planets, though spectral differences are becoming better known with time.[62] Even though they are substellar, the placement of T dwarfs on the H-R diagram is sometimes justified because brown dwarfs are believed to have formed through a starlike nebular condensation, rather than through a planetary accretion-type process. In addition to orbiting single stars, they have been found as part of binary systems and as free-floating objects. Some two dozen brown dwarfs were known by 2002, and since then several hundred more have been verified.[63]

* * *

The discovery of brown dwarfs meant that astronomers were hot on the heels of the real holy grail: planetary systems. The detections were at first indirect, no surprise given the small masses involved. In 1983, the first circumstantial evidence was found: circumstellar disks, now believed to be the remnants of post-planetary formation. In 1992, protoplantary "proplyds" were imaged with the Hubble Space Telescope. The claim of actual planets around solar-type stars was more ambiguous at first, and again intertwined with the search for brown dwarfs. Radial velocity techniques, which relied on an object gravitationally tugging its parent star in the line of sight and resulting in measurable periodic variations in the radial velocity of the star, were being refined at this time to yield greater accuracy. It was using this technique that Harvard astronomer David Latham and his colleagues reported a companion to the star HD 114762 in 1989.

Also applying the radial velocity technique, in 1995 the Swiss team of Michel Mayor and Didier Queloz unambiguously detected the first extrasolar planet around a Sun-like star 51 Pegasi (Figure 4.6). In this case "M sin (i)" was only half a Jupiter mass, twenty times less than Latham's object, and so the inclination factor could not substantially affect the planetary status claim. Their observation was confirmed, and soon supplemented, by the American team consisting of Geoff Marcy and Paul Butler.[64] They became the two primary planet-hunting teams over the following two decades, though they were joined by many others as more discoveries were made. Over the next fifteen years, more than 500 planets were found circling other stars, many of them in systems with more than one planet.[65] The radial velocity technique was supplemented by the transit technique, whereby starlight is observed to dim as it passes across the face of its parent star as viewed from Earth. This is the technique used by the Kepler spacecraft, which as of 2012 had detected more than 1200 candidate planets, most of which were expected to be confirmed.

Many of the 500 planets turned out to be "hot Jupiters," gas giants orbiting very close to their host star. Others were in highly eccentric orbits. A very few

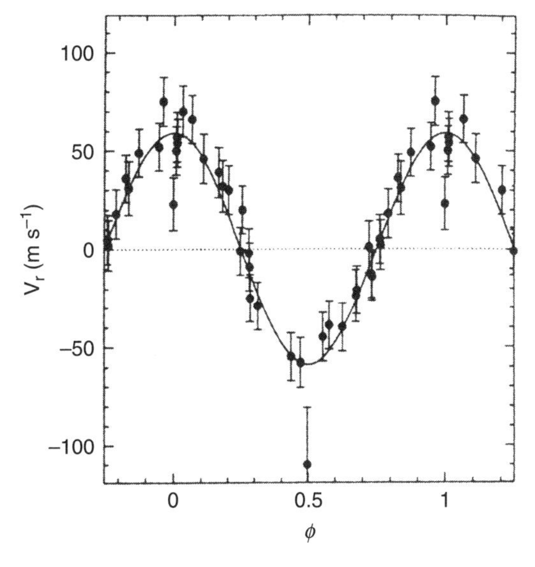

Figure 4.6. Evidence of a planet with about one-half Jupiter's mass around 51 Pegasi. The sinusoid represents a line-of-sight (radial velocity) variation in the motion of the star of ± 59 meters/second as the star is tugged one way and then another over 4.2 days by the inferred planet. Reprinted with permission from Macmillan Publishers, Ltd., Michel Mayor and Didier Queloz, "A Jupiter-mass Companion to a Solar-type Star," *Nature* 378 (1995): 355–9, copyright 1995.

were rocky "super-Earths," perhaps the precursor to finding Earth-sized and Earth-like terrestrial planets. Only a few planets have been directly imaged, notably Fomalhaut b within its debris disk and HR 8799, but more will be imaged and the spectroscopic study of their atmospheres is already becoming a new subfield.[66] Already in 1997 a connection was found between a star's chemical composition, namely its high fraction of elements heavier than helium (its "metallicity"), and the likelihood it will have planets, a finding supported by subsequent exoplanet surveys and by a 2009 survey revealing that 160 metal-poor main sequence stars had no detectable planets.[67] With the exception of the few pulsar planets, virtually all planets detected so far circle main sequence stars, mainly F, G, and K dwarfs and subgiants in the solar neighborhood.

Already attempts are being made to classify extrasolar planets based on their physical characteristics. One study finds that the extrasolar planets discovered thus far across five orders of magnitude in mass can be accommodated into the three classes already known in our own solar system: terrestrial planets, gas giants, and ice giants. Another study, by Italian astronomer S. Marchi, distinguishes five classes based on planetary mass, semimajor axis, eccentricity, stellar mass, and stellar metallicity.[68] Whether the latter are true new classes of objects, or types of a particular Class, remains to be seen and is dependent on the definition of "Class." In any case, these pioneering efforts are likely to expand as more and more planets are discovered around other stars.

4.5 The Discovery of Stellar Systems

We accept today that stars tend not to lead solitary lives, but to congregate. We have seen this already in Chapter 3 with Galileo's discovery of star clusters in 1610 and Herschel's discovery of globular clusters in the context of attempts to separate nebulae from groups of stars in the eighteenth century. But that is only part of the story. On the one hand we now know that stars also congregate into smaller groups of double and multiple stars, and on the other hand they form the huge conglomerations we call galaxies. Astronomers generally have considered these systems at various levels as separate classes of objects, partly because of their penchant to classify, and partly because of the different origins, physical nature, and uses of these systems. The history of their discovery reveals much about the need for precision of language if we are to understand the nature of discovery.

The pioneering double star astronomer Robert Grant Aitken began his classic work on *The Binary Stars* (1918) with the statement, "The first double star was discovered about the year 1650 by the Italian astronomer, Jean Baptiste Riccioli." He was speaking of Mizar, also known as Zeta Ursae Majoris, located in the handle of the Big Dipper. His narrative went on to state that in 1656 Christiaan Huygens saw theta Orionis "resolved into the three principal stars of the group which form the familiar Trapezium," and that three more double stars were discovered before the close of the seventeenth century: gamma Arietis in the Northern Hemisphere, and alpha Crucis and alpha Centauri in the Southern Hemisphere. He pointed out that these discoveries were accidental, made in the course of observations taken for other purposes. Most important, he held that "no suspicion seems to have been entertained by these astronomers or by their contemporaries that the juxtaposition of the two star images in such pairs was other than optical, due to the chance positions of the Earth and the two stars in nearly a straight line." Only in 1767, he emphasized, did John Michell argue based on probabilities that "such double stars, etc., as appear to consist of two or more stars placed near together, do really consist of stars placed near together, and under the influence of some general law." To Michell he gave credit as the first to establish the *probability* of physical systems. He saw the "real beginning" of double star astronomy with the German Jesuit astronomer Christian Mayer, and especially William Herschel's observations during the last quarter of the eighteenth century. And he pointed to Herschel's establishment of physical systems in 1802–1803 as the first proof of physical systems based on observation.[69]

There is much to unpack in this narrative in terms of the analysis of discovery. First, it is now known that Galileo and his colleague Benedetto Castelli were the first to observe a double star, none other than Mizar. At the instigation of his former student and friend, the Benedictine monk Benedetto Castelli, on January 15, 1617, Galileo confirmed Castelli's discovery that the star Mizar, located in the handle of the Big Dipper, was a double star. Moreover, on January 30, Castelli discovered another double star in Monoceros, and ten years later resolved Beta Scorpii.[70] It is not at all clear that Riccioli even claimed to have reported the first double star, an attribution that seems to depend on a single sentence in his *Almagestum Novum* (1651): "There appears to be one star in the middle of the Great Bear's tail, when there are actually two, as the telescope reveals."[71] Even though Galileo and Castelli could not have known these were physically associated stars, like star clusters they constitute the first detection of another new class, the simplest among many stellar "systems" later to be discovered. Similarly, in terms of triple stars, again, it was Galileo who observed triple stars in the Trapezium of Orion in 1617, and Huygens confirmed this a generation later.[72] As it turns out, today the Trapezium is classified as a star cluster, known to contain about 1000 hot young stars about one million years old. Galileo and (as Aitken says) Huygens resolved the three principal stars of theta Orionis, a multiple star system that is the core of the Trapezium.

Then there is the question of class status, which may be parsed here in more ways than one. Did Galileo and Castelli discover double stars in 1610, or Mayer and Herschel almost two centuries later? Did any of them really think it was a new class of objects? And should double stars really be placed in a separate class from multiple stars? After all, "double" is "multiple" in common language. While neither Galileo nor Castelli spoke in terms of a new class, their excitement at the discovery of the new phenomenon of double stars was palpable. By contrast Herschel, with his penchant for natural history, certainly considered them a new class – indeed, he divided them into several classes (we would say "types") based on their separation. It is fair to say that even though Galileo and his student did not use the class terminology for this (or for satellites or for what turned out to be rings), they had the idea they had discovered something new. As to whether double and multiple stars should be placed in separate classes, it is in the end a judgment call, but one influenced by the fact that most astronomers today do so because double stars are much more numerous and can be used in a variety of ways (e.g., stellar masses) that multiples cannot. The terminology and definition of "multiple" is also historically contingent; in his first catalogue of double stars in 1782, Herschel wrote that it contained "not only double-stars, but also those that are treble, double-double, quadruple, double-treble, and multiple."

As in so much in this volume, precision of language is the key in answering the question about who first discovered double stars. Today the term "double star" is sometimes technically used to mean only stars that appear aligned by chance, by contrast to "binary stars," taken to be real gravitationally bound stars. Taken in this way, Galileo and Castelli discovered double stars, and Mayer and Herschel discovered binary stars. The moment of discovery for Galileo and Castelli is clear, because it involved only detection, with no substantial interpretation required: the two stars were clearly close to one another, a situation Galileo thought might be useful for parallax determination. Not so for Mayer and Herschel, who had to go through more steps if they were to infer that stars were gravitationally bound.

It is clear that Herschel was not the discoverer of double stars; Mayer's first catalogues of "Doppelsterne" appeared in 1778 and 1781, and in Herschel's case 42 of 269 doubles in his first catalogue in 1782 had already been observed by others. But when did Herschel first consider them "binary" stars? He certainly did not consider them gravitationally bound as of 1779, the year he began his systematic search for doubles. Herschel was interested in double stars because he (like Galileo) wanted to use them to test a method for measuring stellar parallax – a method that would only work if they were *not* physically associated, in which case the nearer star might move against the more distant fixed background star. At the close of his first catalogue of double stars Herschel explicitly remarks, "I preferred that expression (i.e. double stars) to any other, such as Comes, Companion or Satellite; because, in my opinion, it is much too soon to form any theories about small stars revolving around large ones, and I therefore thought it advisable carefully to avoid any expression that might convey that idea."[73]

After the completion of his second catalogue in 1784, Herschel moved on to nebulae. But when he returned to double stars in 1797, he noticed changes in the relative positions of the components of some of the double stars, and in July 1802, first made the distinction between doubles and physical binaries. Herschel had no way of knowing they were physically associated until he returned late in his career to observe them again, and saw they had physically moved, as he reported in 1802. On June 9, 1803, Herschel published what Aitken called "the fundamental document in the physical theory of double stars," in which he compared his early and later double star observations, concluding "that many of them are not mere double in appearance, but must be allowed to be real binary combinations of two stars, intimately held together by the bonds of mutual attraction."[74]

Therefore it seems clear that Galileo and Castelli discovered double stars in 1617, that Michell inferred the probability of physical systems in 1767, and that Herschel observationally proved and reported physical systems in 1802 and 1803. The latter Herschel termed "binaries." Although technically the distinction remains in some quarters, today the term "double star" in common usage even among astronomers has come to encompass both optical doubles and physical doubles (binaries). The definitive list of such objects, called "The Washington Double Star Catalogue," contains more than 92,000 resolved doubles, of which only 10 percent are proven physical systems. Thus, the distinction has largely disappeared in terms of current astronomical usage.[75] We thus choose to say that Galileo and Castelli discovered the new class known as double stars in 1610, that Michell and Herschel interpreted them as gravitationally bound systems, and that our understanding of their physical nature in terms of the stellar classes that compose them came only later. Galileo and Huygens also discovered multiple star systems at the core of the Trapezium in Orion. Today the Washington Double Star Catalogue includes 6,387 triples, 1774 quadruples, 575 quintuples, and 631 sextuples or more. Where do multiple systems end and sparse open clusters begin? The definition is arbitrary.[76]

These relatively simple stellar systems were only the beginning. We have already seen in Chapter 3 how Galileo discovered open star clusters such as Praesepe, how Herschel distinguished globular clusters in the eighteenth century, and how he classified open star clusters (designated VI, VII, and VIII in his 1786 paper) based on how closely compressed they appeared. Nor was this the end of stellar systems: different classes of star clusters were still being discovered in the twentieth century. The Dutch astronomer Adriaan Blaauw and the Armenian astronomer Viktor Ambartsumian first recognized "stellar associations" in 1946 and 1947. The latter gave them their name and distinguished two types: OB and T associations; the Canadian astronomer Sidney van den Bergh later distinguished an R association for those that illuminate reflection nebulae.[77] Associations are distinguished from open star clusters, which are much older, larger, and more gravitationally bound. Not surprisingly, OB associations contain hot stars of spectral type O and B. OB associations are historically important because they have been used to trace the spiral structure of the Andromeda and Milky Way galaxies, and they continue to be important for addressing fundamental problems of star formation.

Meanwhile, a few years earlier astronomers had originated the idea of stellar "populations," arguably using the idea of stellar systems in a different way than physical proximity. The concept began with the German astronomer Walter Baade while observing the Andromeda Galaxy and two of its companion galaxies at the Mt. Wilson Observatory, located in the San Gabriel Mountains overlooking Pasadena and Los Angeles. In 1943, during the wartime blackouts at the Observatory, Baade was for the first time able to observe details of the Andromeda Galaxy (M 31) and its companions, M 32 and NGC 205. In particular, pushing the limits of his photographic plates, Baade was able to resolve individual stars in these galaxies and distinguish two kinds of populations: the brightest O and B blue stars found in the disk part of the galaxies, and the red giants found in the spheroidal component. Because of the distance of the Andromeda Galaxy, the observations were very difficult, but Baade was confident in his conclusions: "Although the evidence presented in the preceding discussion is still very fragmentary," he wrote in a landmark 1944 paper, "there can be no doubt that, in dealing with galaxies, we have to distinguish two types of stellar populations, one which is represented by the ordinary H-R diagram (type I), the other by the H-R diagram of the globular clusters (type II). Characteristic of the first type are highly luminous O-and B-type stars and open clusters; of the second globular clusters and the short-period Cepheids."[78] The stars of open clusters and the galactic disk are younger, while those at the core of the galaxy and in globular clusters are older. Baade's type I and type II populations quickly became known as Population I and Population II stars. Recently astronomers have used the term "Population III stars" to refer to the very first stars that formed in the universe.

Baade's work had profound implications, not only for Galaxy evolution but also for the scale of the universe. For it turns out there are also two populations of Cepheid variables, those of Population I found in the Galaxy's disk being much more luminous than those of Population II, found in the globular clusters. Hubble had used the period-luminosity relation of Population I Cepheids for his distance determinations to the Andromeda Galaxy, whereas he was actually observing Population II stars (now known as W Virginis stars), which have a different period-luminosity relationship. Because they are intrinsically fainter, Hubble had underestimated Andromeda's distance. By 1952, Baade realized the estimated distance to Andromeda (now known to be 2.5 million light-years), and the size of the universe, should be doubled. The stellar population concept also had deep implications for the composition of the stars.

The term "population" embraces a different concept of "stellar systems," in that they are not gravitationally bound (except in the general sense that all the stars of a galaxy are), but rather share properties of age, location, and chemical composition. The galaxies themselves are the ultimate example of physically bound stellar systems, and to them we now turn.

5

Galaxies, Quasars, and Clusters: Discovery in the Realm of the Galaxies

It will be at once remarked, that the spiral arrangement so strongly developed in 51 Messier, is traceable, more or less distinctly, in several of the sketches ... we are in the habit of calling all objects spirals in which we have detected a curvilinear arrangement not consisting of regular re-entering curves; it is convenient to class them under a common name, though we have not the means of proving that they are similar systems.

William Parsons, Third Earl of Rosse, 1850[1]

Extremely little is known of the nature of nebulae, and no significant classification has yet been suggested; not even a precise definition has been formulated.

Edwin Hubble, 1917[2]

It was on February 5, 1963 that the puzzle was suddenly resolved . . . I noticed that four of the six lines exhibited increasing spacing and strength toward the red . . . I started taking the ratio of the wavelength of each line to that of the nearest Balmer line. The first ratio was 1.16, the second 1.16, the third . . . 1.16! . . . Clearly, a redshift of 0.16 explained all the observed emission lines! The extraordinary implications of a "star" of 13th magnitude having a redshift of 0.16 were immediately clear.

Maarten Schmidt, 1983[3]

Unlike the realms of the planets and stars, long distinguished by the "wanderers" moving among the fixed stars, the realm of the galaxies had to be discovered. Curiously, however, some of its members had been detected long before they were known to be outside our stellar system. As we saw in Chapter 3, a few objects such as Andromeda and the Magellanic Clouds had long been seen with the naked eye, and in the eighteenth century Charles Messier and William Herschel detected numerous fuzzy objects catalogued as "nebulae." But these nebulae were largely believed to be in the realm of the stars, as indeed many of them were. How it was determined that some comprised separate systems of stars far beyond our own is a storied part of the history of astronomy, approached in this chapter through the lens of discovery and its complexities. Despite prescient early guesses and some more scientific inferences, it was only Edwin Hubble in the early twentieth century who provided definitive proof that "extragalactic" objects existed beyond our own Milky Way Galaxy.

But this was only part of the story. Photographically extending Herschel's natural history of the heavens, the parsing of extragalactic nebulae into classes took place well before their true nature was known. Unlike stars, classes of galaxies were not distinguished by spectral lines, but by morphology; like stars it was not clear if morphology represented the deep nature of the object – "the thing itself" – or was just an outward manifestation of the deep nature. Hubble was not the only player involved in the process of separating galactic classes, but he emerged as the long-term winner in the course of community negotiation over nebular and galactic classification. The transformation of the term "non-galactic" (originally meaning outside the plane of our Galaxy) into "extragalactic" was more than a linguistic twist; it represented a community consensus as well as a revolution in astronomy. More than one linguistic and social subtlety was at play, as Hubble himself avoided the word "galaxy," always referring to "the realm of the nebulae," even in his most famous book on the subject by that very title. Only after his death in 1953 did the term "galaxy" become widespread.

Beyond the nebular classes proclaimed or adopted by Hubble and others in the 1920s, the realm of the galaxies held astonishing surprises, heralded a generation later when radio astronomers detected strange "quasi-stellar objects" in the 1950s, objects that turned out to be spewing such immense amounts of radiation that they were designated as entirely new classes of galaxies, collectively known as active galaxies. The further revelation that large numbers of galaxies form structures, ranging from clusters and superclusters to gigantic filaments and voids that stretch across the universe, was another form of discovery that required quite different techniques and modes of thought to uncover. In this chapter we address all these discoveries, and how they came to define the universe as we know it today.

5.1 Discovering a New Realm and Its Classes: Islands in the Universe

We have already seen in Chapter 3 how many of the nebulae catalogued by Messier and Herschel in the eighteenth century actually comprised what we today recognize as six distinct classes of objects in the realm of the stars, as defined by their physical composition: gaseous nebulae including planetary nebulae and hot hydrogen clouds; dusty clouds delineated as dark nebulae and reflection nebulae; and open and globular star clusters actually composed of stars. That a seventh class turned out to be an entirely new realm of objects - galaxies - as far removed from our own Galaxy as the sidereal realm was from the planets, was beyond Herschel's considerable abilities to prove, even with the power of his new telescopes. Natural history could only go so far.

It is true that as early as the mid-eighteenth century, astronomers had speculated that some of the nebulae might lie beyond our own stellar system. But Thomas Wright of Durham's ideas in 1750 were primarily motivated by theology, and the budding philosopher Immanuel Kant's views five years later, though more grounded in observation, were still largely speculation. Both were ignored, though Kant exhibited the spirit, and even the terminology, later used by Herschel: "Here a wide field is open for discovery, for which observation must give the key. The nebulous stars, properly so called, and those about which there is still dispute as to whether they should be so designated, must be examined and tested under the guidance of this theory." Kant's "theory" was that different types of nebulae require different explanations, not much of a theory but a prescient statement if ever there was one. Though ignored at the time, the two eighteenth-century thinkers grasped the basic idea that later became known as the "island universe" theory.[4]

Over the next century and a half, opinion oscillated between the twin poles of near or distant, single system or multiple island universes. The resolution of some nebulae into stars by Lord Rosse in the 1840s tended to support the island universe theory, particularly with the (erroneous) claim of the resolution of the Orion Nebula into stars. On the other hand William Huggins's spectroscopic demonstration in 1864 that some nebulae were composed of luminous gas tended to sway opinion in the other direction. This inside-our-system theory received strong support from the detection in 1885 of the extremely bright "nova" in Andromeda, since if the Andromeda Nebula were remote it meant this star would have to be shining with the equivalent of 50 million suns, inconceivable at the time. By the late 1880s, the island universe theory had fallen completely from favor, and in 1890 the historian Agnes Clerke wrote,

"No competent thinker, with the whole of the available evidence before him, can now, it is safe to say, maintain any single nebula to be a star system of coordinate rank with the Milky Way," an assertion she repeated in 1905.[5]

Meanwhile, the numbers of observed nebulae had increased dramatically as the natural history of the heavens continued. William Herschel's sweeps of the sky with telescopes of his own construction had resulted in some 2500 nebulae being catalogued by the early nineteenth century. By 1864, his son John had doubled the number to 5079 nebulae in his *General Catalogue of Nebulae and Clusters of Stars*, fondly known as the GC. The endeavor could no longer be confined to the Herschel family, and when, in the 1880s, J. L. E. Dreyer combined the GC with other observations to produce the *New General Catalogue* (NGC), the number of objects rose to 7840. Supplemented by two more catalogues, the NGC by 1908 contained more than 13,000 objects. But of these only a few had been resolved into stars.[6]

The biggest advances in understanding the true nature of nebulae came with the application of photography to astronomy. As Allan Sandage has remarked, the faint structural features of galaxies could be detected only when photographic surveys came into general use about 1890, and "these features proved to be decisive in the classification problem because the presence or absence of spiral arms is what divides galaxies into two major groups," the ellipticals and spirals.[7] Even this bipartite division was far from obvious at the turn of the twentieth century, as witness the German astronomer Max Wolf's eclectic forms of nebular objects based on photographs taken at Heidelberg.[8] Wolf's 1909 drawings based on these photographs (Figure 5.1) are reminiscent of Herschel's but considerably more regular in form. More important, Wolf's scheme was the first to display nebular forms in a sequence ranging from amorphous blobs to well-developed spiral arms. But he did not, and could not, tackle the distance problem based solely on natural history methodologies.

As Wolf published his drawings in 1909 in a far-off Germany heading into war, Edwin Hubble (Figure 5.2) was a young American lad of nineteen in the midst of studying mathematics and astronomy at the University of Chicago. There he was influenced most during his undergraduate years by Forest Ray Moulton, who with Chicago geologist T. C. Chamberlin had published their "planetesimal hypothesis" in 1904. Chamberlin and Moulton had drawn attention to the spirals as possible solar systems in formation, based on recent photographs taken by Lick Observatory astronomer James Keeler, who believed they were not only the most abundant form of nebulae, but also likely protosolar systems illustrating Laplace's nebular hypothesis.[9] It was these very spiral nebulae that would prove crucial to Hubble's career, stepping stones to proof that, far from being solar systems inside our stellar system, some nebulae were

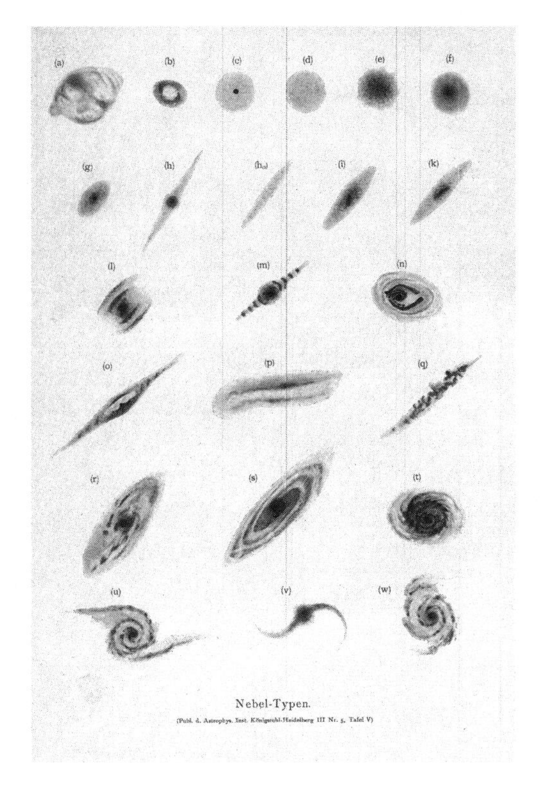

Figure 5.1. Max Wolf's 1909 drawings of nebular forms, many of which turned out to be stellar systems of their own beyond our own Galaxy. Hubble considered Wolf's forms to be only a "temporary filing system," and concentrated on the "e" and "f" types, which turned out to be a major class, elliptical galaxies. Wolf's system was still used extensively until the late 1940s. But did Wolf "discover" new classes of galaxies? Compare with Herschel's nebular forms in Figure 3.2. From Wolf, "Die Klassifizierung der kleinen Nebelflecken" (1909).

actually beyond our own Galaxy. After graduation from Chicago in 1910, Hubble went to Oxford as a Rhodes scholar, from 1910 to 1913 studying Spanish and law and receiving his BA in jurisprudence.[10]

After a brief stint teaching in Indiana, in 1914 Hubble returned to the University of Chicago and its Yerkes Observatory, where he began his life's work. Laboring under the minimal supervision of the nearly blind astronomer Edwin Frost, Hubble produced his thesis "Photographic Investigations of Faint Nebulae" in 1917. A scant seventeen pages, including numerous tables and two plates, the document is nevertheless revealing as a marker of where things stood and what was to come. In the first paragraph of his thesis, Hubble noted that about 17,000 nebulae had been catalogued, and an estimated 150,000 were within reach of existing instruments, namely the Mt. Wilson 60-inch and

Figure 5.2. Edwin Hubble, classifier of "nebulae." In 1924 he demonstrated that most nebulae outside the plane of our Milky Way Galaxy are extragalactic systems of stars. In doing so, he added a new "Kingdom" to astronomy (along with stars and planets), which he divided into four classes. Classes of active galaxies (such as Seyferts in Figure 5.7) would be discovered and delineated later. Hale Observatories, courtesy AIP Emilio Segrè Visual Archives.

100-inch telescopes. In the next two paragraphs he summarized his understanding of the state of knowledge at the time, which was very confusing and therefore worth quoting in full:

> Extremely little is known of the nature of nebulae, and no significant classification has yet been suggested; not even a precise definition has been formulated. The essential features are that they are situated outside our solar system, that they present sensible surfaces, and that they should be unresolved into separate stars. Even then an exception must be granted for possible gaseous nebulae which appear stellar in the telescope, but whose true nature is revealed by the spectroscope. It may well be that they differ in kind and do not form a unidirectional sequence of evolution. Some at least of the great diffuse nebulosities, connected as they are with even naked-eye stars, lie within our stellar system; while others, the great spirals, with their enormous radial velocities and insensible proper motions, apparently lie outside our system. The planetaries, gaseous but well defined, are probably within our sidereal system, but at vast distances

> from the earth. In addition to these classes are the numberless small, faint nebulae, vague markings on the photographic plate, whose very forms are indistinct. They may give gaseous spectra, or continuous; they maybe planetaries or spirals, or they may belong to a different class entirely. They may even be clusters and not nebulae at all. These questions await their answers for instruments more powerful than those we now possess.[11]

This statement is notable, among other reasons, for its clear statement that spirals "apparently lie outside our system." For his dissertation Hubble photographed seven fields with the Yerkes 24-inch reflector during the years 1914–1917. Altogether they showed some 588 nebulae, and Hubble attempted to sort them according to Wolf's classification system, which he said "while admittedly formal, offers an excellent scheme for temporary filing until a significant system shall be constructed." Having done this, Hubble found that "the most striking feature is the great predominance of the classes e and f. These two classes form a continuous sequence from the brightest in the list to the very limits of the plates, where they are but mere faint markings on the films. Eleven are clearly spirals, and the spindles are unexpectedly common. These results are typical ... As far as telescopes of moderate focal length are concerned, the predominant form of nebulae as we know them at present is not the spiral, but is this same 'e, f' class, described as round or nearly so, brightening more or less gradually toward the center, and devoid of detail."[12] Indeed, 429 of the 588 nebulae shown on Hubble's plates belonged to Wolf's 'e' and 'f' classes. As was apparent from Wolf's drawings that Hubble appended to his dissertation, the 'e' and 'f' classes were what Hubble later classed as ellipticals, the terminology still used today. Hubble believed his dissertation was not particularly original, a fair assessment. But it served the important purpose of letting him explore for himself the realm of the nebulae, seeing firsthand what problems needed solving.

After a stint in the U.S. infantry during World War I, in September 1919, Hubble went to Mt. Wilson in the San Gabriel Mountains near Los Angeles. There he had access not only to the 60-inch telescope, with which he did work on reflection nebulae as described in Chapter 3, but also access to the world's largest telescope, the Hooker 100-inch, which had begun regular observations only in 1919, just as Hubble arrived. It was with the latter telescope that Hubble could examine the nebulae outside the plane of the Milky Way Galaxy with the best instrument in the world, seeing fainter and farther than anyone before him.[13]

Given what was to follow, it is somewhat ironic that Hubble's first major paper, published in 1922, was "A General Study of Diffuse Galactic Nebulae," in other words, a study of those nebulae known to be within our Galaxy, not at all

CLASSIFICATION OF NEBULAE

Class	Symbol	Example
I. Galactic nebulae:		
A. Planetaries	P	N.G.C. 7662
B. Diffuse	D	....
1. Predominantly luminous	DL	N.G.C. 6618
2. Predominantly obscure	DO	Barnard 92
3. Conspicuously mixed	DLO	N.G.C. 7023
II. Extra-galactic nebulae:		
A. Regular:		
1. Elliptical (n = 1, 2, ... , 7 indicates the ellipticity of the image without the decimal point)	En	N.G.C. 3379 E0 221 E2 4621 E5 2117 E7
2. Spirals:	Symbol	Example
a) Normal spirals	S	....
(1) Early	Sa	N.G.C. 4594
(2) Intermediate	Sb	2841
(3) Late	Sc	5457
b) Barred spirals	SB	....
(1) Early	SBa	N.G.C. 2859
(2) Intermediate	SBb	3351
(3) Late	SBc	7479
B. Irregular	Irr	N.G.C. 4449

Extra-galactic nebulae too faint to be classified are designated by the symbol "Q."

Figure 5.3. Hubble's 1926 classification of "nebulae" divided them into galactic and extra-galactic. The latter term at first signified outside the plane of our Milky Way Galaxy, and, after 1924, outside of the Galaxy itself. Only after Hubble's death in 1953 did the term "galaxy" come into widespread use. From Hubble (1926), reproduced by permission of the AAS.

the subject of his dissertation or the work for which he was to be come most famous. But it was in this paper that Hubble first ventured, by way of introduction in the paper's inaugural section, his own attempt at a classification system of nebulae, much simpler than Wolf's. "There appears to be a fundamental distinction between galactic and non-galactic nebulae," Hubble noted. "This does not mean that the latter class must be considered as 'outside' our galaxy, but that its members tend to avoid the galactic plane and to concentrate in high galactic latitudes." What is notable here is that the term "non-galactic" first denoted a location outside of the plane of the Milky Way Galaxy, not outside of the Galaxy itself. Hubble also noted a physical difference: galactic nebulae were associated with stars, while non-galactic nebulae were not.[14]

There was therefore no factor of distance involved when Hubble ventured a classification system based on nebular morphology, playing the biologist's role of natural historian. The great bipartite division was between galactic nebulae, consisting of planetary and diffuse nebulae, and non-galactic nebulae, divided into spiral, elongated (spindle and ovate), globular, and irregular morphologies. The following year, in unpublished materials prepared for the International Astronomical Union, Hubble simplified his scheme and defined only three categories of non-galactic nebulae: ellipticals, spirals, and irregulars. This became the terminology employed in his classic paper of 1926, the foundation of the galaxy classification system still used today, the "Hubble sequence" (Figure 5.3). Although the scheme closely followed James Jeans's sequence based on nebular evolution

Figure 5.4. William Parsons, third Earl of Rosse, in 1845 was the first to detect the spiral form of "nebulae," a new class that would prove crucial to cosmology. He employed his six-foot-diameter speculum metal reflector, twice the size of William Herschel's, the largest in the world at the time. AIP Emilio Segrè Visual Archives, E. Scott Barr Collection

theory, Hubble held that "a conscious attempt was made to ignore the theory [of Jeans] and arrange the data purely from an observational point of view."[15]

Hubble's classification system for non-galactic nebulae therefore was in place well before the distances to these objects had been determined, in fact even before it was known whether they were galactic or extragalactic. But more than natural history was needed to solve the problem of non-galactic nebulae; for that, distance techniques were essential. As Hubble himself noted in his retrospective book, *The Realm of the Nebulae* (1936), three landmarks stand out in the road that eventually revealed the nature of the non-galactic nebulae: the first radial velocity measurements of a galaxy in 1912, the discovery of novae in 1917, and his own discovery of Cepheid variable stars in the Andromeda nebula and elsewhere.

In demonstrating the existence of a class of nebulae beyond our own star system (ironically not yet known to be a spiral itself), spiral nebulae were to prove crucial in all three of these landmarks.[16] William Parsons, the third Earl of Rosse in Ireland (Figure 5.4), had first detected the spiral form of nebulae in 1845 when he turned his six-foot diameter speculum metal reflector, twice the size of William Herschel's, to the 51st object in Messier's catalogue. His

Figure 5.5. Lord Rosse's drawing of the spiral nebulae M51, today known as the Whirlpool Galaxy, 1850. Rosse called M51 "the most conspicuous object of that class," among fourteen he had observed beginning in 1845. The class of spiral nebulae would become spiral galaxies after 1925, when Hubble proved they were independent systems of stars outside our Galaxy, though Hubble himself persisted in calling them spiral nebulae.

observations and sketches of M51 (Figure 5.5), now known as the Whirlpool Galaxy, were presented to the Royal Society of London in 1850, with the following unequivocal words: "It will be at once remarked, that the spiral arrangement so strongly developed in 51 Messier, is traceable, more or less distinctly, in several of the sketches."[17] It is notable that Rosse used the word "detected" rather than "discovered" with regard to the spiral form, and in the following spring he "detected" another spiral, M99, and suspected several more from John Herschel's southern catalogue. By the time of the publication of his results in 1850, Rosse and his observing colleagues had detected fourteen spirals, though he cautioned they were "comparatively difficult to be seen, and the full power of the instrument is required to bring out the details . . . 51 Messier is the most conspicuous object of that class." Moreover, he cautioned, "we are in the habit of calling all objects spirals in which we have detected a curvilinear arrangement not consisting of regular re-entering curves; it is convenient to class them under a common name, though we have not the means of proving that they are similar systems." Despite these laudable cautions Rosse was nevertheless the first to detect and declare the spiral form as a new class of nebulae. Today we know that 27 out of 110 objects in Messier's extended catalogue are spiral galaxies.

Rosse himself, however, did not yet know the true nature of his spiral structures as conglomerations of stars. Even in 1887, when Isaac Roberts made the first long-exposure photographs of the Andromeda Nebula and revealed its spiral structure, its distance was so uncertain that Roberts and other astronomers believed it to be a solar system in formation, in line with the nebular hypothesis. And when in 1899 the German spectroscopist Julius Scheiner identified familiar stellar features in the spectrum of the same object, and declared that "the previous suspicion that the spiral nebulae are star clusters is now raised to a certainty," the idea of the Andromeda Nebula being a system of stars was still too radical for most.[18]

The three landmarks that led to the extragalactic distance determinations all employed these spiral nebulae whose form Rosse first detected. In 1912, V. M. Slipher at the Lowell Observatory began his spectroscopic radial velocity observations of spirals, and within two years showed that that spectral lines in fourteen spiral nebulae were redshifted, an observation he interpreted as meaning they are moving away from us. Ejnar Hertzsprung and other astronomers took these velocities, ranging up to 1100 km/second, as evidence that these objects could not be gravitationally bound to our own system of stars: "It seems to me, that with this discovery the great question, if the spirals belong to the system of the Milky Way or not, is answered with great certainty to the end, that they do not."[19]

The second event was the discovery of faint novae in spirals in 1917. Bright novae, known today to be supernovae, had been observed in spirals before, in Andromeda in 1885 and Centaurus in 1895. But it was George W. Ritchey's photographic detection of a nova in the spiral galaxy NGC 6946, as well as Lick Observatory astronomer Heber D. Curtis's discovery of three more in the same year, followed by even more on old photographic plates, that allowed crude distances to be determined to these spirals. By comparing the brightness of novae in spirals to their brightness in our Galaxy, Curtis calculated an average distance to spiral nebulae of 20 million light-years, clearly far outside our Galaxy. By 1917, when Hubble completed his dissertation on the nebulae, the island universe theory was the leading theory of spirals, the one toward which Hubble himself tended.

The island universe theory was not, however, accepted by all. The so-called Great Debate on the scale of the universe, argued between Mt. Wilson astronomer Harlow Shapley and Curtis and held in 1920 in Washington, D.C., under the auspices of the National Academy of Sciences, clearly showed the differences that still existed. Using his nova results and a variety of other methods, Curtis argued that nebulae were island universes. Shapley, on the other hand, was fresh from his discovery that globular clusters are arrayed at considerable distances around our own stellar system, in which we hold an eccentric position. He argued that our Galaxy was larger, and the spirals nearer, than believed, estimating the Galaxy's diameter at 300,000 light-years, ten times the size accepted by most astronomers. If true, it meant the Magellanic Clouds were inside our Galaxy. Moreover, Shapley's colleague at Mt. Wilson, Adriaan van Maanen, had measured internal motions in several spirals including M101, motions which, if true, were much too large to be seen if the spirals were distant island universes. While the island universe theory was still ascendant in 1920, doubts remained.[20]

The third event was the one in which Hubble himself proved central. His method used Cepheid variables instead of novae, and was much more accurate and robust than the nova method. This was because in 1908, Henrietta Leavitt at the Harvard College Observatory had discovered and published a period-luminosity relation for what became known as Cepheid variable stars that she had observed in the Magellanic Clouds. Their periods ranged from 1.25 to 127 days, and the longer the periods, the greater the luminosities. Therefore if the period could be determined, so could the luminosity, and if one had the luminosity, the star's apparent magnitude would yield its true distance. The 1908 paper went largely unnoticed, but Leavitt's 1912 paper gave the results in graphical form and was noticed to great effect, despite its short length – three pages. Shapley used this method in 1918 for some of the nearer globular clusters in his famous work showing our eccentric position in the Galaxy, and Hertzsprung had used it before him. Thus, when in late 1923 Hubble found a Cepheid in the Andromeda Nebula, it afforded an accurate method (at least accurate relative to the time) for measuring the distance to Andromeda. Hubble calculated that distance at about one million light-years, well beyond our own star system. In 1924 Hubble found several more Cepheids in Andromeda and M33.[21]

The first public announcement of Hubble's Cepheid work appeared in the *New York Times* on November 23, 1924. But, still worried about van Maanen's contrary measurements of internal motions of spirals, Hubble's results were not *formally* announced until New Year's Day, 1925, at a joint meeting of the AAS and AAAS in Washington, DC. Even then Hubble himself was not present, but sent his paper to be read.[22] From a total of twelve Cepheids in Andromeda and twenty-two in M33, Hubble announced a distance to the spirals of roughly 285,000 parsecs, placing them even outside Shapley's large version of our Galaxy. Thus, the first day of 1925 marks the official declaration of final proof that spiral nebulae, in particular the Andromeda Nebula M31 and M33, were beyond our Galaxy, and consisted of stars, not gas. Step by step, Hubble could proceed deeper and deeper into space using the Cepheid method; by the end of 1925, he reported eleven Cepheids in the object known as NGC 6822, and showed it also to be beyond our Milky Way Galaxy.[23] Despite a few holdouts due mainly to van Maanen's claim of internal rotation of galaxies (later shown to be erroneous), opposition to the island universe theory collapsed. In addition to the realm of the planets and the realm of the galactic stars, astronomy now had the realm of the galaxies.

In his 1926 paper, significantly titled "Extra-galactic Nebulae," Hubble put the pieces together. It was here that Hubble first published the classification system for galaxies as we now know it (Figure 5.3). He pointed out that it had first been presented in the form of a memorandum distributed to all members of the Commission on Nebulae of the IAU in 1923. It was discussed at a meeting in Cambridge in 1925 and was published in an account of the meeting in *L'Astronomie*. The question of the adoption of the system was left to a subcommittee of the Commission, "with a resolution that the adopted system should be as purely descriptive as possible, and free from any terms suggesting order of physical development." The Commission also indicated a preference for the term "extra-galactic" in place of the necessarily non-committal "non-galactic." The title of Hubble's 1926 paper marked the all-important shift: non-galactic had become extra-galactic, a true landmark in the history of astronomy.

Still, one of the curiosities of the history of astronomy testifying to social factors, at least in nomenclature matters, is that Hubble persisted in calling the spirals extragalactic nebulae, and his popular book on the subject even in 1936 remained *The Realm of the Nebulae*. This terminology was likely due to Harlow Shapley, who had urged Hubble in October 1924 to call his objects galaxies rather than the unwieldy "non-galactic nebulae." Shapley had also urged Slipher, the head of the Commission on Nebulae, to adopt the term "galaxy." Neither did so. As Gale Christianson, Hubble's biographer, put it, "Hubble was in no mood to be dictated to by the man whose universe he had demolished." "Extra-galactic nebulae" was the phrase used by Hubble in his landmark 1926 paper, and by Mt. Wilson astronomers until Hubble's death in 1953, when the term "galaxies" was universally adopted. Alan Sandage, Hubble's successor at Mt. Wilson, confirms that "within a week of Hubble's death in September, 1953, 'galaxy' became the standard name, even at Santa Barbara Street," headquarters for Mt. Wilson staff.[24]

As we shall see in Chapter 8, it is notable that Knut Lundmark also put forth a similar classification system for galaxies in 1926, and Harlow Shapley a very different one in 1927 – not to mention that the British industrialist John H. Reynolds devised a similar classification system in 1920, six years before Hubble's scheme was published. But it was Hubble's that won the day, both for reasons of simplicity and social factors. Thus, just as Herschel and then Huggins were central in proving definitively that nebulous matter exists in the universe, so Edwin Hubble became the central figure in finally proving that galaxies exist beyond our own. As more and more galaxy distances were determined, later in the decade he would propose a velocity-distance relationship: the more distant the object as shown by Cepheids, the faster was it moving away from us as shown by redshifted spectra, a relation that came to be known as "Hubble's law." For most (though not Hubble himself) this was evidence that the universe was expanding, another pillar of modern cosmology.

* * *

With the discovery of the true nature of the spiral nebulae, extragalactic astronomy – and the study of the separate classes of galaxies – could begin in earnest. The extragalactic nature of the spirals had been shown, but how about the other morphological forms in Hubble's classification system, the ellipticals and irregulars? In his landmark 1926 paper Hubble noted that definite distances were known for only six systems including the two Magellanic Clouds, which were then considered irregulars, not spirals. However, he continued, "the similar nature of the countless fainter nebulae has been inferred from the general principle of the uniformity of nature."[25] In other words Hubble was inferring that all the objects in his "non-galactic" nebulae category were actually outside the galaxy. With this leap, "Non-galactic" became "extra-galactic" not only for the spirals but also for ellipticals and irregulars.[26]

As we have seen, elliptical galaxies had long been observed as fuzzy objects inseparable from many other types of nebulae. We now know that Charles Messier's 1781 catalogue of nebulae and star clusters, for example, contained six of them, and four more were later added to the extended catalogue of 110 objects. But they could not be directly proven to be extragalactic because they could not be resolved as stars as was the case for many spirals. It was recognized early on that, in stark contrast to spiral galaxies like our own, ellipticals are almost featureless aggregations distinguished only by their ellipsoidal or spherical morphology. As Hubble wrote even in 1936, "Elliptical nebulae are highly concentrated and show no indications of resolution into stars. The luminosity falls rapidly away from bright, semistellar nuclei to undefined boundaries . . . Small patches of obscuring material are occasionally silhouetted against the luminous background, but otherwise these nebulae present no structural details."[27] Ellipticals were clearly "extragalactic" in Hubble's original sense of outside the plane of our Galaxy, but only by the uniformity of nature could they be declared extragalactic in the sense of being placed outside our Galaxy.

Therefore, assuming ellipticals to be outside our Galaxy, further inferences as to their nature could only be made based on their morphology. In his 1926 classification system Hubble had divided extragalactic nebulae into regular and irregular galaxies, classifying ellipticals and spirals as regulars, those defined as having "rotational symmetry about dominating non-stellar nuclei." These two classes, he estimated, comprised 97 percent of all galaxies. But Hubble went further than the delineation of classes to specify types; depending on their ellipticity he arranged elliptical galaxies from E0 (spherical like M87) to E7 (elongated), shading into what he called "a limiting lenticular figure" at the junction of the spirals. Spiral galaxies were typed as early, intermediate, and late "normal" spirals, as well as early, intermediate, and late "barred" spirals, the latter having a central bar structure not evident in the normal galaxies.

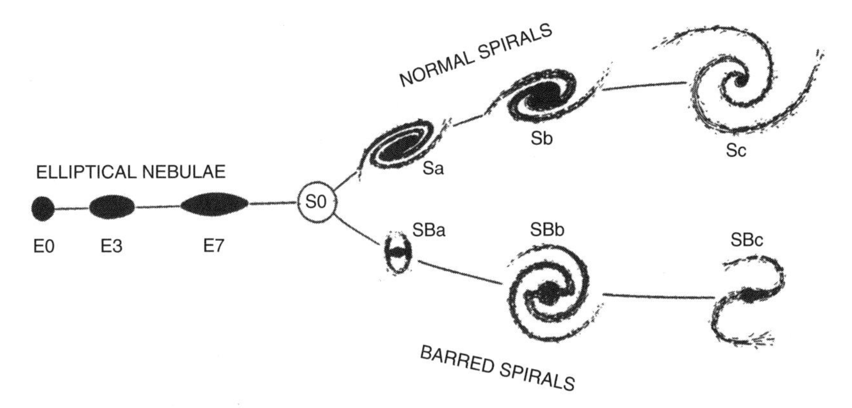

Figure 5.6. Hubble's "tuning fork" diagram of galaxy classification, from *The Realm of the Nebulae,* 1936, p. 45. Hubble referred to the diagram as "the sequence of nebular types . . . a schematic representation of the sequences of classification." For Hubble the diagram represented an evolutionary sequence, moving from left to right. The diagram illustrates the difficulties of delineating classes; Hubble considered E7 "lenticulars" a "transition stage . . . more or less hypothetical," between ellipticals and spirals. The existence of a lenticular class of galaxies remains controversial to this day, when they are more often identified with Hubble's S0 galaxies.

The significance of the "early" and "late" terms is graphically evident in the famous "tuning fork" diagram of galaxy types first published ten years later in *The Realm of the Nebulae* (Figure 5.6), where Hubble placed ellipticals on the handle of the fork because he thought, influenced by Jeans's theory, they were early objects before acquiring their more mature morphologies as spirals. In the 1926 paper, however, he continued to insist on his independence from Jeans, while remarking on the suggestive parallels: "Although deliberate effort was made to find a descriptive classification which should be entirely independent of theoretical considerations, the results are almost identical with the path of development derived by Jeans from purely theoretical investigations. The agreement is very suggestive in view of the wide field covered by the data, and Jeans's theory might have been used both to interpret the observations and to guide research. It should be borne in mind, however, that the basis of the classification is descriptive and entirely independent of any theory." Clearly, in Hubble's mind the tuning fork also represented an evolutionary sequence, although he realized the different forms of the ellipticals could be due to projection effects caused by the viewing angle.

Irregular galaxies originally comprised all those galaxies that did not fit into other classification categories that Hubble defined in 1926, amounting to only about 3 percent. In his book *Realm of the Nebulae* (1936), Hubble, still using his favored term "nebulae" rather than "galaxies," commented that whereas the "regular nebulae" - spirals and ellipticals - "are characterized by rotational symmetry around dominating central nuclei . . . about one nebula in forty is irregular in the sense that one or both characteristics are absent."[28] He recognized the Magellanic Clouds as conspicuous examples, already proven extragalactic using the Cepheid method.

We now know that elliptical galaxies are dominated by older, low-mass, and metal-poor population II stars moving in elongated, randomly oriented orbits, and that ongoing star formation is rare, ellipticals having mostly depleted their interstellar matter. Although they are typically surrounded by numerous globular clusters, because they are structureless they seem simpler than other classes of galaxies. And because of their older stars, ellipticals are still sometimes referred to as "early-type" galaxies, as opposed to "late-type" spirals, harking back to Hubble's original designation but for different reasons. Round elliptical galaxies are believed to be the oldest ellipticals, forming from galaxy mergers 2 to 3 billion years ago and exhibiting no star formation. While small ellipticals are the most common galaxies in the universe, particularly in clusters of galaxies, ellipticals in general represent only 10–15 percent of galaxies in the local universe. And we know that in contrast to spirals, they were less common in the early universe than now.

Today irregulars are defined as galaxies that either have little structure or lack any structure whatsoever, and, having a much larger sample than in Hubble's day, about one-third of all galaxies are classified as irregular. They are more common at higher redshift, as one moves back to earlier epochs in the universe. Irregular galaxies vary in mass from 100,000 to 10 million solar masses, and in size from a thousand to tens of thousands of light-years. They are often rich in gas and dust, and may represent ellipticals or spirals that have been distorted over time by interactions with other galaxies. Irregular galaxies are well represented among the nearest galaxies in our Local Group. The Magellanic Clouds were for a long time considered the nearest irregular galaxies to our own. Based on Harlow Shapley's work on Cepheids, Hubble considered the Large Magellanic Cloud (LMC) to be at about 85,000 light-years, and the Small Magellanic Cloud (SMC) at about 95,000 light-years from the Milky Way. We now know them to be at about 160,000 and 200,000 light-years, respectively. And only recently they have been shown to have some spiral characteristics, a paean to the difficulties of classification.

Thus, the delineation of the basic classes of galaxies, and their placement in a separate realm beyond the stars, was essentially complete already in 1926. Hubble's classification system did not change much in the next ten years. In his 1936 book *Realm of the Nebulae*, he still identified lenticulars with the most flattened form of ellipticals, known as E7, and hypothesized what he called an S0 class between ellipticals and spirals. It is this S0 class we today refer to as lenticulars - an interesting case of adjusting a classification system as more data became available.

That is not to say that disputes about classes have all been resolved. Even today some dispute whether lenticulars should be considered a separate class of galaxies. At a different level Hubble's prototype of an E7 elliptical, NGC 3115 (also known as the Spindle Galaxy), is now recognized as an early S0 system because new observations clearly reveal a thin disk. In fact, three of twelve of Hubble's prototypes have been reclassified since 1936 based on better evidence.[29] Moreover, it is notable that when Hubble proposed an intermediate S0 class of galaxies, none had been observed. But eventually they were observed, and the types, if not the classes themselves, adjusted accordingly, as in Gérard de Vaucouleurs's 1959 extension of the Hubble sequence of galaxies, where lenticulars are further typed as barred (SB0) or unbarred (SA0). De Vaucouleurs later considered all S0 galaxies to be "lenticulars." Nevertheless, even today it is uncertain whether galaxies such as M84 and M86 should be classified as highly elongated E7 ellipticals or S0 lenticulars. Such are the problems any classification system encounters, whether biological or astronomical. In the case of astronomers, however, their objects are located millions of light-years distant.

Similarly, while Hubble originally classed the Magellanic Clouds as "typical irregular nebulae - highly resolved, with no nuclei and no conspicuous evidence of rotational symmetry," de Vaucouleurs recognized rotational symmetry and some spiral structure. He suggested they were in fact part of the spiral sequence, and his classification system added the Sm and Im stages to account for these characteristics in the Magellanic Clouds as well as in other similar galaxies. In his system they are thus typed as Irr/SB (s)m, the "B" indicating characteristics of a barred spiral, and the "s" indicating an s-shaped spiral emerging directly from a central bulge or the ends of a bar.[30]

The de Vaucouleurs system and others greatly refined Hubble's classification work, leaving the original classes intact, the ambiguous lenticulars notwithstanding. But quite aside from these developments, great surprises were in store in the form of the discovery of galaxies so different from the "normal" varieties that they are considered to be an entirely new family of classes, known as active galaxies.

5.2 Where the Action Is: Quasars, Blazars, and Black Holes

Even before Hubble's death in 1953 there were hints of unusual activity in some galaxies, dating all the way back to H. D. Curtis's 1918 observation of a "jet" that appeared to be associated with the nucleus of M87. Unusual activity began to be noticed more substantially with a few spirals in the 1940s, spread to the ellipticals in the 1950s and, with the discovery of quasars, eventually involved all morphological classes of galaxies in the Hubble sequence. Gradually astronomers realized they were dealing with a qualitatively new family of galaxies, far different from normal galaxies in terms of the activity in their central nuclei and their resultant energy output, and therefore eventually termed "active galaxies." As they were discovered to the increasing astonishment of astronomers over three decades, these objects were separated into four basic classes based on their spectra and energy output, presumably a reflection of their physical nature: Seyfert galaxies, radio galaxies, quasars, and blazars. The search for an explanation of the energy source of all this activity eventually led through a circuitous route to black holes, an idea so bizarre that one may question whether they are actually "objects" in the usual sense of the word rather than "singularities" as they are often described.

Today a unified theory of active galaxies holds that some, and perhaps all, classes of active galaxies may fundamentally be one and the same kind of object, appearing as different phenomena only because they present different viewing angles as seen from Earth. There is as yet, however, no consensus on this unifying theory, a subject of continuing research. Thus, active galaxies offer lessons not only in how totally unexpected new classes of objects were discovered in the realm of the galaxies based on features other than morphology, and what those features were that astronomers used to designate classes, but also how class designations sometimes require adjustment with new knowledge. Moreover, for the first time a full understanding of these classes involved observations across a wide range of the electromagnetic spectrum, adding an important new dimension to the process of discovery and classification. Indeed, the opening of the electromagnetic spectrum, including the deployment of spacecraft above the Earth's atmosphere made possible by the Space Age, proved crucial for the discovery and mature understanding of these classes.

The first indication of something unusual in the realm of the galaxies came in 1943, when astronomer Carl Seyfert, who had graduated from Harvard under Harlow Shapley seven years earlier and was working for two years at Mt. Wilson, observed that a very small proportion of "spiral nebulae" had bright optical nuclei. More specifically, these galaxies exhibited "high excitation nuclear emission lines superposed on a normal G-type spectrum," that is to say, superposed on a normal solar-type spectrum with its continuous spectrum and Fraunhofer absorption lines.[31] As early as 1899, the German spectroscopist Julius Scheiner had recognized absorption lines in the spectrum of the Andromeda Nebula (M31), indicating it might consist of stars since that spectrum was analogous to those produced by the Sun and other stars. The conclusion was too radical for the time, and remained so even when Edward Fath at Lick Observatory confirmed the result for M31 and six other spirals in 1908. As we have seen, conclusive proof of the extragalactic nature of M31 awaited Hubble's detection of Cepheids in 1923, with results announced on the first day of 1925.

But Fath had also noticed something else: not only absorption lines but also six emission lines were present in one of his six spirals, known as NGC 1068. And not only emission lines, but the same "high excitation" emission lines as had been found in planetary nebulae heated by their central star.[32] As in the case of Huggins's landmark 1864 observation of planetary nebulae, this was an indication of a hot luminous gas, requiring a source of extreme energy. The conclusions seemed suspect, and no one knew how to interpret such a result, since clearly much more than a single star or group of stars was involved for extragalactic objects at this distance. By 1939, spectra of many spirals obtained by V. M. Slipher, Hubble, and Nicholas Mayall showed that about 50 percent of them had one emission line either in the nucleus or the arms. But very few of them had six emission lines indicating extreme activity in the nucleus, lines that were also broader than normal, usually a sign of Doppler motion (in this case up to 8500 km/second). In his 1943 paper Seyfert not only confirmed Fath's result for NGC 1068, but also found other galaxies with such broad emission lines, "most of them intermediate-type spirals with ill-defined amorphous arms, their most consistent characteristic being an exceedingly luminous stellar or semi-stellar nucleus which contains a relatively large percentage of the total light of the system." Seyfert identified twelve galaxies in "this unusual class of object," beginning with the prototype spiral M77 (NGC 1068), now known to be located at a distance of some 60 million light-years.[33]

Significantly, Fath's 1908 detection of these peculiar properties in NGC 1068 was considered so suspect and anomalous that the observation was not taken seriously; the object was declared the prototype of a new class only when Seyfert reported other galaxies with similar properties in 1943. Also significantly, most of Seyfert's observations were made at the 60-inch reflector of the Mt. Wilson Observatory, where he held a National Research Council Fellowship from 1940 to 1942, and where Edwin Hubble was his principal advisor (accounting for his use of Hubble's favored term "spiral nebulae" rather than "spiral galaxies"). With Seyfert's work at Mt. Wilson, there is thus a passing connection between Hubble's reconnaissance of "normal" galaxies in the first half of the century and the new era of active galaxies that would dawn in the second half. But as Don Osterbrock has pointed out, there was very little physical interpretation in Seyfert's paper because very little was known about the processes involved in

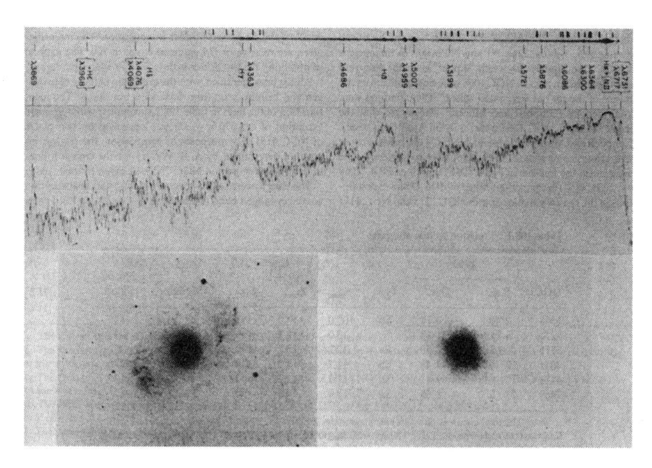

Figure 5.7. Evidence for the first class of active galaxies, known as Seyfert galaxies. This image from Seyfert's 1943 article indicates how difficult the detection of a new class of objects can be. Here the image shows (top to bottom) the spectrum, the microphotometer tracing of the spectral lines, and direct images of one of the six galaxies, NGC 4151. The two most prominent emission lines that Seyfert considered anomalous are at wavelengths 4959 and 5007, seen at top right of center. The spectrum was taken with the Mt. Wilson 60-inch reflector, and the images with the Mt. Wilson 100-inch. The image at bottom left shows the weak amorphous arms of the spiral, while the image at bottom right zooms in on the nucleus. Plate I from Seyfert (1943). Reproduced by permission of the AAS.

even normal spirals with emission lines: "Seyfert's paper defined the class, and that was all he could do at the time."[34]

Seyfert's achievement, however, should not be underestimated. His declaration of a new class was quite subtle: separating those spirals with six high excitation and broad emission lines in the nucleus from those showing only one emission line. His declaration was made, moreover, only with a small sample of twelve often faint galaxies showing these spectral phenomena. But the evidence that presented itself in the form of microphotometer tracings of the emission lines told a story that demanded explanation, at least to his discriminating eyes (Figure 5.7). Seyfert's declaration proved correct: broad emission lines in spirals with bright semi-stellar nuclei are the two characteristics that still define what are now called Seyfert galaxies, even though they are now divided into two types. But Seyfert's paper hardly caused a stir; it was not referred to for sixteen years after its publication.[35]

The plot thickened a few years later with the discovery of what we now call "radio galaxies," though they were not immediately recognized as being active galaxies. The discovery of radio galaxies as a new class of object owes its existence, as the name implies, to the rise of radio astronomy. Although Karl Jansky first detected the galactic background radio noise in 1932, and Grote Reber produced maps of radio emission in the 1940s, the British physicist J. Stanley Hey and his colleagues are credited with the discovery of the first discrete radio source, later known as Cygnus A, in 1946, an extension of their wartime radar work. As Hey recalled, "We could locate the position of the source only to within about 2 degrees, and there was no obvious optical clue to its identification with visible stars. Nevertheless, we concluded that only discrete sources could produce such fluctuations." The signal was coming from the direction of the constellation Cygnus, and they believed the source must be a similar but more distant source than the Sun, which they had already observed in radio emission.[36] Only in 1952 did Walter Baade and Rudolph Minkowski identify Cygnus A with an optical source, an extremely faint 18th magnitude galaxy, and only in 1954 did they publish their results.[37]

The energy output implications of such a distant object were astonishing. The *New York Times* described Cygnus A as a radio station at a distance of 600 million million million miles, with a power of 400,000,000,000,000,000,000, 000,000,000,000 kilowatts. British cosmologist William H. McCrea recalled that astronomers "had to practise hard before breakfast at believing" in a radio luminosity about a million times greater than that of our Galaxy.[38] Cygnus A is now known to be an elliptical galaxy and one of the brightest radio sources in the sky. Because of its early discovery, it is often considered a prototype of the radio galaxy class. Its huge physical structure spans 300,000 light-years, and it emits some 10 million times more energy than normal galaxies. Clearly this was a different class of object than more radio-quiet ellipticals. But because radio galaxies were observed in the radio spectrum and Seyfert's in the optical, as yet no connection was made between the two. Unlike Seyfert's, the anomaly was not made evident by spectral lines, but simply by the strength of the source and the power that must be present to generate it.

It took some time for astronomers to realize how common the astonishing phenomenon of radio galaxies really was, and to separate them from other radio-emitting objects. Radio positions were not well determined in the early years of the field; optical identifications were difficult because the error box might contain thousands of optical objects. Nevertheless, over a period of years some identifications were made, so astronomers could begin to see just how extraordinary a phenomenon they were dealing with. In 1947, the year after Hey and his British colleagues reported the first discrete radio source, John Bolton and his Australian colleagues discovered more discrete radio sources, and in 1949 they suggested optical identifications for three of them: Taurus A, Virgo A, and Centaurus A, the A indicating the brightest radio object in each constellation. They identified Taurus A with the Crab Nebula within our

Galaxy, and the other two with the "nebulous objects" M87 and NGC 5128, respectively. But they were reluctant to identify the latter two as extragalactic, not only because "there is little definite evidence to decide whether they are true extra-galactic nebulae or diffuse nebulosities within our own galaxy," but also because their identification as radio sources "would tend to favor the latter alternative, for the possibility of an unusual object in our own galaxy seems greater than a large accumulation of such objects at a great distance."[39] The extragalactic nature of both was confirmed in due course.

We now know that radio galaxies are almost always large elliptical galaxies, by contrast to the Seyfert class, which are typically spirals. Virgo A, discovered optically already in 1781, turns out to be a giant elliptical galaxy, the largest near Earth and the dominant galaxy in the what we now know as Virgo Cluster. Centaurus A (NGC 5128) is now classified as an elliptical or lenticular galaxy, and, at a distance of 12 million light-years, the nearest radio galaxy and one of the largest and brightest in the sky. John Herschel first called attention to this object as peculiar from its appearance in the nineteenth century, but it took radio astronomy to demonstrate just how peculiar it was.

In a way reminiscent of Hubble's morphological separation of different classes of nebulae in our own Galaxy, radio astronomers were eventually able to identify different classes of radio objects, not all of them extragalactic. Radio surveys in the early 1950s, such as the first Cambridge survey and the Mills survey in Australia, identified several hundred radio objects, of which only ten or so were identified with optical counterparts by 1953. The leading theory for several years was that the majority of them were a new class of object, very nearby, dark stars termed "radio stars." But in what astronomer and historian Woodruff T. Sullivan III called the "bible" of optical identifications as of 1954, Baade and Minkowski identified four major categories of optically identified radio sources: peculiar emission nebulosities within our Galaxy such as Cassiopeia A and Puppis A; peculiar extragalactic nebulae including Virgo A, Perseus A, Centaurus A, and Cygnus A; normal galaxies with much lower radio output than the second category; and supernova remnants such as Taurus A.[40]

As Sullivan has shown in detail, these categories of radio objects took time to separate, and in fact their first and fourth categories eventually proved to be one and the same – related to supernova in our own Galaxy. The early identification of Taurus A with the Crab Nebula had already shown that not all discrete radio sources were radio galaxies, but the second half of the twentieth century demonstrated that radio galaxies, with their prodigious energy outputs, were quite common. Radio astronomy had opened a new window on the universe, and the universe was forever changed, or at least our perception of it. Radio galaxies and related objects are now known to number in the millions, featuring an array of morphological characteristics including lobes, plumes, and jets, characteristics that turned out to be clues to the energy source lying at their centers.

The existence of such prodigious amounts of energy begged the question of just what that energy-producing mechanism was. Already in the 1950s, the favored theorized mechanism was synchrotron radiation, caused by electrons spiraling close to the speed of light in a magnetic field. Although controversial among radio astronomers until the late 1950s, this process is now known to be the most widespread type of non-thermal radio generation, that is, radiation not arising from thermal radiation at different temperatures. The mechanism was plausible, but the energy source causing electrons to spiral at such relativistic speeds remained an enigma, even more so when Geoffrey Burbidge pointed out in 1959 that they must contain enormous energy equivalent to transforming one million to 100 million solar masses into energy. As one astronomer in the field later wryly noted, "This conclusion did not raise a strong interest in the community as it should have done." Nevertheless, in the same year Burbidge and his colleagues, as well as the Dutch astronomer Lodewijk Woltjer, finally revived Seyfert's 1943 paper on his optically anomalous galaxies, noting that their nuclear activity could bear some relation to that of radio galaxies. They came to opposite conclusions, however, Burbidge arguing that the plasma giving rise to the emission lines should be ejected from the galactic nucleus, Woltjer finding it should be gravitationally confined by a massive body. It was the beginning of a long and heated controversy.[41]

* * *

The situation would become even stranger in the coming decade. Fourteen years after J. S. Hey's discovery of Cygnus A as the first discrete astronomical radio source in 1946, and eight years after its identification as the first radio galaxy in 1952, a dramatic series of events began that would lead to another class of object with an outpouring of radiation far more powerful than even Cygnus A. As we have seen, with the birth of radio astronomy strong discrete radio sources had been detected since the 1940s, and hundreds had been observed by the early 1960s with radio surveys of the sky such as the Third Cambridge (3C) Catalogue. They were believed to be "radio stars," but because their positions had not been accurately determined, most of their optical counterparts had not been identified, making distance determinations impossible.

The saga began with one of the objects from the 3C Catalogue. In 1960 Thomas A. Matthews, who had received his PhD in 1956 at Harvard under Bart Bok, obtained an accurate radio position for 3C 48, using Caltech's new Owens Valley Radio Telescope in California. The object at first was believed to be extragalactic, but when Allan Sandage, Hubble's protégé and successor at Mt. Wilson, took an optical image of the object in September of that year, it appeared stellar in nature. And when he obtained the first spectrum the following month, it showed broad emission lines. Though a very peculiar object, it was generally believed by 1961 to be one of the relatively nearby "radio stars" in our Galaxy.

Meanwhile Caltech astronomer Maarten Schmidt, taking up the mantle of Rudolph Minkowski after the latter's retirement from Mt. Wilson in 1960, continued Minkowski's work on optical counterparts of radio sources. He obtained possible optical identifications from Matthews, and began to make spectroscopic observations with the idea of determining the distance and nature of the objects. Some were misidentifications, but in 1962 he obtained spectra of two objects with peculiar spectra. At about the same time as this was transpiring, Cyril Hazard in Australia obtained accurate radio positions of the object known as 3C 273, using a lunar occultation technique with the new Parkes Radio Telescope. The object had been catalogued already in 1959, but neither its radio nor optical position was well known. Lacking large optical telescopes to follow up, their Australian colleague John Bolton forwarded the position to Schmidt in August. Matthews soon found that the radio source coincided with a 13th magnitude star and with a "nebular wisp or jet." Schmidt believed the wisp and the star were likely unrelated, but obtained a spectrum of the relatively bright star in late December 1962, using the 200-inch Palomar telescope. It appeared to be a blue star, but the spectrum was completely anomalous, featuring unknown broad emission lines. "It was clear that 3C 273 belonged to the class of 3C 48," Schmidt later wrote.[42] But the nature of both remained unknown. And that jet, it turned out, would be physically associated with the "star," a clue understood only later.

Schmidt has given us his account of the moment of discovery of what turned out to be the newest, most famous, and most astonishing class of active galaxy: "It was on February 5, 1963 that the puzzle was suddenly resolved. Cyril Hazard had written up the occultation results for publication in *Nature* and suggested that the identification results be published in an adjacent article. It was in the process of writing the article that I took another look at the spectra. I noticed that four of the six lines exhibited increasing spacing and strength toward the red. I attempted (not necessarily for any good reason) to construct an energy-level diagram based on these lines, then made an error which seemed to deny the regular pattern. I remember being slightly irritated by that, because it was clear the lines were regularly spaced – and to check on that I started taking the ratio of the wavelength of each line to that of the nearest Balmer line. The first ratio was 1.16, the second 1.16, the third . . . 1.16!" Schmidt measured some of the other lines and suddenly the answer dawned on him: "Clearly, a redshift of 0.16 explained all the observed emission lines! The extraordinary implications of a 'star' of 13th magnitude having a redshift of 0.16 were immediately clear"[43] (Figure 5.8).

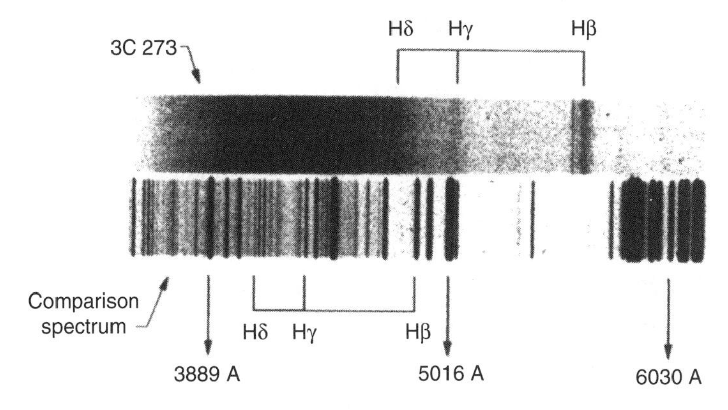

Figure 5.8. Evidence for quasars – the spectrum of 3C 273 (top), showing a redshift in the Balmer lines of hydrogen amounting to 16 percent with respect to the "at rest" comparison spectrum (bottom), placing the object at cosmological distances. The spectra are negatives, in which the bright lines appear dark. Courtesy of Maarten Schmidt, "The Discovery of Quasars," *Proceedings of the American Philosophical Society* 155 (2011): 142–6.

In his moment of discovery Schmidt realized that some of the emission lines of 3C 273 were actually well-known "Balmer" hydrogen lines with an extremely large redshift, some 16 percent, meaning they were not only beyond our Galaxy, but at the exceedingly large distance of almost 2 billion light-years. With this Rosetta Stone decipherment, Schmidt called his Caltech colleague Jesse Greenstein to his office where, "within minutes," they realized that the lines of 3C 48, discovered earlier by John Bolton, had been redshifted by 0.37, amounting to a distance of 3.5 billion light-years. These results were published in four consecutive articles in *Nature* in 1963, including Schmidt's on 3C 273, "A Star-like Object with Large Red-shift." Early the following year Schmidt and Matthews reported two more similar objects, 3C 47 and 3C 147. This paper was the first to use the term "quasi-stellar," rather than the more unwieldy "star-like object" in the title of the first paper. [44] By 1964, the Chinese astronomer Hong-Yee Chiu suggested that the term "quasi-stellar objects" be shortened to "quasar"; this term caught on, though "quasi-stellar" was still used throughout the 1960s and beyond, despite the fact that quasars were completely unrelated to stars.

The interpretation of "quasi-stellar objects," "QSOs," or "quasars," however, was not so straightforward. In his published article in 1963 Schmidt himself pointed out that an alternative explanation was a "gravitational redshift" caused by a massive star located inside the Galaxy. Based on a variety of considerations he concluded that "the explanation in terms of an extragalactic origin seems most direct and least objectionable." This conclusion was strengthened in a long and still historically fascinating joint article by Greenstein and Schmidt on 3C

48 and 3C 373 published the following year.[45] If the extragalactic interpretation was true, the prodigious amounts of energy at these distances made them a new class of object. This interpretation was made even more difficult by another extraordinary fact reported by the astronomers Harlan Smith and Dorrit Hoffleit in 1963. Examining Harvard's historical plate collection, they found that 3C 273 had varied in brightness over the past eighty years with periods as short as one year. This implied an object only a few light-years in diameter, even if the disturbing effect causing the variation propagated at the speed of light. Yet if the object was a galaxy, it should typically be more than 100 light-years in size. It was hard to see what kind of an object could generate this phenomenon.[46]

Quasar hunting became a favorite sport of astronomers, and their popular appeal is evident in the fact that Schmidt appeared on the cover of *Time* magazine for March 11, 1966. But the energy source still remained an enigma, Greenstein and Schmidt explicitly declaring in their 1964 paper, "We deliberately have not attempted to discuss the origin of these large energies, nor do we discuss the numerous other physical problems concerned with suggested mechanisms in the quasi-stellar objects." As Greenstein recalled thirty-five years later, "we did leave out the words 'black hole,' but we vigorously emphasized the occurrence both of renewal and of gravitation as the ultimate energy source."[47] Their work triggered an outburst of activity, leading observers and theorists to discuss the possibilities of gravitational collapse. In 1964 and 1965, Edwin Salpeter and Yakov Borisivich Zel'dovich independently suggested "accretion of interstellar matter by massive objects" as the source of energy.[48]

A few well-known astronomers, including Margaret and Geoffrey Burbidge, Fred Hoyle, and Halton Arp, continued to believe that quasars were not at cosmological distances.[49] Though the survivors held this view even into the early twenty-first century, after the detection in 1964 of the cosmic microwave background as a remnant of the Big Bang, there were increasingly few proponents of this renegade view. As we shall see below, the standard interpretation came to be that quasars are the cores of extremely distant active galaxies, the most luminous of all classes of active galaxies discovered before or since. Their extreme distance of billions of light-years makes them the most energetic objects in the universe aside from short-lived explosive events like supernovae and gamma ray bursts. Their vast amounts of radiation, and sometimes jets of material, originate from their centers, and their brightness may vary over time scales from months to hours. Today, different quasars have been observed in radio, infrared, optical, ultraviolet, X-ray, and gamma-ray regions of the spectrum, and with a variety of ground-based and space-based telescopes, including all four of NASA's Great Observatories: Hubble, Compton, Chandra, and Spitzer, as well as Fermi.

Quasars also stand as an example of the importance of new technology to the process of discovery, in this case the whole new regime of the radio spectrum. Schmidt pointed out that "Quasars are hard to find in the optical sky; there are as many as 3 million stars brighter than the brightest quasar, 3C 273. The situation is radically different at radio wavelengths. In the 3C catalogue 3C 273 is the sixth strongest source above galactic latitude 15 degrees. In hindsight, then, it is clear why radio astronomy was destined to lead us to the first quasars. If radio astronomy had developed much later, X-ray astronomy would have played the same role for the same reasons."[50] As it turned out, X-ray astronomy was destined to make its own landmark discoveries in the understanding of active galaxies.

* * *

Just when astronomers thought galactic astronomy couldn't get any stranger, in 1972, a decade after the discovery of quasars, Peter Strittmatter at the University of Arizona and his colleagues declared yet another class of objects of what they called the "BL Lacertae type." "Recent investigations have suggested the existence of a new class of astronomical objects with the following characteristics," they wrote in the *Astrophysical Journal*. Those characteristics included rapid variations in intensity at radio, infrared, and visual wavelengths, a high proportion of emission at infrared wavelengths, absence of emission lines in their spectra, and strong and rapidly varying polarization at visual and radio wavelengths.[51] They wrote that the objects were compact sources and very similar to quasars, but exhibited no spectral lines, making a determination of their distances difficult. Reminiscent of statements made when radio galaxies and quasars were discovered, they found it "hard to understand within any reasonable physical model how a very small source of radiation can produce such a large observed flux unless the objects are not far away." But because of their similarity to quasars, even without the broad emission lines, they declared BL Lacertae objects likely to be extragalactic.

Two years later, in 1974, astronomers identified BL Lacertae itself with a galaxy rather than a star. In concluding their otherwise staid paper, astronomers J. B. Oke and James Gunn wrote, "the same phenomena and approximately the same level of outrageousness exist for BL Lac as for higher-redshift QSOs if they are placed at the distances suggested by their redshifts."[52] With astronomers' penchant for thrift, their name was soon shortened to "BL Lacs," and by 1978 Edward Spiegel dubbed them "blazars," a term first used in the title of a paper in 1984 and that has caught on since that time.[53] More than 780 blazars had been discovered by 2012 by the Fermi telescope alone, though these blazars include much more than classical BL Lacs – an example of expanding the boundaries of a particular class.

The rise of the blazars provides a final illustration in the realm of active galaxies of the extended nature of discovery and the difficulties of declaring a new class of object. The objects that came to be called blazars were first believed to be irregular variable stars in our galaxy; they changed in brightness over periods of days or years, but with no pattern. BL Lac itself, discovered in 1929, varied from 13th to 16th magnitude with optical fluctuations of almost a magnitude in a few days and radio variations over a month. But in 1968, John Schmitt of the David Dunlap Observatory identified the BL Lac "variable star" as a powerful radio source, and pointed out that its optical properties, radio polarization, and unusual microwave spectrum "make it outstandingly interesting."[54] Only when they were inferred to be extragalactic in 1972 were BL Lacs given the status of a new class of object. It was the last (thus far) among the classes of active galaxies to be discovered, classes that may be collapsed into one if unification theories are successful.

* * *

The detection and extragalactic interpretation of the four classes of active galaxies, today referred to as active galactic nuclei, or AGNs, was only the beginning of the story. What was lacking was an understanding of their nature. Even a basic understanding of their true nature required some knowledge of their energy source, and there was no lack of speculation. Chain reactions of supernovae, galactic flares, stellar collisions, and especially supermassive stars were all discussed. As astronomer Susan Collin has pointed out, the Burbidges, Hoyle, and Willy Fowler had all been very active in stellar studies, and when it came to active galactic nuclei they naturally sought out a stellar explanation. However, all these possibilities were eventually dismissed, due to their shortcomings involving instabilities, inefficiencies, and other technical reasons.[55]

Eventually the true answer was to come in the form of the concept of black holes, infinitely dense regions of space with gravitational fields so intense that even light cannot escape. With black holes we enter a world bizarre even to astronomers accustomed to the extreme productions of nature such as white dwarfs and neutron stars. It is a world in which matter is compressed into an infinitely small volume known as the singularity, a world populated by objects with so-called "event horizons," the distance at which it is impossible to escape the gravitational pull of the black hole. Yet, it was a world that theorists had foreseen, though not on the galactic scale, long before active galactic nuclei were found in the last half of the twentieth century. Astronomers found the idea too bizarre to believe.

In fact the idea of black holes has a long, if sketchy and sporadic, history dating back to the eighteenth-century British natural philosopher John Michell, the French mathematician Pierre-Simon de Laplace, and the German astronomer Karl Schwarzschild.[56] Schwarzschild, father to the same Martin Schwarzschild we last saw in Chapter 4 in connection with the theory of subgiant stars, described the gravitational field of a point mass in mathematical terms in 1916, the year of his death at the young age of forty-two from a painful disease contracted on the Russian front during World War II. As it turns out, he was the first to discover the solutions of Einstein's equations of general relativity that describe non-rotating black holes.

A more concrete part of the story stemmed from the discovery of the extremely compact white dwarf stars in 1910, as discussed in the previous chapter. We recall that in 1931 Chandrasekhar showed that the compaction of a white dwarf could not continue forever due to electron pressure. He also calculated the upper mass of a white dwarf, now known as the Chandrasekhar limit of 1.4 solar masses, beyond which a star would be forced to collapse even further beyond the white dwarf stage.[57] In the Chandrasekhar-Eddington imbroglio that followed, Chandra embraced relativistic degeneracy as the basis of his work, while Eddington believed it was a mathematical result with no astrophysical meaning.

Chandrasekhar's calculations begged a question: what exactly would happen when a star more massive than 1.4 solar masses collapsed? The answer came in part when Walter Baade at Mt. Wilson and his colleague Fritz Zwicky at Caltech proposed a new and more powerful class of exploding objects they termed "supernovae" in 1934. In the same paper they proposed the idea that "neutron stars," supported by neutron pressure rather than the electron pressure of white dwarfs, might be collapsed objects produced by supernovae explosions of massive stars. Chandrasekhar himself began to glimpse the truth about white dwarfs and more massive stars as endpoints of stellar evolution when he wrote in 1935, "for a star of small mass the natural white dwarf stage is an initial step towards complete extinction. A star of large mass [greater than the upper limit for white dwarfs] cannot pass into the white dwarf stage, and one is left speculating on other possibilities."[58]

Taking the argument one step beyond neutron stars, what would happen if a neutron star, like a white dwarf, had a maximum mass beyond which even the neutron pressure could not support the star? What could be more dense than a neutron star? In 1939, J. Robert Oppenheimer and his graduate student, George Volkoff, building on Chandrasekhar's work, provided an answer: "stellar matter after the exhaustion of thermonuclear sources of energy will, if massive enough, contract indefinitely, although more and more slowly, never reaching true equilibrium." Later in the year, with his student Hartland Snyder, Oppenheimer followed up on the theoretical implications of this mind-boggling scenario, still finding that for a sufficiently massive star, "unless fission due to rotation, the radiation of mass, or the blowing off of mass by radiation, reduce the star's mass to the order of that of the sun, this contraction will continue indefinitely." In other words, a star above a certain mass would collapse completely to an infinitely dense region, what we today term a singularity, or "black hole."[59]

As of 1940, then, most of the work on gravitationally collapsed objects had been in connection with the collapse of stars to form the observed white dwarfs, the theorized neutron stars, and beyond that something so bizarre it dare not be named. Neutron stars were finally discovered in the form of pulsars in 1967, but more massive collapsed objects had never been observed, even by their indirect effect. Nevertheless, with the discovery of quasars and other active galaxies in the 1960s, the now hot topic of collapsed objects was reluctantly transferred to the galactic realm and became a very general and non-rigorous explanation for their observed energies.

As mentioned above, it was Salpeter and Zel'Dovich who in the mid-1960s first suggested "accretion of interstellar matter by massive objects" as the energy source for active galaxies. But as one astronomer active in the field recalled, "nobody took the idea seriously. Black holes ... were rapidly quite popular among theoretical physicists, and a large literature was devoted to them, but they were considered as a utopia by astronomers, and not associated with the release of energy in AGN."[60] It was a Princeton physicist, John Archibald Wheeler, who christened such gravitationally collapsed objects "black holes" in 1967 during a talk at NASA's Goddard Institute for Space Studies. A possible turning point in their acceptance by astronomers might have come with Donald Lynden-Bell's 1969 paper, "Galactic Nuclei as Collapsed Old Quasars," a concept he pushed at a Vatican conference on "Nuclei of Galaxies" the following year. But the real turning point seems to have come with the paper of the British astronomer Martin Rees at a 1977 meeting in Copenhagen, after which astronomers increasingly accepted the idea. So increasingly did popular culture, as evidenced by the Disney movie *The Black Hole* (1979). And by this time pulsars had shown that extremely compact neutron stars were a reality.[61]

The question was how black holes could be observationally proven, a problem not synonymous with the detection of active galactic nuclei themselves, the phenomenon whose energy source required proof. More direct observational evidence of the mechanism was required, and it came only grudgingly and with the help of the opening of yet another realm of the electromagnetic spectrum – the very short wavelength X-ray realm.[62] Because X-rays do not make it through the Earth's atmosphere, the field began haltingly with rocket flights, long before spacecraft could provide more stable platforms above the Earth's atmosphere. In 1964, a team led by Herbert Friedman, using an Aerobee rocket flight from White Sands Missile Range in New Mexico, had discovered a bright X-ray source they labeled Cygnus X-1. After satellite observations in 1971 and ground-based spectra constrained the nature of the object, in 1972 Louise Webster and Paul Murdin of Royal Greenwich Observatory theorized it must be an X-ray binary inside our own Galaxy. [63] They and others suspected it harbored a stellar mass black hole after measurements of the orbiting blue giant component indicated the presence of an object of at least ten solar masses, too massive for even a neutron star. By 1974, after a massive effort, one astronomer put the odds at 80 percent that Cyg X-1 harbored a black hole. But there was no unequivocal black hole signature, and physicist Stephen Hawking bet physicist Kip Thorne that it was not a black hole. Hawking conceded to Thorne only in 1990.[64]

Ironically, stronger proof of black holes would come not from the stellar but the galactic variety, which, unlike stellar black holes, were never predicted to exist but were much more energetic. As we have seen above, and as Thorne has emphasized, their discovery was "serendipity in its purest form," even though the discovery of active galaxies was by no means synonymous with the discovery of black holes.[65] It was Lynden-Bell's paper of 1969 that argued convincingly that gas streams falling toward a black hole would spiral inward, producing what he called an "accretion disk." By the mid-1970s, Roger Blandford, Martin Rees, and Donald Lynden-Bell, all at Cambridge University, showed that such a rapidly spinning accretion disk would produce oppositely pointed jets of matter streaming from the center of the accretion disk and its black hole – the very type of object seen in 3C 273.[66]

The actual discovery of accretion disks and jets associated with other active galaxies fit the theory of black holes and was good circumstantial evidence. In 1984, CalTech astronomer John Tonry reported evidence for a "central mass concentration" of 5 million solar masses in the elliptical galaxy M32, one of the companions of the Andromeda Galaxy. A "black hole" was one of his candidates for an explanation.[67] Others made similar suggestions for other galaxies. But even in the early 1990s not all were quite convinced. Referring to the jet structures exhibited in some of these claims, Kip Thorne wrote in 1994, "These black-hole-based explanations for quasars and radio galaxies are so successful that it is tempting to assert they must be right, and a galaxy's jets must be a unique signature crying out to us 'I come from a black hole!' However, astrophysicists are a bit cautious. They would like a more ironclad case." Thorne pointed out that alternative explanations did exist, including a "rapidly spinning, magnetized, supermassive star, one weighing millions or billions of times as much as the Sun – a type of star that has never been seen by astronomers, but that theory suggests might form at the centers of galaxies."

But evidence of something more exotic kept piling up, especially with the eagle eye of the newly launched Hubble Space Telescope. In 1994, Hubble astronomers found circumstantial evidence for black holes in the giant elliptical galaxy M87, 50 million light-years distant, as well as in other galaxies in the form of an increase in the number of stars around galaxy centers, possibly due to the gravitational pull of black holes. With the installation of Hubble's corrective optics, known as COSTAR, during its first servicing mission in 1993, astronomers were able to measure spectroscopically the motion of the gas around the putative black hole using the Faint Object Spectrograph, and thereby determine its mass. By 1994, Hubble astronomers claimed "seemingly conclusive evidence" for a black hole of about at 3 billion solar masses at the center of M87, based on velocity measurements of orbiting gas.[68]

Later in 1994 radio astronomers confirmed a 40-million solar mass black hole in the spiral galaxy NGC 4258, and numerous other supermassive black holes have been "observed" in this way since then. By 1997, Hubble astronomers were suggesting that most galaxies contain black holes, some in more quiescent form in normal galaxies. Once galaxies become quiescent, they suggested, black holes are revealed only by the glow of the surrounding gases heated by the black hole's gravitational energy, or by the motions of nearby stars as they swirl around it. Consistent with this idea, but nevertheless shocking because so close to home, in 2008 astronomers confirmed a black hole of 4 million solar masses in the center of our own Milky Way Galaxy; at a distance of 28,900 light-years, it is the nearest supermassive black hole.

Thus, several decades passed before it was widely accepted that Active Galactic Nuclei (AGNs) – those powerful galactic centers first seen as Seyferts, then radio galaxies, quasars, and blazars – are powered by black holes. Because most galaxies are believed to harbor a supermassive black hole, and because 100 billion galaxies exist that can be seen from Earth, at least 100 billion supermassive black holes may exist in the observable universe. Because of their size supermassive black holes were the first to be observed, followed by stellar black holes. It has been estimated that one out of every thousand stars in our galaxy is massive enough to end its life as a black hole, in which case the Milky Way Galaxy may contain 100 million stellar-mass black holes. About a dozen have been identified, the nearest at a distance of 1600 light-years.

Intermediate black holes, weighing in at a few thousand solar masses, have remained the most elusive. Although in 2000 the ROSAT and Chandra X-ray Observatories identified extremely bright X-ray sources that could be interpreted as intermediate black holes in star-forming galaxies such as M82, the Hubble Space Telescope clinched the discovery in 2002 with the announcement of intermediate black holes in the globular cluster M15 at a distance of 32,000 light-years in Pegasus, and in the giant globular cluster G1, located at a distance of 2.2 million light-years away in the neighboring Andromeda Galaxy.[69] Those black holes were 4000 and 20,000 solar masses, respectively. In both cases the black hole masses were identified by the speed of stars orbiting them. This announcement culminated a thirty-year search for intermediate black holes. Always indulging their penchant for classification, astronomers now talk about three "classes" of black holes based on their mass: stellar, intermediate, and supermassive.[70]

* * *

Whether or not supermassive black holes should be considered a class of object in the galactic realm distinct from stellar black holes remains an open question, as indeed is the more general question of whether any type of black hole is in reality an "object." What now seems virtually certain is that the supermassive variety power active galactic nuclei and that more quiescent black holes form the central core of many normal galaxies as well. Perhaps more to the point is the question whether active galactic nuclei should be considered as four distinct classes of galaxies. Today most astronomers do consider them so, even in the midst of attempts to unify them under a coherent scheme. The data are still confusing, and undoubtedly incomplete, not uncommon characteristics in attempts at classification. But the criteria for classification are revealing, often harkening back to the original demarcations at the time of discovery.

It is no longer possible to classify active galaxies in terms of energy output across a particular segment of the spectrum. While first discovered based on their radio emissions, 90 percent of quasars are now known to be radio quiet. Radio galaxies and blazars are always radio loud. Seyferts, originally discovered in the optical, are now known to radiate in the infrared, ultraviolet, X-ray, and gamma ray regime, in addition to the optical wavelengths, and they are less powerful and lower luminosity than quasars. They are in general much closer to us than quasars or blazars, normally tens of millions of light-years rather than hundreds of millions or billions of light-years.

Then there is the matter of host galaxies in terms of the Hubble sequence. Unlike Seyferts, which are associated mainly with spiral galaxies, and radio galaxies found mainly in giant ellipticals, the Hubble Space Telescope has shown that quasars reside in a variety of host galaxy types, including spirals, ellipticals, and even colliding galaxies. Astronomer John Bachall encapsulated the uncertainties that remained more than three decades after the discovery of quasars when he commented regarding their discovery in so many types of galaxies that "If we thought we had a complete theory of quasars before, now we know we don't. No coherent, single pattern of quasar behavior emerges. The basic assumption was that there was only one kind of host galaxy, or catastrophic event, which feeds a quasar. In reality we do not have a simple picture - we have a mess."[71] Nevertheless, all quasars are active galaxies because of the vast amount of radiation they emit in all wavelengths, including some that emit high-energy gamma rays. Because quasars and other classes of active galactic nuclei are found only at great distances, they must become dormant as galaxies age, leaving behind a quiescent normal galaxy to live out its life. The exceptions are the radio-quiet Seyfert galaxies, which are much closer.

Then there is the question of the relation of quasars to BL Lacs, like radio galaxies usually associated with elliptical galaxies. BL Lacs have most of the characteristics of quasars except for the lack of broad emission lines, rapid and large variability, and somewhat less luminosity. In fact, many early quasars now turn out to be blazars, including the first one discovered, 3C 273. Why blazars vary in optical and gamma ray brightness over periods from minutes to days is still mysterious. The EGRET experiment on the Compton Gamma Ray Observatory in the 1990s detected 66 blazars; 233 blazars were known by 1995, and as of 2003 a few hundred were known. The Compton's spacecraft's successor, the Fermi Gamma Ray Telescope (formerly the GLAST) mission, has now observed almost a thousand. A blazar may well be an active galactic nucleus viewed head-on, made visible by its relativistic jet pointed directly at us. This orientation may enhance the continuum radiation while depressing any spectral lines.

Active galaxies thus well illustrate the problem of what it means to be a class. Most likely united in their class status by their source of energy in the form of black holes, they differ in many observable characteristics. Yet, with the possible exception of Seyferts, they may differ from each other only in the viewing angle as seen from Earth, in which case their status as distinct classes may be in jeopardy. In short, active galaxies in the form of Seyferts, radio galaxies, quasars, and blazars continue to exercise the imaginations of astronomers. Nowhere have their powers of discovery, interpretation, understanding, and classification been more tested than in the realm of the galaxies, first reconnoitered by Edwin Hubble, elaborated by ground-based and space telescopes, and studied by numerous astronomers in between. From Hubble to Hubble the realm of the galaxies has revealed its secrets in often roundabout ways of discovery, with more undoubtedly to come.

5.3 The Discovery of Structure: Clusters, Superclusters, Filaments, and Voids

In the same way that double stars and star clusters had to be discovered in the realm of the stars, so did double galaxies and larger clustering systems have to be demonstrated in the realm of the galaxies. This time, however, the distances involved were incomparably larger, straining not only contemporary ideas about the cohering power of gravitation, but also the very concepts of "object" and "structure." This time there were good reasons to believe that such clustering might not take place over such vast stretches of space and time. And this time the very structure of the universe was at stake, the cluster controversy for the first time providing empirical data to an enterprise that had heretofore been largely a speculative era of cosmology. What began tentatively with the discovery of galaxy clustering eventually revealed in the 1950s a new phenomenon not found in the realm of the stars: clusters of clusters. This "hierarchical clustering" revealed in turn the large-scale structure of the universe in the 1980s and beyond. By the end of the twentieth century extragalactic discovery had enabled a preliminary reconnaissance of the universe, a universe that only a century earlier had plausibly been argued to consist only of our own Galaxy. The process of discovery for hierarchical clustering stretched the very concept of discovery in all three of its components: detection, interpretation, and understanding.

The first evidence for gravitational clustering at what we now know are galactic distances came in the form of double galaxies. And, not surprisingly, given the telescopic power at his command, it came from William Herschel. Already in his "Catalogue of one thousand new nebulae and clusters of stars" published in 1786 based on his sweeps of the sky beginning in 1783, Herschel had observed and remarked on nebulae that were very close together, a practice he continued in his catalogues of 1789 and 1802.[72] And, only a few years after he declared double stars to be true physical "binary" systems (Section 4.5), he did the same for double nebulae. In his remarkable paper of 1811 on the construction of the heavens – the same one we discussed in Chapter 3 in relation to separating star clusters from true nebulosity – Herschel discusses the nature of 139 double nebulae (labeled 5 and 6 in Figure 3.2). "With regard to their being double nebulae, it may be objected that this double appearance may be a deception; and indeed if this were a double star, instead of a double nebula, there might be some room for such a surmise," Herschel wrote. Because of the large number of stars, he argued, optical alignment of double stars was more likely than for the less numerous nebulae. And he had an important physical reason: "Add to this their great resemblance in size, in faintness, in nucleus, and in their nebulous appearance; from all which I believe it must be evident that their nebulosity has originally belonged to one common stock." Moreover, Herschel also distinguished larger groups in his observations: "20 treble, 5 quadruple and 1 sextuple nebulae," all he believed originating from the breakup of "former extensive nebulosities." Although Herschel had no way of knowing these were what we today call "galaxies," most of them were in fact galaxies, and Herschel was seen by his successors as first detecting double galaxies as well as what we would today call galaxy "groups" such as our Local Group.[73]

Herschel, however, had no idea of the extent to which this clustering would be carried. Already in the eighteenth century telescopic observers had noticed a concentration of "nebulae" in certain regions of the sky. In the third edition of his catalogue of nebulae and clusters published in 1781, Charles Messier wrote that "The constellation of Virgo, & especially the northern Wing, is one of the constellations which encloses the most Nebulae."[74] Although Messier knew neither the true nature of the nebulae nor their distances, today we recognize that 15 out of the 110 objects in Messier's extended catalogue are actually galaxies in the Virgo Cluster. But like double stars and double galaxies, the distinction between optically aligned systems and physically bound systems was all-important. And even Herschel dared not suggest these were physically bound systems.

The discovery that large groups of galaxies tend to form coherent gravitationally bound groups awaited large telescopes that could detect more and fainter galaxies. Surprisingly, however, it did not require knowledge of the absolute distance of the galaxies, a time-consuming process involving Hubble's Cepheid variables and eventually redshifts obtained through spectroscopy and Hubble's remarkable "velocity-distance relation" of 1929. In any case, Cepheids could only be detected in nearby galaxies, and redshifts were still subject to interpretation. Rather, relative galaxy distances were inferred in a more roundabout way. Assuming all galaxies had the same luminosity, distances could be determined by apparent faintness. But, as in stars so in galaxies: there were giant galaxies and dwarf galaxies and everything in between, in other words, individual galaxies existed over a variety of luminosities. In Hubble's time this difficulty was circumvented by taking the average magnitude of a large group of galaxies in a given volume, and deriving the "luminosity function" of that volume of galaxies. On the reasonable but still bold assumption that the luminosity function remained constant for all distances in all directions, in a statistical sense "apparent faintness would measure relative distances, although absolute distances were unknown." Using this technique, the large-scale distribution of galaxies was found to be uniform, but the small-scale distribution was not, more galaxies appearing in some volumes of space than in others. And relative distances of these different galaxy-laden volumes were roughly indicated by their relative dimensions, the smaller "clusters" assumed also to be the most distant.[75]

With such nascent but necessary distance techniques – anchored by the few nearby galaxies in which the Cepheid distance technique could be applied – Hubble emphasized in his classic 1936 volume *The Realm of the Nebulae* that "Surveys of the sky show that nebulae are scattered singly and in groups of various sizes up to the occasional great clusters." Significantly, Hubble considered this perfectly sensible by analogy with the realm of the stars: "The small-scale distribution resembles that of stars in the stellar system. Analogies among the nebulae [galaxies] are readily found with individual stars, doubles, triples, multiples, sparse clusters, and open clusters. Globular clusters alone seem to have no counterpart in the realm of the nebulae."[76] Double galaxies such as the Magellanic Clouds and triple systems such as the Andromeda Galaxy with its two companions were well known. The trick would be to find the analogous larger clusters of galaxies.

The observational confirmation of larger galaxy groups was given credence with the discovery that the Milky Way Galaxy is part of a "Local Group" of galaxies. Hubble began detailed studies of nearby galaxies in 1925, and by 1936 he spoke of nine definite and three possible members of the Local Group. Walter Baade appears to have been the first to use the term "Local Group" in a paper the year before, where he noted, "our galaxy and the nearest extra-galactic systems form what may be termed a local group of nebulae." In an entire chapter devoted to the subject in *The Realm of the Nebulae*, Hubble related what was known about each of the nine certain members in addition to our own "galactic system": the Magellanic Clouds, Andromeda (M31) with its companions M32 and NGC 205, and M 33, NGC 6822 and IC 1613, as well as the three possible members. Of the nine certain members, four were irregulars, two spirals and two ellipticals, our own galaxy type still being unknown. Hubble characterized our local group as "a typical, small group of nebulae which is isolated in the general field." Such groups, he found, were very common.[77]

Much rarer, however, were great clusters of galaxies, averaging hundreds of members. The idea of such large clusters of galaxies was born in the 1920s, shortly after the extragalactic nature of some of the nebulae became known. Already in 1926 Harlow Shapley and his young colleague at Harvard, Adelaide Ames, wrote a brief paper on their "investigations of a cluster or cloud of a hundred bright nebulae of the spiral family," the same family in Virgo that Messier had spied 150 years earlier in much less detail. With Hubble's proof that spirals are actually extragalactic, they pointed out, "this work becomes a study of a higher system of stellar systems – a research, one might say, on a cloud of galaxies." They pointed out that the 24-inch Bruce telescope at Harvard's Southern station in Arequippa, Peru – ideally suited for such studies because a single image covered a wide field of 30 square degrees – had recently revealed several other "regions of high nebular frequency." But in particular they found the area of the sky in Coma and Virgo "remarkable for bright nebulae of the spheroidal, oval, spindle, and truly spiral forms."

The straightforward reasoning of Shapley and Ames in terms of physical, rather than apparent, clustering is notable. First, "No better illustration that these various forms are all sub-divisions of the same family could be desired, for in a definitely circumscribed area of approximately one hundred square degrees there are 103 of these objects much alike in brightness and, for the majority, fairly comparable in angular dimensions." There was only one more step: "From an inspection of the distribution, luminosity, and dimensions, one is led immediately to the conclusion that the cluster is a physical system – a group of all sorts of spirals associated in space." Among the 103 objects were seven Messier objects known to be spirals and four "globular nebulae," known today as ellipticals. Assuming galaxy luminosities comparable to those Hubble had already determined by the Cepheid method as a first approximation, they estimated the distance to what they called "the Coma-Virgo group" to be ten million light-years.[78]

By 1931, Hubble and Milton Humason listed eight such galaxy clusters in a follow-up to their famous paper on the velocity-distance relation published twenty-six months earlier. One-by-one they discussed the known data on these clusters – consisting of Virgo, Pegasus, Pisces, Cancer, Perseus, Coma, Ursa Majoris, and Leo. With the exception of Pisces, which Hubble termed "not a cluster, but a group of some 25 elliptical nebulae," all the clusters harbored hundreds of galaxies, each in a circumscribed volume of space ranging from 10 by 12 degrees for Virgo to 1 or 2 degrees for most of the rest. Virgo was obviously the first among equals, being the largest and nearest, and containing 16 of the 34 extragalactic objects in Messier's list, by Hubble and Humason's count. Others had been discovered well before Hubble and Humason's synthesis – the Coma cluster and Perseus cluster having been seen on Max Wolf's photographic surveys and reported in 1901 and 1905 respectively. In fact, the German astronomer Heinrich D'Arrest had visually observed the twenty-five brightest members of the Coma Cluster as early as 1865, and H. D. Curtis counted more than 300 on his Crossley reflector plates of the region in 1918 – altogether making difficult the idea of a "discoverer" of the Coma cluster. Other clusters had been discovered much more recently – the Ursa Major cluster being described by Baade in 1928 and the Leo cluster being "called to our attention" by W. H. Christie, who discovered it on plates taken with the 60-inch (Mt. Wilson) reflector in December 1929. In most of these cases Hubble uses the "discovery" terminology, obviously referring to it in its most rudimentary form of detection rather than interpretation or understanding, since some were discovered even before they were known to be extragalactic. As Shapley and Ames had put it, the study of a collection of nebulae had now become a study of higher systems of galaxies.[79]

More than that – and the entire point of Hubble and Humason's landmark paper – individual galaxies and clusters had also become tools in the velocity-distance relation. Clusters served as data for the most distant galaxies known, and thus the extension of the velocity-distance relation. Whereas

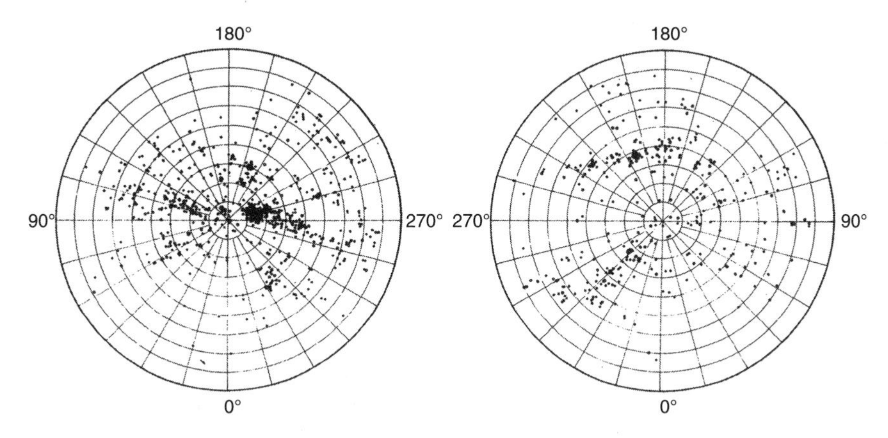

Figure 5.9. Early indications of galaxy clustering, according to Harlow Shapley and Adelaide Ames, from The Shapley-Ames Catalogue, 1932. The plots show the distribution of nebulae brighter than 13th magnitude in the northern (left) and southern (right) galactic hemispheres.

Hubble's initial 1929 paper on the velocity-distance relation had been limited by V. M. Slipher's observation of redshifts of nearby galaxies extending to five bright members of the Virgo cluster, Hubble's program now "contemplated the use of much greater distances, far beyond the limits at which stars can be detected, and hence included the several brightest members of as many clusters or groups as possible."[80]

The year 1932 saw a landmark publication by Shapley and Ames, the same year in which the latter tragically died in an accidental drowning at age thirty-two. In producing a catalogue of 1249 bright galaxies complete down to 13th magnitude, they advanced the identification of even more clusters. To their eyes, the plot of the 1025 galaxies brighter than 13th magnitude (Figure 5.9) clearly indicated "strong clustering in the northern galactic hemisphere, and the general unevenness in distribution" of galaxies. The Shapley-Ames catalogue served as the starting point for many other studies of galaxies, including clustering, and was still being revised fifty years later.[81]

The distinction between groups and clusters, however, was by no means clear even in the aftermath of these surveys. When in 1933 Harlow Shapley published a paper on the average density of matter in twenty-five "groups of galaxies," he included both what we would now call groups and larger clusters. Shapley remarked that "A dozen or more groups of galaxies are now known in which the members exceed two hundred in number. Some of these systems are spheroidal in form, centrally concentrated and of rather definite boundary. The majority of the groups, however, are of irregular form, with indefinite boundaries, and small in membership." And he added the analogy that Hubble would also later use: "Such groups are to the metagalactic system as the open star clusters are to the stellar system."[82]

All these studies begged a series of questions: how to define a cluster, how to distinguish a group from a cluster, and why distinguish the two anyway? In *Realm of the Nebulae* Hubble emphasized, "The nomenclature of the clusters is still arbitrary, and in these discussions the term 'cluster' will be restricted to the great clusters alone. The term 'group' will be used for all the lesser organizations." Ever tactically cautious in his strategic boldness, Hubble spoke of about twenty galaxy clusters, with "fragmentary evidence" that more existed. Despite the fragmentary nature of the evidence, in his mind and in the minds of other astronomers, there was reason enough to separate groups from clusters as qualitatively different classes of objects. In Hubble's formulation it began with the disparity in numbers (averaging 500 versus a dozen), continued with appreciable, though "not very conspicuous," concentration toward the center resembling open rather than globular star clusters, and ended with the observation that elliptical nebulae predominated – all in contrast to much smaller "groups" of galaxies. Hubble speculated that over time clusters might dissipate into the general field.[83]

One who did not distinguish qualitatively different properties based on sizes of clusters was the Swiss Caltech astronomer Fritz Zwicky, a colorful, irascible, and brilliant figure who came up with not one but three novel methods for determining masses of galaxies and clusters. In a 1937 paper remarkable for its prescience on a number of concepts, Zwicky argued that luminosities gave only a lower limit for masses, and that internal rotations such as claimed by van Maanen did not work at all. Zwicky conceived of galaxies having a central core "whose internal viscosity due to the gravitational interactions of its component masses is so high as to cause it to rotate like a solid body." Applying the concept to the Coma cluster, Zwicky found from the orbital motions of its constituent galaxies a mass for the cluster that implied "dark matter" must be holding it together in an amount much greater than the luminous matter that was visible. This was the first use of the so-called virial theorem method, whereby random motions of the members of a cluster are used to yield the mass of the entire cluster. In the case of the Coma cluster, Zwicky found a mass of 45 billion solar masses, a considerable underestimate but also a considerable improvement over any mass calculated before. A second distance determination method called for the use of a "gravitational lens" effect, one of the earliest references to a concept now in widespread use. A third method applied statistical mechanics to the system of nebulae, conceived of as a system similar to molecules in a gas.[84] In a final tour-de-force – and perhaps a poke at Hubble's estimates based on the distance-velocity relation – Zwicky concluded his article by suggesting that an age of one billion years for the universe "is hardly adequate" considering the relatively stationary distribution of nebulae in clusters, a concept he elaborated in 1939.[85]

In his 1937 article Zwicky adopted the terminology of the "Coma cluster" as one of several "great clusters" detected. But Zwicky had a different idea of "cluster" than most of his colleagues, as is evident when he suggested that Hubble's separation of nebulae into "relatively few *cluster nebulae* and a majority of *field nebulae*" might be fundamentally wrong. Instead, "from the preliminary counts reported here it would rather follow that practically all nebulae must be thought of as being grouped in clusters – a result which is in accord with the theoretical considerations." In other words, Zwicky believed a galaxy cluster was analogous to an isothermal gravitational gas sphere that extends indefinitely "until its extension is stopped through the formation of independent clusters in the regions surrounding it."[86] This was practically the opposite of Hubble's idea of more well-defined clusters dissipating into space, but as we shall see below, Zwicky's concept helps to understand his later opposition to the very idea of "superclusters." In his 1938 paper "On the Clustering of Nebulae" and in subsequent work, Zwicky held to the idea that "practically all nebulae are bunched in more or less regular clusters and clouds of nebulae." Noting that about twenty clusters were located within the sphere of radius 40 million light-years, Zwicky pushed the idea of extended "cluster cells," each containing 2000 to 4000 galaxies. He estimated that some 30,000 clusters might be contained in the 500 million light-year limit of the 100-inch telescope.[87]

In Hubble's more conventional sense, as late as 1949 only a few dozen clusters of galaxies were known. But this changed dramatically in the 1950s with the advent of large-scale photographic programs, including Lick Observatory's proper motion survey and, most especially, the National Geographic Society–Palomar Observatory Sky Survey, which revealed tens of thousands of galaxy aggregates. California astronomer George Abell determined the 2700 richest of these in his catalogue of clusters published in 1958, a number that was extended to more than 4000 objects in 1989 after Abell's death.[88]

All the while, the ever more numerous galaxy clusters were employed to probe the extent of the universe ever further into deep space using the velocity-distance relation. In 1956, Humason, Nicholas Mayall, and Allan Sandage used 474 of the brightest members of clusters of galaxies to determine a Hubble recession factor, now called the Hubble constant, of 180 kilometers per second per megaparsec.[89] Much controversy ensued over the next fifty years about the true value of the Hubble constant, with most determinations found to be between 50 and 100, corresponding to an age of the universe between 10 and 20 billion years – quite different from the one billion Zwicky had mocked. In 1994, a team led by

Wendy Freedman used the Hubble Space Telescope to determine a distance of 17 megaparsecs (about 56 million light-years) to the Virgo Cluster spiral galaxy known as M100. The team determined a Hubble constant of 72, corresponding to an age of the universe of 13 billion years, with an uncertainty of a billion years. The currently accepted value, obtained with the help of spacecraft such as WMAP and Planck, is between 67 and 73 kilometers per second per megaparsec, or 13.7 billion years old, with an uncertainty of only a few hundred million years. Hubble's velocity-distance relation had done quite well.[90]

Already by the 1930s it was clear that clusters were a common phenomenon in the universe, and very useful for probing its structure. In the conclusion to his 1938 paper, Zwicky wrote that "Clusters of nebulae are the largest known characteristic aggregations of matter, and their investigation provides the last stepping stone for the investigation of the accessible fraction of the universe as a whole." He viewed his Coma cluster counts as having furnished the first and only proof that Newton's law of gravitation describes the interactions among nebulae in clusters. In fact, galaxy clusters were only the beginning of the hierarchy that gradually revealed itself, accompanied by considerable controversy.

* * *

The discovery of a higher order of clustering came as a surprise, and was slower to be accepted than the idea of lower-order clustering. As usual there were precursors. The German natural philosopher Johann Heinrich Lambert, a contemporary of Immanuel Kant and Thomas Wright of Durham, had proposed a hierarchy of systems in the universe all connected by gravity. More seriously, the Swedish astronomer Carl Charlier in 1908 and 1922 had proposed a kind of hierarchical clustering in the sense of nested universes – spurred on as a way of resolving Olbers's paradox, whereby the sky should not be dark at night in an infinite universe with infinite numbers of stars. In a hierarchical universe, Charlier reasoned, the sky would remain dark because each cluster of larger size has a lower density, and the light reaching us would not fill the sky. Charlier's argument was flawed; the idea of hierarchical clustering was not revived for thirty years, and even then it was not widely believed for another thirty years. Its chief modern proponent, the American astronomer Gérard de Vaucouleurs, offered two reasons why the idea "nearly vanished" after Charlier's proposal: Hubble and Humason's redshift data were seen as a better way to resolve Olbers's paradox, and the "working hypothesis" of an isotropic and homogeneous universe, introduced as a simplifying assumption by theoretical cosmologists, "was soon accepted as fact and elevated to the status of dogma."[91] De Vaucouleurs's implied critical attitude toward how science sometimes works was no fluke, but a long-held feeling based on his own experience.

Aside from these early suggestions, our "local" neighborhood was naturally the first to support the idea with solid observational evidence. It was only in 1953 – the year of Hubble's death – that de Vaucouleurs suggested the Virgo cluster might be a dominant member of a large aggregate of galaxies that formed what he called a "supergalaxy." Among his evidence was the spatial distribution of galaxies in the Shapley-Ames catalogue, claims based on radial velocities of a differential rotation of the "inner metagalaxy," and the strong flattening of the system, often a feature of rotating systems. By 1958, he referred to the supergalaxy as the "Local Supercluster." As he put it in this paper, "Preliminary evidence has been presented . . . that the majority of the brighter galaxies in the Harvard Survey [Shapley and Ames, 1932] and a good many of the fainter ones listed in the NGC, whether members of recognized groups or clusters or so-called 'field nebulae,' belong to a flattened super-cluster of galaxies, the 'Local Supergalaxy,' of which the Local Group of galaxies is merely an outlying condensation."[92] He estimated the diameter of the supercluster at 60–100 million light-years, depending on the still uncertain distance to the Virgo cluster – a cluster that so dominates the Local Supercluster that it is sometimes called the Virgo Supercluster. By 1974, de Vaucouleurs had identified fifty-four clusters forming the Local Supercluster.[93]

The general idea of superclusters, however, remained controversial for at least a quarter century after de Vaucouleurs's initial proposal in 1953. While George Abell agreed in his 1957 catalogue of 2712 rich clusters that "hyperclusters" did in fact exist, some astronomers, including Zwicky, found no such higher order clustering. Many others dismissed superclusters as accidental alignments or sheer speculation well into the 1970s. In the 1970s, however, Princeton University astronomer James Peebles and his colleagues, using galaxy counts and a tool called the "galaxy correlation function," showed that galaxy clustering appeared to occur on a very wide range of scales, from small groups to superclusters. Using the Lick Observatory counts of galaxies first carried out by Donald Shane and C. A. Wirtanen, they produced a famous "Map of a Million Galaxies," widely published both in science publications and in popular culture venues such as posters and Stewart Brand's *Whole Earth Catalogue* as a way of graphically depicting a perspective on Earth's place in the universe.[94] The map showed a very inhomogeneous distribution of galaxies, but even Peebles was skeptical of the reality of the true nature of the inhomogeneity because of the tendency of the human eye to find patterns.

Even more skeptical was Zwicky, who vigorously denied higher-order clustering throughout his career into the 1970s, partly as a question of nomenclature since he considered clusters to consist of anything ranging from small groups of galaxies to those spanning 150 million light-years. In any case, higher-order clustering is different from one large cluster, and in fact is one of the distinguishing features of a cluster from a supercluster. Concentrating on one particular area of the sky as a test case, Zwicky wrote that "there are no indications whatsoever for the existence of any systematic clustering of clusters of galaxies."[95] What was needed was definitive redshift data, and shortly after Zwicky's death in 1974, the Estonian astronomer Jaan Einasto and his colleagues used just such data to show beyond doubt that structure did indeed exist beyond galaxy clusters.[96] Such redshift data, which gave three-dimensional information including distance, finally proved the existence of superclusters.

Why did the "discovery" of superclusters take so long? De Vaucouleurs characterized the discovery of the Local Supercluster as "a good example of how long it takes to convince the scientific community of the reality of some simple facts which although obvious for many years to a few unprejudiced researchers, were persistently denied or ignored by the majority." He held it up as "yet another example of the damaging effect of prejudice on the progress of science."[97] The problematic concept of an "unprejudiced" astronomer aside, de Vaucouleurs seems himself to have oversimplified the concept of discovery. It was not a simple matter of looking and seeing, as he well knew. He pointed out that William and John Herschel, and even the popularizer R. A. Proctor after mapping the nebulae in Herschel's *General Catalogue* (1864), had noticed the concentration of nebulae in what we now believe comprises the Local Supercluster. De Vaucouleurs himself gave credit for discovering clustering "roughly along the median (galactic) longitude 100 degrees" to J. H. Reynolds in the 1920s. Lundmark noticed the same thing in 1927. But neither appreciated the significance of this clustering, and neither followed up.[98]

Nor did the more mainstream astronomers interpret their data as hierarchical clustering. Although from de Vaucouleurs's point of view the plots of the Shapley-Ames galaxies brighter than 13th magnitude "strikingly confirmed the phenomenon discovered by Reynolds," those two astronomers "merely commented on the elongated cloud of galaxies stretching across Virgo and Centaurus, which they described as 'an extension of the Coma-Virgo supergalaxy'." And Zwicky himself saw the same thing, but did not interpret it as hierarchical clustering. De Vaucouleurs thus characterized his own work in the 1950s as a "rediscovery" of the Local Supercluster. To him, referring to the Milky Way of galaxies all around the sky in both hemispheres, even if faintly populated in the southern galactic hemisphere, "it seemed obvious that unless improbable chance encounters had accidentally brought together nearly exactly in the same plane a number of independent clusters and clouds of galaxies, this phenomenon indicated the existence of a large, flattened supersystem of galaxies including our own Galaxy and Local Group in peripheral location reminiscent of the outlying position of the Sun in the Galaxy." In characterizing his work as a "rediscovery," de Vaucouleurs perhaps did not give enough credit to the structure of discovery as consisting not only of detection and interpretation - undertaken in spades by his predecessors - but also understanding.[99]

The existence of superclusters is now widely accepted, including de Vaucouleurs's "supergalactic" coordinate system recognized in 1953 by analogy with the galactic coordinate system.[100] Through the relentless reach of gravitation, overcoming some of the Hubble flow due to the expansion of the universe, superclusters apparently cohere over a span a hundred million light-years, ten times more extensive in size than clusters. Whereas the Virgo Cluster with its 2000 galaxies spans about 8 million light-years, the "local" Virgo Supercluster, with its 100 galaxy groups and clusters, covers 110 million light-years. About 50 superclusters are known, 17 of them within a billion light-years of our Local Supercluster. Within 200 million light-years is the Hydra Supercluster, about 100 million light-years long. Stretching between 150 and 400 million light-years away is the Pavo-Indus Supercluster, not particularly rich but identifiable nonetheless. The Perseus-Pisces Supercluster is one of the largest structures in the universe, at a distance of about 250 million light-years. Next out is the Coma Supercluster, located at about 350 million light-years distant, some five or six times farther away than the Virgo Cluster. The Shapley Supercluster - named after Shapley when Somak Raychaudhury rediscovered it in 1989 and referred to it as the "Shapley Concentration," a concentration that Shapley had noted already in 1930 - lies at 650 million light-years. It has been estimated that about 10 percent of the motion of our Local Group of galaxies is due to the pull of the Shapley Supercluster. Such is the power of gravity, the weakest of the four fundamental forces in the universe.[101]

While Hubble had clearly accepted clusters of galaxies, he had not lived long enough to see the supercluster controversy blossom in the form of hierarchical clustering. Summarizing results in his 1936 book, he had noted "the small-scale distribution is irregular, but on a large scale the distribution is approximately uniform. No gradients are found. Everywhere and in all directions, the observable region is much the same." This concept held true even with the discovery of clusters and superclusters: over large enough areas one could still conceive of a uniform distribution of matter. Surprisingly, that too was to change at the end of the twentieth century with the discovery of large-scale structure.[102]

* * *

The existence of structures larger than superclusters, on the scale of hundreds of millions to billions of light-years, awaited surveys of many more galaxy redshifts than had been accumulated even by the late 1970s. Even though Edwin

Hubble in the 1920s waffled over whether redshifts in the absorption or emission lines of galactic spectra were actually a measure of recessional velocity, and that these velocities were in turn a measure of distance, others did not hesitate to interpret the correlation in this way, and Hubble's velocity-distance relationship published in 1929 became one of the landmarks of twentieth-century cosmology – a prime example of the discovery of a new phenomenon by contrast to a new class of object (see Chapter 7). In any case, lists of galactic redshift measurements were gradually and painstakingly built up over the decades. Some 600 were known by 1956, 2700 by 1976, 5000 by 1980, and 30,000 by 1989. During the sixty-year period from 1930 to 1990, measurement time for a single spectrum went from several hours with a spectrograph on the Mt. Wilson 100-inch telescope to a few minutes with digital detectors and image intensifiers. Still, as the authors of the first large redshift survey emphasized in recounting this history and announcing the first definitive large-scale structure, these advances allowed the mapping of only one ten-thousandth of the volume of the visible universe, equivalent to the surface of Rhode Island compared to the surface of the Earth.[103]

Even so, an early hint of large-scale low-density "voids" came in 1981, with the discovery of the "Boötes Void," a region 250 million light-years in diameter where the density of galaxies was observed to be less than 20 percent of the average. This was the largest of several other isolated regions previously noticed, and one of several dozen voids or "supervoids" since discovered. Another hint came in 1982 with the discovery of long sheets, walls, or "filaments" of galaxies.[104]

But the first large redshift survey to reveal a pattern in the large-scale structure of the universe was undertaken by the Harvard Center for Astrophysics (CfA), led by Margaret Geller and John Huchra. Their survey, begun in the early 1980s, consisted of thousands of galaxies in a 6 by 117 degree slice through the Coma cluster out to 600 million light-years. When their graduate student, Valerie de Lapparent, first plotted 1057 of these galaxies in the summer of 1985 as part of her dissertation, the "slice map" revealed chains and sheets of galaxies separated by giant empty regions they called voids (Figure 5.10). "Several features of the results are striking," they wrote in their paper published the following year. "The distribution of galaxies in the redshift survey slice looks like a slice through the suds in the kitchen sink; it appears that the galaxies are on the surfaces of bubble-like structures with diameter 25–50 h^{-1} Mpc. This topology poses serious challenges for current models for the formation of large-scale structure."[105] In the lingo of astronomers, 50 h^{-1}, where h is one-hundredth of the Hubble constant now believed to be between 67 and 73 (km/sec)/megaparsec, is equivalent to about 225 million light-years.

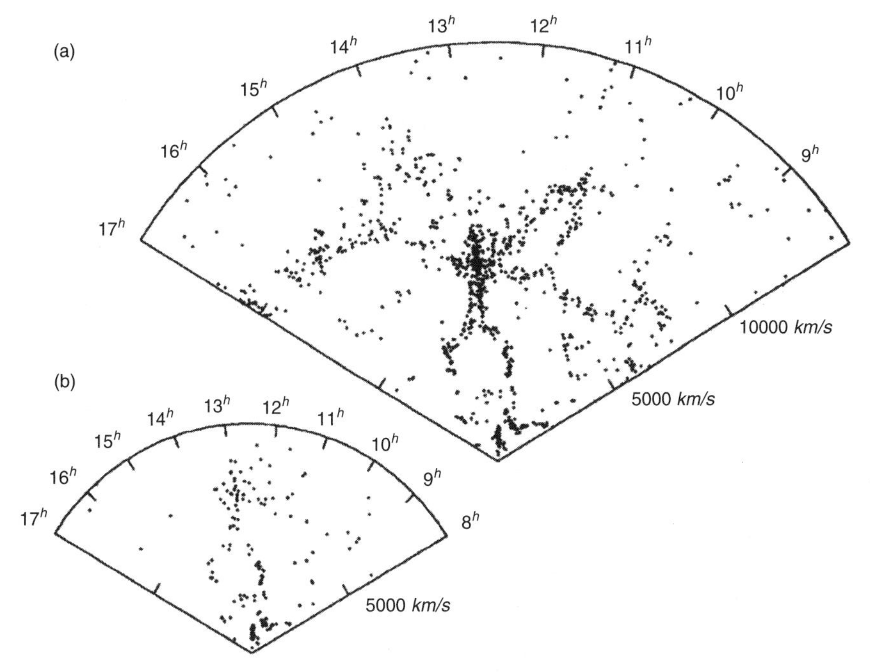

Figure 5.10. A slice of the universe, evidence for large-scale structure of filaments and voids, 1986. The top represents a map of observed velocity of 1061 objects plotted vs. right ascension in the declination wedge between 26.5 and 32.5 degrees. The scale along the right side is 15,000 kilometers per second at the top to 0 at the bottom. The smaller wedge, given for comparison, plots a smaller sample of 182 objects, demonstrating how larger samples are important for seeing large-scale structure. From Lapparent, Geller, and Huchra, "A Slice of the Universe," *ApJ* 302 (1986): L2.

This result was truly astonishing. "Many astronomers had imagined roughly spherical galaxy clusters floating amongst randomly scattered field galaxies, like meatballs in sauce," two galactic astronomers wrote looking back from the vantage point of the year 2000. "Instead, they saw galaxies concentrated into enormous walls and streamers, surrounding huge voids that appear largely empty."[106] These structures have variously been called "sheets," "walls," "bubbles," and "holes," but the terms most used are filaments and voids.

The Harvard results were only the beginning. The following year astronomers reported a thin "filament" of galaxies with a narrow redshift range, known as the Perseus-Pisces filament.[107] And when the Harvard map was completed in 1989 incorporating 14,000 bright galaxies, it revealed a structure known as the "Great Wall," a sheet of galaxies 500 million light-years long, 200 million light-years wide, and 15 million light-years thick, surrounded by great voids

200 million light-years in diameter.[108] In 1986, J. Richard Gott and his colleagues at Princeton showed that the large-scale structure was "sponge-like," with the galaxies representing the material of the sponge and the voids representing the holes in the sponge. Others prefer to describe it as bubble-like. Ever-deeper surveys continued to show the same structure at increasing distances. In 1996, the Las Campanas Redshift Survey, employing data from 26,418 galaxies, showed the cellular structure extending at least four times further than the original Harvard survey. In 2001 astronomers at the Anglo-Australian Telescope announced that their survey of redshifts of 180,000 galaxies, the so-called "2dF" survey, showed the same cell structure out to five times the distance of the Huchra-Geller survey.[109]

By 2003, the Sloan Digital Sky Survey had measured redshifts of a million galaxies out to 2.5 billion light-years. Like the nearer Harvard survey, the Sloan survey found voids and filaments, in this case out to 2 billion light-years, including a 1.37 billion-light-year structure dubbed the "Sloan Great Wall."[110] Three times farther away and 80 percent longer than the Great Wall discovered by Geller and Huchra in 1989, it is the largest observed structure in the universe. Other well-studied filaments include the Sculptor Wall, the Centaurus Wall, and the Coma Wall. Several dozen voids have also been identified, ranging from the "Local Void" some 200 million light-years in diameter, to the "Giant Void" more than a billion light-years in diameter, the largest void in the northern galactic hemisphere.[111]

It is striking that, like its smallest structures, the large-scale structure of the universe is predominantly determined by the work of gravity. In this case those structures likely have cosmological significance, reflecting the enhanced density regions in the early universe as measured over the last several decades by the COBE and WMAP spacecraft. There are other cosmological queries raised by this structure. For example, it remains unclear how well this large-scale structure of the universe matches the invisible dark matter, believed to constitute by far the majority of matter in the universe.

With filaments and voids we stretch the definition of "object" to mean a gravitationally bound "coherent structure," as indeed we have been for clusters and superclusters, as for star clusters in the stellar realm and Trans-Neptunian systems in the planetary realm. If we accept such coherent structures as objects, by their very nature filaments and voids constitute the largest class of objects in the universe in terms of their size.

Glancing backward, we can see how far we have come in the course of one century. As the twentieth century began it was highly uncertain whether extragalactic objects even existed, a reputable opinion being that our Galaxy constituted the entire universe. By the 1920s, Hubble had classified galaxies into three or four distinct groups, determined that they were indeed outside our Galaxy, and discerned the first few members of the Local Group, which today we recognize as constituting at least forty members, spanning more than 6 million light-years. By the 1930s, the galaxy Atlas of Shapley and Ames showed that the Local Group is, in turn, part of the Virgo Cluster containing more than 100 galaxies, today known to be some 2000 galaxies, spanning more than 10 million light-years. And by the 1950s – even as active galaxies were being discovered – Gerard De Vaucouleurs had (at least to his own satisfaction) good evidence of the Local Supercluster, containing at least 50 galaxy clusters and spanning 110 million light-years, of which our Local Group is only an outlying member.[112] Thirty years later, the first large-scale structure of the universe was discovered. Altogether, these events comprise a century of discovery often seen as completing the Copernican revolution, successively demoting humanity from any central position whatsoever.

Part III Patterns of Discovery

6

The Structure of Discovery

If both observation and conceptualization, fact and assimilation to theory, are inseparably linked in discovery, then discovery is a process and must take time.

Thomas Kuhn, 1962[1]

The idea of discovery is conceptually too complex for any "average" historical, psychological, or sociological analysis. Put more controversially, settling on the meaning of "discovery" is too important to our understanding of science to be abandoned to scientific discoverers or to psychologists or to sociologists or to historians of science. Conceptual analysis is important enough to be pursued by conceptual analysts, and not just given over to fact-gatherers.

Norwood Russell Hanson, 1967[2]

Epistemologically the problem [of the nature of discovery] is insoluble from an individualistic point of view. If any discovery is to be made accessible to investigation, the *social point of view* must be adopted; that is, the discovery must be regarded as a *social event*.

Ludwik Fleck, 1935[3]

Discovery is not an atomized contribution to knowledge that others need merely recognize and accept, but rather represents a retrospective characterization of a complex process of transformative negotiation, characterization that simultaneously formalizes the essential character of the discovery and confers upon it the stamp of objectivity as, by implication, an aspect of the physical world that was there waiting to be "discovered."

Kenneth Caneva, 2005[4]

Using the Pluto affair as our starting point, in Chapters 1 through 5 we have discussed the discovery of several dozen classes of astronomical objects, a significant subset of the whole as defined in Appendix 1. In the realm of the planets we have told both the canonical and more nuanced stories of the first detection of moons and rings by Galileo, of asteroids by Piazzi, of the terrestrial, gas giant, and ice giant classes of planets, as well as of the Trans-Neptunian objects in the Kuiper Belt and perhaps even the inner edges of the Oort Cloud. In the realm of the stars we have witnessed not only the discovery of six separate classes of nebulae, but also the discovery of the stars themselves: giants and dwarfs, subgiants, supergiants, and subdwarfs, the extreme white dwarf and neutron stars, returning by way of the elusive brown dwarfs to the subject of planets once again, this time the planets of other stars. And in the realm of the galaxies we have visited "normal" galaxies in the form of ellipticals, lenticulars, spirals, and irregulars; active galaxies such as Seyferts, quasars, and blazars with their powerful central black holes; and structures of galaxies ranging from clusters and superclusters to enormous filaments and voids stretching hundreds of millions of light-years – structures so large as to strain the very concept of "object."

Having laid out these narratives of discovery in all their glory, in the next two chapters we turn to the question of patterns in the discovery of these classes, as well as classes of astronomical objects not yet discussed, which at first glance may appear rather different from the familiar classes already treated. In doing so we recognize that the discovery of new classes of objects is a distinct category of discovery. Indeed, historians and philosophers of science have long recognized that the term "discovery" is used to designate a variety of processes in a variety of contexts. In a delightful article submitted to the *Journal of Philosophy* the day before his death at age forty-two when his private plane crashed into a hillside in dense fog, the renegade philosopher of science Norwood Russell Hanson laid out "An Anatomy of Discovery," including the logic of things, events, and processes properly said to be "discoverable," and a taxonomy of discovery activity. In particular Hanson saw a tripartite division between the discovery of "*an* X," some localizable object such as a planet, satellite, or asteroid; the discovery "*of* X," such as an elementary particle unseen as an individual entity or a universal process such as radioactivity, the quantum of action, or the Compton effect; and the discovery "*that* X," in other words, that there is no ether through which waves propagate, that electrons have both particle and wave-like properties, and that DDT can kill insects.[5]

In terms of discovery activities Hanson found there were "trip over" discoveries – those made in the absence of any theoretical or psychological expectation – such as Galileo's telescopic discoveries, the discovery of radioactivity, and the discovery of the positron; "back-into" discoveries – in which all theoretical expectation is against a particular discovery, but it is discovered anyway – such as the Michelson-Morley experiment meant to check the properties of the ether wind only to find there was none; and "puzzle out" discoveries – those with full theoretical and psychological expectation – such as Hertz's discovery of radio waves based on Maxwell's theoretical expectations. Hanson believed Pluto's discovery based on prediction also fell into the latter category, but we now know that story is considerably more complicated.[6]

Such parsing is just the kind of thing philosophers love doing, and often excel at, to the bewilderment of "normal" people. Yet, Hanson undoubtedly purposely penned "An Anatomy of Discovery," rather than "*The* Anatomy of Discovery" because he realized the term discovery can be sliced and diced in even more ways. Naturally, philosophers have done just that. Most recently, Greek historian and philosopher of science Theodore Arabatzis distinguished at least seven such uses of the term: the discovery of phenomena through controlled experiment; the discovery of entities accessible to immediate inspection; the discovery of objects not accessible to unaided observation; the discovery of entities that are unobservable in principle; the discovery of new properties of well-established entities; the discovery of new principles/laws; and the discovery of new theories. The list is not exhaustive, and beyond these uses there are many "subclasses" of discovery itself, among the most prominent being those predicted from theory and those that are not, two of Hanson's categories and a favorite distinction of philosopher of science Thomas Kuhn, who revolutionized the field in the 1960s.[7]

Philosophers of science have therefore provided a framework within which to explore discovery in more historical detail, even while exploring the framework itself. Some of their categories will indeed prove useful in the discussion that follows. Clearly, using Hanson's lingo, in analyzing new classes of astronomical objects we are concerned with the discovery of "*an X*" – in particular the prototype of "many Xs," the first in a class of localizable objects – rather than the discovery of processes or properties. And since we are dealing with the discovery of new classes of objects, one would think we were dealing primarily with "trip-over" discoveries, not foreshadowed by theoretical or psychological expectation, though that remains to be explored. A more common name for these "trip-overs" is "serendipitous" discoveries, widely known to occur in astronomy; whether serendipitous and trip-over are coextensive domains also remains to be explored.[8]

Again, in terms of the categories of Arabatzis, most of the discoveries we have discussed in this volume fall into the category of "objects not accessible to unaided observation," since they required the telescope and its various detectors, spectrometers, and photometers for their most basic detection and more complex analysis. There are, however, some objects that fall into his other categories: comets, meteors, and "new stars" (novae) did not require telescopes for their "discovery" in the loose sense of the word, for they were seen, if not understood, in antiquity. Moreover, energetic particles such as cosmic rays and the solar wind are collections of exceedingly small "objects" unobservable in principle, and detected only by their effects. And at the other end of the size spectrum, some of the structures we designate classes, such as clusters, superclusters, and filaments of galaxies, defy the normal definition of "object."

In addition, quite aside from *objects* discovered with or without the telescope, many astronomical *phenomena* have been discovered. Some of these fall into Hanson's category of a universal process, including the expansion (and later acceleration) of the universe, a "universal" process if ever there was one. Others, such as interstellar magnetic fields, and the cosmic microwave background (usually taken as proof of the Big Bang), are properties of the universe. The discovery of new classes of objects might in principle be quite different from other types of discovery as enumerated by Hanson, Arabatzis, and Kuhn. They might also differ from other subsets of discoveries such as new members of an already determined class, where the observer already "knows what to expect" or "knows what to look for." We therefore cannot claim universal application for any structure of discovery that we might find in the case of new classes of astronomical objects. But this is perhaps compensated by the possibility that we will understand more deeply the nature of discovery for this particularly important category of discovery, a precondition for comparing different structures of discovery.

In short, in this chapter we attempt to tease out the internal structure of discovery for new classes of astronomical objects, which remain our chief focus. In the next chapter we examine how this structure might differ from the structure of other types of discovery. We begin by analyzing what Thomas Kuhn long ago called "the structure of scientific discovery," now situated specifically in the context of the discovery of the new classes of celestial objects we have already discussed. We then look at the microstructure of discovery, including technological, conceptual, and social roles. And finally we discuss how discoveries end, ushering in the new era of "post-discovery."

In all of this we attempt to be sensitive to the concerns and conclusions of historians, sociologists, and philosophers of science. Curiously, the latter have written considerably more on the subject than historians – perhaps taking to heart Hanson's injunction that the subject should be left to conceptual analysts, by which he meant philosophers of science such as himself! Hanson, who began his study of discovery with his book *Patterns of Discovery* in 1958, followed his comment with one of the most perceptual analyses of discovery that exists even fifty years later. But following Hanson's dictum that discovery is what science is all about and the cutting edge of scientific knowledge, surely the historical facts can only strengthen any conceptual analysis. We shall indeed find that even in one of its limited categories - new classes of astronomical objects - discovery is a complex process when viewed under the microscope, so to speak, extended in time as Kuhn argued, and multifaceted by its very nature, grounded in reality yet influenced by social and institutional circumstance, and infinitely fascinating in its variety even where patterns appear. But in the midst of complexity there is indeed structure.

6.1 Macrostructure: Detection, Interpretation, and Understanding

Beginning with the earliest telescopic discoveries, we have noticed a pattern that we now investigate in more detail with an eye toward probing its generality as a defining characteristic of discovery: that a discovery is not an event at a discrete moment of time, that it has a structure, and this structure consists most often of detection, interpretation, and multiple stages of understanding (Figure 6.1). As we shall see, discovery is sometimes preceded by a "pre-discovery" phase, and (less surprisingly) is always followed by a "post-discovery" phase, both of which delimit the structure of discovery itself in both time and space. As we shall see in Chapter 8, classification also plays its role in discovery, sometimes problematically if attempted before understanding has reached a mature level, as notoriously in the case of Pluto. On the other hand, and counterintuitively, classification has historically been both a pre-discovery and a post-discovery activity, corresponding to phenomenology in the first case and "the thing itself" in the second.

As is evident in the opening quote to this chapter, the general idea that discovery has a structure is an insight of Thomas Kuhn, whose book *The Structure of Scientific Revolutions* incorporated this extended concept of discovery fifty years ago. Kuhn's structure famously involved three stages: anomaly, an extended period of exploration and struggle with anomaly, and (sometimes) an adjustment to "paradigm theory" and worldview that resolves the anomaly and may result in a scientific revolution.[9] This tripartite structure, he argued, applied only to non-predicted substances, objects, and phenomena such as oxygen, the planet Uranus, and X-rays. Kuhn's controversial treatment, however, was skewed in the service of his idea of scientific revolutions, surely a specialized case of discovery, since most discoveries do not result in such revolutions. Nevertheless, his more general idea of discovery as an extended process has been largely accepted by those who have studied the nature of discovery

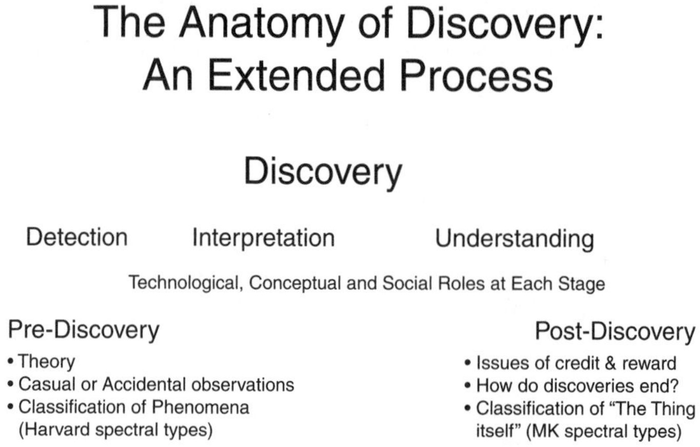

Figure 6.1. The extended structure of discovery, showing its three stages of detection, interpretation, and basic understanding, as well as pre-discovery and post-discovery phases. Classification can occur both as a pre-discovery and post-discovery activity. Discovery also has a microstructure consisting of conceptual, technical, and social roles.

over the last fifty years. The details of the structure, however, remain open to discussion, and the structure presented here would seem to follow more closely what most astronomers actually do, and see themselves as doing, in an exercise better characterized as normal rather than revolutionary science. Indeed, in his 1981 study *Cosmic Discovery*, astronomer Martin Harwit asserted that "in astronomy the major discoveries seem to differ qualitatively from the revolutions that Kuhn describes."[10]

We begin our examination of the structure of this extended process by juxtaposing the discovery of a significant sample of the new classes of telescopic objects discussed earlier, and listed in Appendix 2 along with other classes in chronological order of first detection (or inference or declaration in some cases such as stellar classes and dwarf planets). These classes, taken from all three realms of the planets, stars, and galaxies, are moons, rings, and asteroids; globular clusters, galactic clusters, nebulae, and various classes of stars; and some of those classes belonging to both normal and active galaxies as well as clusters of galaxies. We first turn our metaphorical microscope on the structure of their discovery at low resolution, proceeding from the seemingly "simplest" to the more "complex" discoveries, which in general follow a chronological ordering. We will see that while no discovery is simple, some are simpler than others when viewed over the extended period of their discovery. And not surprisingly, the more complex the instrumentation and the more convoluted the chain of reasoning, the more complex the structure of discovery.

No discovery appears simpler than the very first observations of Galileo. In textbook accounts, Galileo pointed his telescope at Jupiter and "saw" its moons. But as we recall from Chapter 2, Galileo's discovery of the moons of Jupiter was not consummated in a single night. His detection was made on January 7, 1601, when he observed "three little stars" just to the east of Jupiter. He thought they were fixed stars, even though he was intrigued that they were arranged along a straight line and parallel to the ecliptic, and were brighter than others. The next night Galileo saw the three "fixed stars" on the west side of the planet. Not yet thinking of the possibility of the mutual motions of these objects, he was at a loss as to how to explain this. Two nights later, only two stars were visible to the east of Jupiter. Only then did it dawn on him that "the observed change was not in Jupiter but in the said stars." And on the following night, January 11, he arrived at the conclusion, "entirely beyond doubt," that "in the heavens there are three stars wandering around Jupiter like Venus and Mercury around the Sun." Subsequent observations showed four orbiting objects.

Here we have, in the space of five days, and in a single mind, the progression from detection to interpretation to a basic understanding of what the nascent Italian astronomer was dealing with. The moons of Jupiter were an entirely unexpected discovery, Galileo's five-day mental journey demonstrating his great surprise at finding orbital motion around planets. It is revealing that Galileo first drew the analogy with Mercury and Venus rather than our own Moon (in the sense of their orbital motion around the Sun), but it is understandable that he did so with the validation of the Copernican system in mind. And he soon used the Moon for the same argument, arguing that if some doubted the possibility of the attendance of one Moon around the Earth as it circled the Sun, Jupiter now displayed four "stars" wandering around it even as Jupiter made its circuit of the Sun in twelve years. Thus, within five days from January 7 to January 11, with the aid of the analogy of the Moon circling the Earth even while the Earth and planets circled the Sun, Galileo detected, interpreted, and understood – "discovered" – a new class of objects. And in short order, Galileo quickly deployed his discovery in the service of the Copernican theory.

Not everyone agreed, notably those still in an Aristotelian frame of mind. In particular the prominent Jesuit professor Christopher Scheiner, who in March 1611 had observed sunspots with a 30-power telescope of his own construction, did his best to preserve the old cosmology by arguing that the Medician stars of Jupiter resembled nothing so much as the transient sunspots; he interpreted Galileo's observations as flights of tiny starlets not necessarily associated with Jupiter at all. Scheiner would go on to become the expert on sunspots, but in an interesting aside, Galileo and Scheiner became involved in a priority dispute over the discovery of sunspots, Galileo claiming to have seen them at about the same time in the spring of 1611. As Heilbron has remarked relevant to our theme, "According to Galileo's game rules, it did not matter when Scheiner first saw the spots or that he established some of their properties before Galileo did. That gave him no priority. Why not? Because he misinterpreted what he saw. The lion's share of the discovery of the spots belonged to their first correct interpreter."[11] The extended structure of discovery, even in the case of the property of an object rather than the object itself, allows for a more nuanced view of what actually happened.

The structure of discovery is even more apparent, but much more drawn out, in the case of the rings of Saturn, which, unlike the moons of Jupiter, had no precedent. Again, it is often said that Galileo discovered the rings of Saturn when he turned his telescope on that planet on July 30, 1610. But we recall Galileo's perplexed reaction: "[T]he star of Saturn is not a single star, but a composite of three, which almost touch each other, never change or move relative to each other; and are arranged in a row along the zodiac, the middle one being three times larger than the two lateral ones."[12] There is no doubt that Galileo detected something new on that date; in fact he detected what we now know to be the rings of Saturn. But it is equally evident that Galileo had no idea what he had detected. As Albert van Helden has shown in detail, during his lifetime Galileo treated these bodies as "collateral globes," as satellites such as those he had discovered around Jupiter six months earlier, though strange satellites to be sure. This remained the dominant concept for more than thirty years. Galileo had gone through the phases of detection and interpretation, but as yet had no correct understanding of the most basic nature of the object. And by his own rules applied against Scheiner in the case of sunspots, Galileo should not be credited with the discovery of the rings of Saturn, since his interpretation was not correct. Although one commonly sees in both textbooks and historical treatments that Galileo "discovered" the rings of Saturn, it is historically more accurate to say that Galileo contributed significantly to the *process* of the discovery of Saturn's rings.

Only in March 1655, when Christiaan Huygens turned his larger telescope toward Saturn, did he explain what Galileo had seen as handles or "anses," rings that changed appearance. He came to this conclusion by an extended process of reasoning he does not fully reveal in the *Systema Saturnium*. Thus, although the "interpretation" phase had certainly begun already with Galileo, only forty-five years later did Huygens give what we now consider the correct interpretation. Was it Huygens, then, who discovered the rings of Saturn in 1655 rather than Galileo in 1610? Not really, it could be argued, on two counts: the interpretation was not widely accepted for several decades, when his prediction of the appearances and disappearances of the rings proved accurate, and, perhaps more seriously, lacking knowledge of dynamics Huygens had no idea how such rings could physically exist and persist around Saturn. That understanding came only in the mid-nineteenth century, when James Clerk Maxwell provided a basic dynamical explanation. And an understanding of the true composition of the Saturnian rings awaited the Voyager spacecraft, with many mysteries remaining even today. In short, it does not seem fair to credit Galileo, Huygens, Maxwell, or the Voyager scientists individually with the discovery of the rings of Saturn. Rather, discovery in the case of Saturn's rings appears to be an extended collective process of detection and interpretation, as well as basic and more mature understanding. Moreover, it does not seem proper to credit the Voyager scientists with the discovery of the Saturnian rings in any sense, begging the question, "when does a discovery end?" We return to this problem in the last section.

Incredibly, aside from the planetary classes themselves discussed in Chapter 2, it was almost two centuries before the discovery of another new class of objects in the solar system. Yet again an obvious pattern emerges, for the traditional story is that minor planets (dubbed "asteroids" by William Herschel) were discovered on New Years Day, 1801, when Piazzi observed an 8th magnitude star in the constellation Taurus. But it was certainly not on that night that he discovered the first asteroid, for the object appeared to be just another point of light indistinguishable from the rest of the stars. Not until the next night did he see something more interesting: the object had moved 4 arcminutes to the north and west. This movement put it in the realm of the wanderers, and Piazzi knew he had seen something different from a normal star. By "the evening of the third, my suspicion was converted into certainty, being assured it was not a fixed star." But what exactly was it? At first Piazzi thought it was a comet. But by January 24 he had doubts: "I have announced this star as a comet; but the fact that the star is not accompanied by any nebulosity and that its movement is so slow and rather uniform, has caused me many times to seriously consider that perhaps it might be something better than a comet. I would be very careful, however, about making this conjecture public."[13] The "something better" was still not an asteroid, for no such object had been conceived. Rather, twenty years after Herschel's discovery of Uranus, Piazzi had in mind a planet.

And now an element of theory enters the picture of discovery for the first time, for the object was lost in the glare of the Sun in mid-February, and only after Carl Friedrich Gauss's predictions based on a new orbit determination did the German astronomer Baron Franz Xavier von Zach definitely recover the object exactly one year after its first detection, January 1, 1802. "Finally, the new primary planet of our solar system has again been discovered and found, like a starfish on the beach," von Zach wrote.[14] This new element in the thought process of discovery is notable: Von Zach was confident of its planetary nature not only because it was moving and lacked the nebulosity of a comet, but also because Gauss had determined an orbit that placed it at a distance in accordance with the prediction made by what was known as Bode's law. But the object remained an anomaly. If it was between Jupiter and Saturn, why was it so dim? Such dimness implied smallness, a possibility but not what was expected. Oddness turned to chagrin two months later when the German astronomer Heinrich Wilhelm Olbers discovered another object, quickly determined to be in the same orbit, and when William Herschel determined both objects were extremely small, estimated at only 162 and 70 miles in diameter. It was this unprecedented situation that led Herschel to declare a new class of object by early May 1802: "we ought to distinguish them by a new name, denoting a species of celestial bodies hitherto unknown to us . . . From . . . their asteroidical appearance . . . I shall take my name, and call them Asteroids."[15] But not all astronomers agreed on this declaration of a new class, and for fifty years many astronomers, including official government almanacs, considered these objects to be planets. Only when their number increased significantly was their designation as minor planet or asteroid widely accepted, a classification issue that we will address further in Chapter 8.

Thus, in the case of minor planets, we have a rather more extended process of discovery than in the case of the moons of Jupiter. We once again see detection, followed by interpretation beginning the next day, perhaps a basic understanding that small "minor planets" existed within a few months, but acceptance of asteroids as a new class of objects only after many decades. As with Saturn's rings, we can now say we have a mature understanding of asteroids, knowing something of their origin and structure, having had close spacecraft encounters with many, and even having landed on several. The vexing question of the demarcation of "mature understanding," however, remains open.

* * *

Meanwhile, as is evident from the chronological listing in Appendix 2, new classes of objects had been discovered for the first time in the realm of the stars, namely diffuse nebular objects (1610 or 1618), "galactic" clusters of stars (1610), and, much later, globular clusters of stars (1785–1789). Here another complication emerges in delineating the structure of discovery, for a few exemplars of each had been seen as naked-eye objects even before the telescope. Yet no one would say they were "discovered" before the telescope. Rather, *their discovery is bound up with the problem of when they were distinguished as different kinds of objects*. Arguably, Galileo first detected star clusters when he observed the profusion of stars in the Pleiades, while the question of the discovery of diffuse nebulae and globular clusters is tied up with the general problem of the

Figure 6.2. Charles Messier: discoverer or pre-discoverer? Arguably neither La Caille nor Messier detected globular clusters as a new class of object, in part because they did not consider them to be composed of stars, but just another nebular morphology. In this interpretation they would be considered pre-discoverers. The problem of who first detects a new class of object is common in astronomy. This portrait, likely made around 1771 just before the first edition of Messier's catalogue of nebulae and star clusters was published, is held at the Paris Observatory and attributed to Nicolas Ansiaume (1729–1786). A similar Messier portrait, held at the Louvre, is often ascribed to the French artist Alexandre-François Desportes, but Messier was only twelve when Desportes died in 1743, so the ascription cannot be correct and the painting is likely by the same artist. This image is from a photo by Owen Gingerich.

nature of the nebulous objects, which came to a head in the late eighteenth century. What we now think of as bright globulars, such as M22 in Sagittarius and Omega Centauri in the Southern Hemisphere, could be seen with the naked eye as faint fuzzy objects and were well-known in the late seventeenth century. But they were fuzzy objects not distinguished from other nebulae, a distinction made only in the course of the eighteenth century. As we saw in Chapter 3, Nicolas-Louis de La Caille published forty-two nebulae and clusters of the Southern Hemisphere in 1755, while in the Northern Hemisphere Charles Messier (Figure 6.2) published his *Catalogue des Nébuleuses & des amas d'Étoiles* [catalogue of nebulae and star clusters], in three editions between 1774 and 1781, implying by its very title that nebulae and star clusters are two different classes of objects. The catalog of 1781, the last edition published in Messier's life, in fact contains a total of twenty-eight "round nebulae" as he referred to them, and one (M4 in Scorpius) that he claimed to have resolved into stars. The question was whether all "round nebulae," or indeed all nebulae in general, are resolvable into stars, given a large enough telescope, or whether they were two separate classes of object. That question was largely answered by William Herschel around 1790, and definitively with the spectroscope by William Huggins in 1864. Herschel discovered thirty-seven new globular clusters during his sweeps of the sky, was the first to resolve virtually all of them into stars, and coined the term "globular cluster" in the discussion adjacent to his second catalogue of 1000 deep sky objects published in 1789.

These objects present a dilemma in terms of the first element in the triple structure of discovery. Clearly, there is no identifiable first person to "detect" bright naked-eye globular clusters. "Detection" implies a conscious effort at observation – we might say by a skilled observer, but that would rule out Galileo, who honed his skills as he went along – whether purposeful or serendipitous.[16] Rather, *casual sightings of such objects with the naked eye would appear to constitute a pre-discovery phase*; as we shall see in the next chapter, this is also true of other naked-eye classes of objects, and even most non-naked eye objects if the idea of pre-discovery is extended to embrace theory and other activities. Did La Caille and Messier, then, detect globular clusters *as a new class*? They were certainly among the first to "detect" what we now know to be globular clusters in a conscious way, and based on their shape Messier had already distinguished them from other nebulae in his *Catalogue* in 1774. Though it is a close call, and though Messier even claimed that at least one round nebula was resolvable into stars, arguably only Herschel demonstrated this empirically with his superior telescopes, detecting globular clusters *as a new class*. We begin to see how dicey it is not only to declare who really "discovered" a new class of objects, but even who "detected" it as a new class.

Similarly, what are today known as "galactic clusters," much younger and smaller associations of stars such as the Pleiades, were first observed in prehistoric times. The mention of the Pleiades in the Bible, Homer, and Hesiod also points to a pre-discovery stage, for again, casually noticing something in the sky, and even putting it to use for time and calendrical purposes, does not qualify as the discovery of a new class of objects. And although Galileo had no conception of a "cluster" of stars when he discussed the Pleiades in his *Sidereus Nuncius*, he did know enough to recognize them as a separate class. Such objects were clearly composed of stars, though not until the middle of the eighteenth century were the stars comprising the bright clusters widely believed to be physically associated. Messier listed thirty-three of them in his catalogue. Following their pre-discovery phase, the detection was thus made by Galileo; the interpretation and basic understanding began with Messier, and was confirmed by Herschel. In this case the demarcation between pre-discovery and detection is clearer than with globular clusters, since the pre-discovery observations were casual, not purposeful.

Distinguishing what we now know to be diffuse nebulae was considerably more difficult than resolving galactic and globular clusters, for a nebula composed of something other than stars was truly a new class of objects in the heavens, not just a grouping of already known objects. In a very early telescopic observation Nicolas Claude Fabri de Peiresc is often credited with detecting the Orion Nebula in 1610, and more likely Johann Baptist Cysat in 1618; as far as we know neither ventured an interpretation, nor did anyone else for more than a century (with the possible exception of Edmond Halley) until the advent of William Herschel's improved telescopes. As we saw in Chapter 3, for the decade prior to 1784 Herschel believed in true nebulosity based on apparent changes in the Orion Nebula. From 1784 to 1790 he was skeptical of true nebulosity, only to be convinced of it again in 1790 when he made his observation of "a most singular phenomenon," a "star which is involved in a shining fluid, of a nature totally unknown to us." This object he termed a planetary nebula, a class of object he had observed since 1782 without distinguishing the central star that the nebula surrounded.

In the case of planetary nebulae, with their notoriously misleading name based on their planetary appearance in Herschel's telescope, in the course of eight years Herschel made both the detection and interpretation that yielded basic understanding of this new class of object. The understanding was "basic" in the sense that he believed the object to be a "shining fluid," but had no idea of its composition, much less its place in cosmic evolution, a subject in which he was greatly interested. Only by inference could he transfer this "shining fluid" composition to diffuse nebulae. Even as he did so, doubts remained, only to be resolved by the early spectroscopic observations of Huggins in 1864 when he turned his telescope toward the planetary nebula in Draco. By contrast with planetary nebulae, too faint to be detected without the telescope, in the case of diffuse nebulae, there was a pre-discovery stage, followed by usual phases of detection, interpretation, and basic understanding. The place of both diffuse and planetary nebulae in cosmic evolution – constituting a more mature understanding – was not known until the twentieth century.

With the nebulae we also begin to see that just as there is a fine structure for detection, so is there a fine structure for interpretation. This is particularly evident in the case of Huggins's pioneering spectroscopy to determine the gaseous nature of nebulae. Behind that determination was the nascent history of spectroscopy itself: Newton's dispersion of light into a spectrum of colors with the prism; Wollaston and Fraunhofer's observation of dark lines in the solar spectrum; the network of laboratory experiments that identified those lines with certain elements; and above all, in the case of the nebulae, the laboratory observation that a cloud of warm gas will emit light only at specific wavelengths characteristic of its composition, giving rise to bright lines at those wavelengths against a black background, a phenomenon that came to be known as an "emission spectrum." The spectrum of the planetary nebulae that Huggins observed on August 29, 1864 consisted of bright lines without the continuous spectrum characteristic of stars, proving it was gaseous rather than stellar in nature. Huggins's excitement was still palpable when he recalled many years later, "The riddle of the nebulae was solved. The answer, which had come to us in the light itself, read: Not an aggregation of stars, but a luminous gas. Stars after the order of our own sun, and of the brighter stars, would give a different spectrum; the light of this nebula had clearly been emitted by a luminous gas."[17]

As Huggins's biographer Barbara Becker has shown, this reconstruction after many years likely portrays this as more of a Eureka moment than it actually was; a certain amount of interpretation had to waft through Huggins's mind. Moreover, a good deal of interpolation went into identifying the gaseous nature of planetary nebulae with the gaseous nature of the diffuse nebulae. True, in the four years after his first discovery Huggins examined the spectra of about seventy nebulae, including diffuse nebulae such as found in Orion. But he found that only one-third of them displayed bright-line spectra as opposed to the continuous spectra displayed by star clusters; the gaseous interpretation was not one that could be conferred on nebulous-appearing objects *en masse*; each individual object had to be observed to determine the class to which it belonged. Thus, with the entry of spectra into the realm of discovery, we see the increasing complexity of its interpretation phase. If Galileo did not "simply" turn his telescope toward Jupiter and "see" its moons, even less did Huggins turn his telescope, with its spectroscope, toward the planetary nebula in Draco and "see" a gas cloud. The chain of reasoning in the interpretation phase had extended even more, complicated in this case by the fact that the spectrum did not match any known terrestrial substance.

Nor would this be an isolated characteristic of discovery, for spectroscopy would be one of its essential components for many more classes of objects, not least the stellar classes themselves. Indeed, spectroscopy at first was the *only* discriminator for stellar classes in the sense of stellar spectral types, since aside from their magnitudes, it was the *only* observable characteristic for the vast majority of stars. As we saw in Chapter 4, only when their distances, and thus their luminosities, were known with some accuracy and in sufficient numbers to construct the

Hertzsprung-Russell diagram in the years spanning 1905–1914, could the more sophisticated and physically meaningful classes of giant stars be consciously and graphically separated from the class of dwarf stars. And just as the H-R diagram was two-dimensional, so was the complexity of the interpretation of giants and dwarfs multiplied: just as the interpretative history of spectroscopy was behind the one dimension of the diagram, so also the vagaries of distance determination (essential for luminosities) were behind the other.

Here we have a more serious complication to the structure of discovery, for with the two-dimensional nature of the H-R diagram the detection phase is replaced by *inference*: Hertzsprung and Russell did not so much detect giants and dwarfs as deduce their existence based on their spectra and distance (and thus luminosity), in the process employing interpretation, while leaving plenty of room for further interpretation and basic understanding. A more mature understanding was not reached until the work of Megnad Saha's ionization theory demonstrated what had been long suspected: that the spectral color sequence was actually a temperature sequence. It is notable that in this case a certain amount of interpretation had to occur even before the inference of new classes. Even in a more straightforward "detection" case of a new class such as the Jovian satellites, a form of eye-brain interpretation must be involved in the process of detection before the more explicit post-detection interpretation begins.[18]

The discoveries of the supergiants (1917), subgiants (1930), and subdwarfs (1939) follow a similar pattern as the giants and dwarfs. Because of their relative rarity they were only detected or inferred as new classes after many more stellar distances and luminosities had been determined. For example, Adams and Kohlschütter's monumental volume *Spectroscopic Absolute Magnitudes and Distances of 4179 Stars*, published in 1935, not only gave clear delineations of the giant and dwarf classes, but also intimations of a new class called subgiants. Interpretation depended on singling out only ninety of the more than 4000 stars in the catalogue in the G and K spectral classes that were "somewhat fainter than normal giants," and the initial interpretation was tentative, Adams and Kohlschütter saying only that "there is some spectroscopic evidence to support the suggestion" that Gustav Stromberg had made as early as 1917. Most doubts were removed the following year when trigonometric parallaxes yielded more accurate distances and luminosities and astronomers could better distinguish subgiants. This gave astronomers the most basic understanding that subdwarfs existed as a separate class. A fuller understanding of the physical mechanisms for subdwarfs, as well as all other classes of stars, awaited knowledge of the nuclear reactions inside stars, modeling for each of the classes, and an understanding of a physically based evolutionary sequence that turned out to be very different than Russell had proposed even for dwarfs and giants.

The realm of the galaxies exhibits similar triple macrostructure in the discovery process. As with the nebulae, galaxies have a long pre-discovery phase: as we saw in Chapter 5, like globular clusters, galactic clusters, and diffuse nebulae, numerous galaxies were present already in Messier's late-eighteenth-century catalogue. Between 1780 and 1860, thousands were catalogued based on visual observations by the Herschels and others. In the 1890s, photography revealed a variety of morphologies aside from the already well-known spirals. But there was no way such natural history methodologies could determine their nature as we understand them today, which, as with the classes of stars, required a knowledge of their distances. With the galaxies, however, parallaxes were out of the question for distance determinations; no parallax had ever been determined for a Messier object, and the question was whether they were inside or outside our Galaxy. Some, such as the diffuse nebulae and globular clusters, were associated with our Galaxy, but the definitive proof of some of the "nebulae" as extragalactic came only with Hubble's work on Cepheids in the early 1920s. The use of Cepheids as distance indicators made use of the period-luminosity relation; the period was the observable, the luminosity was inferred from distance and apparent magnitude. Thus, in a pattern of discovery similar to the stars, Cepheid periods replaced parallax as the distance indicator to the galaxies.

This distance determination revealed the extragalactic nature of spirals and the "irregular" Magellanic Clouds, but not the apparently starless elliptical galaxies. The extragalactic nature of the latter had to be inferred from the "uniformity of nature." Even worse, the distance determinations to the galaxies did not reveal the nature of the classes of galaxies as we conceive them today, only the fact that they were truly extragalactic. Classes had been distinguished only by morphology, inferring that shape implied a substantial real difference in the nature of the object, perhaps related to an evolutionary sequence as Hubble suspected in the case of his "tuning fork" diagram. But as with the stars, Hubble's sequence turned out not to be the correct evolutionary sequence, and in fact even today the evolutionary connections among galaxies is not well established.

We need not belabor the point in the case of active galaxies and clusters of galaxies. Aside from the Seyfert galaxies, declared a class in 1943 on the basis of optical observations, the detection of active galaxies, including radio galaxies (1952), quasars (1963), and blazars (1972), required the opening of the electromagnetic spectrum, in particular radio astronomy. We have already seen the details of their discovery in Chapter 5. In each case detection and interpretation as a new class were relatively swift once the new technology was at hand, optical identifications had been made, and at least a few members of the class confirmed. Understanding of their physical nature and the energy mechanisms driving these powerhouse galaxies took considerably longer, dawning only with the theoretical prediction and eventual discovery of those singularities we know today as black holes. It is also notable that, in the case of all classes of active galaxies there was a pre-discovery phase during which the object had been seen but not distinguished as a new class. The peculiar properties of the prototype Seyfert galaxy were observed already in 1908, but it was not declared a new class until 1943, when Seyfert named it as one of twelve galaxies with similar properties. The strong discrete radio sources that turned out to be radio galaxies, quasars, and even blazars had been observed by the hundreds from the 1940s to the early radio surveys of the 1960s. BL Lac itself had first been detected at optical wavelengths in 1929 and interpreted as a peculiar variable star. Thus, as we shall see in more detail in the next chapter, pre-discovery is often the rule rather than the exception for new classes of objects, even those discovered with the telescope.

The discovery of clusters of galaxies follows a similar pattern: pre-discovery by Messier in the eighteenth century of at least one region with higher numbers of nebulae; the detection/inference of our Local Group of galaxies in the 1920s by Hubble, of the Coma and Virgo clusters by Shapley and Ames in 1926, and of eight clusters by Hubble and Humason as of 1931. The recognition of this class of object came relatively quickly following new technology, in this case the improvement of telescopes and methods for distance determination, especially redshifts via spectroscopy. The interesting twist in the case of clusters is the question raised about not only the definition, but also the very nature of an "object." Where did clustering end? As we saw in the last chapter, de Vaucouleurs's declaration of "superclusters" was never accepted by Fritz Zwicky. More fundamentally, can a structure stretching across millions of light-years really be termed an object? Astronomers have generally come to answer in the affirmative, ending with the ultimate objects of filaments of galaxies, a powerful demonstration of the grasp of gravity, the "weakest" force in the universe compared to the other three fundamental forces.

In summary, based on this sample of classes, it seems clear that the discovery of new classes of astronomical objects is an extended process consisting of detection, interpretation, and understanding, all of which may be preceded by a pre-discovery phase. Discovery begins with the detection of something totally novel, often with no theoretical or paradigmatic framework whatsoever, except the general background of scientific data and principles.[19] The detection phase is subject to all the cautions that philosophers of science have foreseen with the problematic nature of observation, including the property of often being theory-laden, in the sense of metaphysical presuppositions rather than a well-formed physical theory. Interpretation, too, can be a complex process, involving not only what is seen but what is inferred, assuming principles inherent in techniques such as spectroscopy and in methods such as distance determination. Often only after long intervals does discovery finally proceed to the various forms of understanding, which has its own problems. We have seen strong cases of what we have called pre-discovery, not only for naked-eye objects, but also for those requiring a telescope. And we have seen reason to question the very nature of the idea of an object in the case of some of the more extended classes.

Clearly, the triple macrostructure of discovery raises further questions, having to do with the details of each stage, in other words, with the microstructure of discovery. As we shall see in the next section, and as we might have expected, microstructure is rich with detail, both human and technical. And as we shall see in the last section, pre-discovery must be counterweighted by post-discovery, otherwise any particular discovery would never end.

6.2 Microstructure: Real People and Their Social Roles in Discovery

The triple detection-interpretation-understanding structure we have documented above offers a useful pattern among most, if not all, discoveries of new classes of astronomical objects. But even if it gives substance to the concept of extended discovery, and without the controversial and rarely occurring concept of revolutions in science, is this pattern so general as to be meaningless? One way to answer this question is to ask whether we can conceive of a pattern of discovery that does not include all three phases, or includes more or different ones. The answer is yes, even within our limited sample: some objects replace the "detection" phase with "inference" (stellar classes) or "declaration" (dwarf planets) of new classes; some such as nebulae and galaxies have a history of pre-discovery; others such as clusters of galaxies strain the concept of one or more phases or even the concept of "object"; still others have a considerably more complex history in any one of the three phases, especially interpretation involving long chains of reasoning. And certainly the duration of each phase varies among discoveries. Another way of approaching the question is to ask if any scientific discoveries actually do not follow this pattern. Again the answer is yes – as can be seen in the case of the discovery of new astronomical phenomena discussed in the next chapter.

But perhaps the best way to answer the question of over-generality is to emphasize that this tripartite structure is only the macrostructure of discovery. A study of the microstructure of each component would likely be even more revealing, uncovering particular forms of detection, interpretation, and understanding, as well as the problems associated with each, such as the problematic nature of observation.[20] Even more importantly, the conceptual elements we have emphasized so far inevitably had technological, social, and psychological components, revealing even more about the nature of discovery. The technological elements have been apparent throughout our narrative histories: Tycho Brahe and his quadrant aside, almost none of the discoveries we have described could have taken place without the telescope in one of its myriad forms (or, as in the case of cosmic rays, other forms of detectors). From Galileo to the Hubble Space Telescope and its ground-based counterparts, discoveries have been dependent not only on telescope size, but also on related factors such as sensitivity, resolution, detectors, and analysis technology such as spectroscopes and photometers. This is such a rich area to explore that we will devote most of Chapter 9 to telescopes as "engines of discovery." In this section, then, we will focus on the other factor of no less importance, social roles in discovery.

As we have tried to emphasize, and as anyone who has worked in a scientific institution knows all too well, discoveries are not a disembodied series of events but human activities situated in a specific context of place and time. In fact there are a variety of contexts, ranging from national and local to institutional and individual. It should be no surprise, then, that when we turn our microscope on high power to the process of discovery, it inevitably resolves into the psychology and sociology of discovery. There has been no lack of attempts to find the patterns of this psychology, and similarly no lack of attempts to show the social basis of scientific discovery. Long before Kuhn and others rode the tide of social history, Ludwik Fleck, Robert K. Merton, Michael Polanyi and others had demonstrated the importance of the social role in science, Fleck making the specific point in 1935 that "discovery must be regarded as a social event."[21] And for decades after Kuhn, in a tide of postmodernism only now beginning to run its course, the field of science studies demonstrated the power of context at many levels in not only influencing, but determining, the practice and content of science.[22] While nothing could be less surprising than to find human beings at the basis of discovery, nothing could be more controversial than the conclusions that some eventually drew from this obvious fact: to insist that science, and thus discovery – its engine of creativity – is socially constructed to such an extent that those discoveries have no basis in reality. Strange as it may seem, this process of throwing out the baby with the bathwater is precisely the point of contention of the most radical proponents of postmodernism in the so-called science and culture wars.

We make no such point here, but we do insist on a social role in scientific discovery, if not a social basis for scientific discovery as the sociologist Augustine Branigan put it in his book by that title.[23] Quite aside from the fact that nothing *counts* as a discovery unless a relevant community so deems it, such social roles are important on at least three levels; the environment and psychology of the individual, the specific social networks in which scientific work often takes place, and the society or societies in which science is undertaken, now increasingly globalized. The importance of the immediate environment, for example, has been demonstrated in a variety of books having to do with laboratory practices and other aspects of science.[24] Although astronomical discovery generally takes place in observatories rather than laboratories, the case has recently been made that the cultural, economic, and scientific factors at play in the "observatory sciences" are in fact analogous to the "laboratory sciences" recently analyzed by historians. In this formulation – in which the observatory sciences include not only astronomy but also cartography, geodesy, meteorology, physics, and statistics – social factors must play a similar role as they have been demonstrated to play in laboratory sciences. This is equally true of observatory techniques and methods, including the concepts of precision, statistics, and social networks, all placed in contexts ranging from local utility to imperial enterprise.[25]

More generally, the importance of social roles has been emphasized with respect to entire societies at various points throughout history. Why did the Greeks invent Western science? Why did the Islamic civilization have its own scientific Enlightenment 1500 years later, only to be dead-ended after several centuries, giving way (though not causally) to the rise of science in Europe during the Renaissance? Clearly, large-scale social factors are involved in such fundamental questions. Galileo could not have made his discoveries in the Islamic world in the seventeenth century. Conversely, social, not technical, reasons explain why societies beyond Europe did not utilize the telescope for scientific discovery. A recent study concluded that the arrival of the telescope in Ottoman lands in the late 1620s had no impact on Muslim astronomy, a situation that persisted for two centuries: "This surprising delay in adopting the telescope in Muslim regions, given their long preoccupation with astronomical inquiries, especially for timekeeping and for computing directions to Mecca, left the Middle East in an underdeveloped state of scientific inquiry. These contrasting paths of scientific development in the Muslim world and in Europe reveal the very different levels of scientific curiosity that prevailed in the West and in Muslim countries for centuries to come."[26]

Similarly, the telescope arrived in China in the closing decades of the Ming dynasty, when science was "stagnant . . . with no significant developments of which to speak." A recent study of the impact of the telescope in China concluded that during the seventeenth century, "More telescopes were brought to China, but usually they did not have much influence beyond the court and circle of high officials. Although the telescope aroused some curiosity and marvelous remarks among the Chinese – the emperors and some high officials wrote poems about telescopic observations – no new astronomical research problems were raised along with the introduction of the new instrument to China."[27] Fast-forwarding 400 years, Hubble likely would not have made his discoveries in Europe in the early twentieth century, not just for reasons of technology, but also for social reasons. Transporting the 100-inch Hooker telescope from Mt. Wilson to Europe would not guarantee the same discoveries would have been made: the essential help of Humason, the funding, freedom of research, and social network of Mt. Wilson, and a dozen other individual, institutional, and social factors all played a role in uncovering the extragalactic universe and the velocity-distance relation. This is an enormous subject touched on here only to make the point that discovery does have a microstructure, and one that matters just as much, if not more, than the epiphenomenal macrostructure.

While this is not the place to undertake such grandiose themes as cultural factors in the invention of science and in the practice and content of science as a whole, we focus in the remainder of this section on the relative roles of social factors in the process of the discovery of representative new classes of objects. Once again, there is no better starting point than the still-fresh case of Pluto, both in the discovery of the object itself in 1930, and in its declaration seventy-six years later as a new class of object. If we look at the detection stage, it was driven first of all by the theory, informed by empirical anomalies in the orbits of Uranus and Neptune, that a trans-Neptunian planet should exist. That is a conceptual element. Second, it was driven by the obsessive personality and the abundant funding of Percival Lowell, a social element that allowed the purchase of the 13-inch telescope, the proximate agent of discovery. Third, it was enabled by the persistence of one individual, Clyde Tombaugh, whose meticulousness at both the telescope and the blink machine allowed him to pick out one object moving among thousands, after eliminating known asteroids. When the discovery was made, despite some indications based on albedo assumptions that it was a small object, it was designated a planet because it most closely fit the consensual definition of a planet, and no other category existed for such an object except possibly a minor planet, then thought to exist only between Mars and Jupiter. Only after James Christy's discovery of Pluto's moon Charon in 1978 was it possible to determine Pluto's true mass, about the size of our Moon. Even then, it took the discovery of numerous Trans-Neptunian objects, some believed to be larger than Pluto, to trigger the official discussion that led to the designation of Pluto as the prototype of a new class of object – dwarf planets. As is evident in Chapter 1, that designation was clearly a social negotiation among different communities of astronomers, notably planetary and dynamical astronomers.

Nevertheless, while social and institutional factors are involved from the very beginning in the discovery of Pluto and its moons, without which the discovery would not have been made, the fact of its existence represented a reality, not a social construction. By contrast, the process of declaring an object a new class is dominated by social factors, and is clearly a contingent activity that is definitely a social construction – as we shall see in more detail in Chapter 8. The dominant social role in declaring Pluto a dwarf planet was very public and all-too-apparent, but is it unusual? Not at all. The same happened for asteroids, though in a much less public way, when Herschel urged Piazzi that his new object should be a new class, in part because it was better to have discovered a new class rather than a new member of a class. And, as David DeVorkin has shown, the acceptance of the entire Harvard classification system for stars was a process of social negotiation. For the later MKK classification of stars it was more a case of usefulness rather than official adoption, but again rival institutional feelings and personal politics played their role in shaping "usefulness." And in the case of the Hubble sequence of galaxies it was a combination of both negotiation and usefulness.

Similarly, while it is difficult to make the case that distinguishing a class of terrestrial planets from a class of gas giant planets is a social construction, for more borderline cases like ice giant planets, the class may well be socially constructed, even once the physical details are well known. By contrast, the rings of Saturn, while certainly an extended discovery as discussed above, did not have to be negotiated as a new class after it was realized these were no normal satellites but something completely novel. This suggests a relationship between conceptual and social roles, as mediated by technological capabilities: *the less is known conceptually about an object or class of objects, the more social construction plays a central role.*

The 1960s offer two very vivid examples of social roles in astronomical discovery: quasars in 1963 and pulsars in 1967, the first shown to be extragalactic objects and the latter first discovered within our own Galaxy. Again, the technological factor is obvious: quasars could not have been discovered without the advance of radio telescopes and optical telescopes to make the optical identification, and pulsars could not have been found without the array of dipole antennas with short time constant constructed to observe a very different phenomenon, interplanetary scintillation. Social factors are less apparent, but no less important. Nothing like quasars had ever been predicted. Maarten Schmidt had to puzzle over spectra redshifted to such an extent that he could not recognize the spectrum until, on February 5, 1963, he finally identified the pattern as the Balmer lines of hydrogen redshifted by 16 percent, placing 3C 273 at the extraordinary distance of 2 billion light-years. The object was clearly in a new class never before seen. And while it might seem that Schmidt was the main player, the discovery was, in fact, a combination of many factors: Schmidt's remarkable insight, yes, but also the culmination of both radio and optical observations made by teams in Australia and the United States; Cyril Hazard's pinpointing of the position of the radio source was particularly crucial. The identification of more such objects in the class, and working out the nature of the objects, required even more collaboration. The discovery of quasars was therefore hardly a solitary individual event.[28] By contrast, pulsars had been predicted as neutron stars already in the 1930s, but the prediction had been forgotten when Jocelyn Bell and her mentor Anthony Hewish pored over the radio scruff on their chart recorders. They contemplated "Little Green Men" and other possibilities before Tommy Gold fingered the "pulsars" as the neutron stars predicted almost a half-century earlier. In the case of both quasars and pulsars, there was a strong element of social interaction that is sometimes the essential ingredient in discovery, as opposed to the lone discoveries of observers like Galileo and Herschel.

Black holes provide a particularly vivid example of discovery by social agreement, at least during the lengthy interpretation phase from the mid-1960s, when "nobody took the idea seriously," to the mid-1990s when Hubble astronomers claimed "seemingly conclusive evidence" for a black hole of about 3 billion solar masses at the center of M87. During that time astronomers gradually came to accept these exotic objects, even without definitive observational proof. While direct detection was out of the question by the very nature of the object, indirect observations in the form of jets and accretion material were made. Yet, as Kip Thorne pointed out, other scientific explanations were possible, including extremely massive stars. With a few notable exceptions, an almost inevitable "accretion of ideas" seemed to take hold as a critical mass of scientific opinion began to favor black holes in the mid-1970s, some twenty years before they were more or less definitively proven. One particularly revealing scientific cartoon by cosmologist Richard McCray depicted scientists at first as "beyond the accretion radius" while others plunged headlong toward it, some of them subject to irrational forces, both in their rush to be the first to make a discovery and in the aftermath when a fashionable idea has become an article of faith[29] (Figure 6.3).

The examples could be multiplied, but likely with the same conclusion: In its finest microstructure, discovery resolves into institutional, social, and even psychological components, for many discoveries could not have been made without the requisite institutional funding, social environment, and individual psychological temperament of the discoverer. Discovery is thus socially constructed in the sense that all of these factors are involved; indeed, one wonders how anyone ever thought they could not be, least of all the scientists involved.

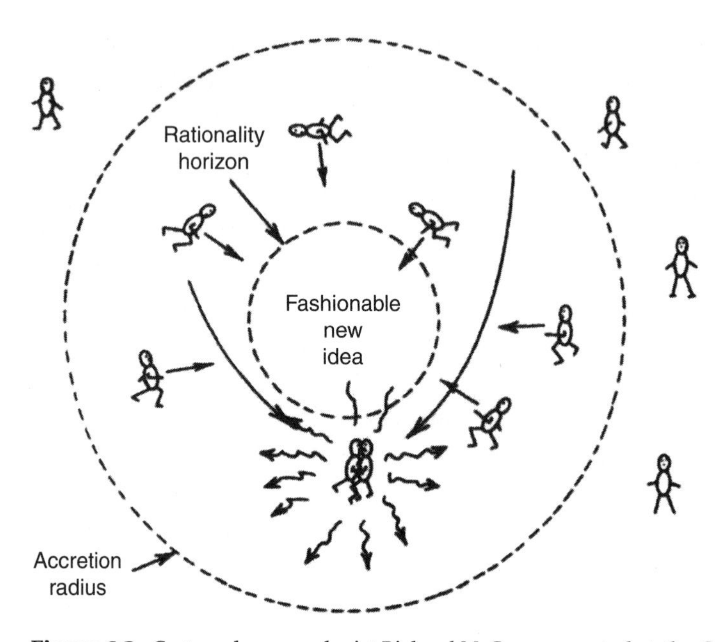

Figure 6.3. Cartoon by cosmologist Richard McCray presented at the Cambridge summer school in 1977 illustrating social influences in science. Titled "response of astrophysicists to a fashionable new idea," McCray wrote that "Beyond the accretion radius [outer circle] astrophysicists are sufficiently busy with other concerns to not be influenced by this fashionable new idea . . . But others, within the radius, begin a headlong lunge toward it. In their rush to be first, they almost invariably miss the central point, and fly off on some tangent . . . In the vicinity of the idea, communication must finally occur, but it does so in violent collisions . . . Some individuals may have crossed the rationality horizon [inner circle] beyond which the fashionable idea has become an article of faith. These unfortunate souls never escape. Examples of this latter phenomenon are also familiar to all of us." The diagram was originally used to illustrate spherically symmetric accretion of interstellar gas onto a massive black hole. At the time there was a controversy about whether active galactic nuclei were at cosmological distances, with Geoffrey Burbidge and Fred Hoyle advocating the non-cosmological view. With the two cartoon figures undergoing the violent crash, McCray had in mind Geoffrey Burbidge and John Bahcall, the latter being very vociferous in challenging Burbidge and Hoyle. By Permission of Richard McCray. From McCray, "Spherical Accretion onto Supermassive Black Holes", in *Active Galactic Nuclei*, C. Hazard and S. Mitton, eds. (Cambridge: Cambridge University Press, 1978), pp. 227–39.

Moreover, the delineation of classes is largely a social activity, though the continual interrogation of Nature that we have seen in the process of discovery constrains reality, the well-known difficulties of observation notwithstanding. This claim for the social construction of discovery is a recognition that, in the absence of proper conditions, many discoveries might not have been made at all – or declared as such by the scientific community.

6.3 How Discoveries End: Demarcating Post-Discovery

The inclusion of an "understanding" component in the structure of discovery begs an important question: how do discoveries end? It does not seem to be the case, in common parlance at least, that we are still "discovering" the rings of Saturn themselves, even if we are still discovering fascinating things *about* them and even though our ever-increasing knowledge remains incomplete. If discovery ends with basic understanding, how does one determine, beyond arbitrary definition, when one has reached basic understanding? To put the question in more pragmatic terms: once again, did Galileo, Huygens, Maxwell, or the Voyager spacecraft "discover" the rings of Saturn? Did Peiresc/Cysat, Herschel, Huggins, or Russell discover gaseous diffuse nebulae? And did Messier, Kant, Rosse, or Hubble discover galaxies? In the opinion of Hanson, Galileo in some sense must be counted as the discoverer of Saturn's rings, since he first detected the phenomenon, though not rings as such, but as handle-like appendages.[30] Similarly, in 1610, Peiresc (or 1618, Cysat) detected the diffuse cloud known now known as the Orion Nebula; with more evidence Herschel adduced such diffuse objects were indeed nebulosities rather than an unresolved group of stars; Huggins proved this spectroscopically in 1864, and a number of astronomers in the 1920s finally identified these kinds of objects as composed mostly of ionized hydrogen, what we now call H II regions. And so on down the line, in the pattern we have documented in the first section.

With numerous exemplars, we are now in a position to shed some light on these questions, at least for astronomy. At the macrostructure level, the problem of discovery is advanced first of all by using more precise language. If discovery, as we have found, begins with detection, continues with data-gathering and its interpretation, and ends with a basic understanding of the object discovered, then the role of the discoverer in each of these phases must be made clear. Galileo detected handle-like appendages around Saturn, Huygens interpreted these appendages as rings, and Maxwell demonstrated how this could be theoretically possible based on dynamical considerations. Similarly (after a pre-discovery phase of casual naked-eye observations), either Peiresc or Cysat first observed diffuse nebulae telescopically, Herschel suspected their gaseous nature, Huggins proved this nature spectroscopically, and Russell and others showed how this could be theoretically true. And again, after a pre-discovery phase in which La Caille and Messier (among others) detected fuzzy "nebulae," and Kant suspected at least some were "island universes" analogous to our Milky Way, Rosse delineated the first of their classes by observing spiral structure, and Hubble definitively proved they were extragalactic, the *sine qua non* for any basic understanding of the true nature of galaxies.

Although such precision provides clarity, some might justifiably claim we have finessed the subject. Skeptics, not to mention scientists with a vested interest, will still want to know, who is the "true discoverer"? This, after all, is a practical problem of great interest to the putative discoverers themselves, and addressed by the Nobel Prize Committee every year, with considerable consequence to careers, funding, and institutions. We can perhaps best approach an answer through the conclusions of the few philosophers, sociologists, and historians who have taken discovery as their subject outside the narrow confines of the well-worn (and here mostly irrelevant) debate over "context of discovery versus context of justification." That distinction, which traces back at least to Hans Reichenbach in the 1930s, has to do with the dynamics of scientific creativity and is not the heart of our subject here. The heart of our problem is the nature and anatomy of discovery itself, not as a problem of creativity, but as a problem of natural history. And despite their extended structure, it would seem only logical that discoveries should have a beginning and an end.

In both advocating an extended structure of discovery and demarcating the end of discovery, the best solution to these questions would seem to be an embrace of the idea of "collective discovery." As historian Ken Caneva put it, this is the concept that "scientific discovery – seen not in terms of the creativity of individuals but as a collective construction cum retrospective characterization – must be understood as an extended process by which a diversity of individuals develop and come collectively to accept as true a statement about the physical world."[31] Such a view, which readily absorbs the idea of extended structure, not only provides a more sophisticated picture of what discovery actually entails; it also avoids what are usually counterproductive arguments about the true discoverer. But it does not illuminate the problem of the end of discovery.

For that we must turn to the historian of science Peter Galison, who has undertaken the subject in his book *How Experiments End*, a history of microphysics that reaches the same conclusion about the extended nature of discovery as we do, but in the case of elementary particles rather than astronomical objects. Having detailed the history of the muon, for example, he asks whether the muon was discovered by eighteenth-century natural philosophers who noticed how electroscopes spontaneously lost their charge, a phenomenon *now* ascribed to sea-level muons, the main constituent of cosmic rays at that level. Or was it in 1929, when Walther Bothe and Werner Kolhörster's detectors indicated the passage of particles that *in retrospect* we call muons? Or was it in 1936, when John Carlson and Robert Oppenheimer suggested a particle of intermediate mass as the penetrating component of cosmic rays? Or was it the following year, when Carl Anderson and Seth Neddermeyer presented data showing the cosmic rays shower particle fit quantum theory, implying, again in retrospect, that the penetrating particles were not electrons and had to be a new particle? Or was it a month later, in April 1937, when Jabez Street and Edward Stevenson demonstrated the energy difference between shower particles and penetrating particles of cosmic rays? These are precisely the kinds of questions we are asking. Galison concludes that "the concept of a single moment of discovery, while perhaps valuable for prize committees and physics textbooks, corresponds to little or nothing in the historical record," a finding that provides further independent support for the idea of extended and collective discovery.[32]

How, then, do experiments end? In Galison's view, they end with "persuasive evidence," but again such evidence does not come at a discrete moment of time. "Instead of looking for a 'moment of discovery,'" he writes, "we should envision the ending of the muon experiments as a progressively refined articulation of a set of phenomena. In a sense the experiment had to end several times. At each stage of the process, a new characteristic could be ascribed to the cosmic rays: they discharged electroscopes; the discharge rate varied in a certain fashion with depth in matter; the shower particles were more easily absorbed than the single particles." Moreover, the final experiments on this subject rested on a large number of earlier experiments. And not only that; as Galison points out, theory also played a complex role in the discovery of the muon, since the quantum theory of electrodynamics helped isolate the experimental techniques that might decide the subject.[33]

We could hardly ask for a better description of what we have found in the case of discoveries of new classes of astronomical objects. They, too, may rest on large numbers of earlier observations and discoveries. They, too, often end as a progressively refined articulation of the nature of the object detected, then interpreted and supplemented by additional observations, until a basic enough understanding is reached to conclude that the object indeed constitutes a new class. And astronomical discoveries, too, may have been influenced by theory, though (as we shall see in Section 9.3) not nearly to the extent as is true of particle physics, and more often in the interpretation and understanding phases than in the detection phase. Thus, stellar classes ranging from dwarfs and giants to supergiants had a large number of prior observations of their spectra and distances; the data were refined, interpreted, and reinterpreted in what was believed to be a temperature sequence; and the understanding of stars was influenced by theories of how stars worked (though not yet satisfactory theories), before they were declared new classes over a period of decades. Similarly, centuries passed from the beginning of the telescope until the several classes of known nebulae were teased out from among those objects that were believed to be resolved or unresolved star clusters. One of these classes turned out to be extragalactic, and once this was realized it was determined that extragalactic objects had their own classes based on morphology, and later on physical differences. It should be noted, however, that neither in the case of particle physics nor astronomical objects was an evolutionary understanding necessary to the discovery of a new class. *Discoveries thus end with a basic enough understanding of an object for it to be declared a new class, but before the mature understanding as defined by knowing an object's place in an evolutionary scheme.*

This parallel between how experiments end and how discoveries end should come as no surprise, for what else are experiments but controlled observations, and observations are the engines of discovery, *sine qua non*. As one might expect, there are differences in the details of how discoveries end, but the basic pattern seems to hold. Put another way, discoveries end with the basic understanding *of* the object, rather than *about* the object, Hanson's distinction between discovery *of* an object rather than *that* an object. The detection of an object that turns out to be Saturn's rings, the interpretation of that object as Saturn's rings, and the basic understanding of the physical possibility of rings in terms of dynamics all fall within the realm of the discovery *of* the object. The discovery of the composition of Saturn's rings, its braids, twists, and other structures, are discoveries *about* the object. In other words, the span of discovery is the difference between discovery of the object itself and the discovery of its properties - discoveries both, but at two very different levels. It is important to note that this dichotomy may not be applicable to the discovery of unobservable, or indirectly observable, objects such as cosmic rays and the solar wind, where the discovery may be accomplished by the discovery of its properties, in the same way the electron was discovered by its mass-to-charge ratio.

The end of a particular discovery implies a *post-discovery period*. That period encompasses not only the determination of the detailed properties of an object, but also its reception among scientists and the public, and its place in any evolutionary scheme. Caneva's work showed that "what a scientist is typically credited with having discovered often differs significantly from the way in which the scientist himself characterized his work."[34] That, too, is a fascinating part of the post-discovery process, but beyond the scope of our discussion. Instead, we turn to the varieties of discovery, with the idea of illuminating and providing context for our discussion so far.

7

The Varieties of Discovery

> I do believe that one can discern general themes in the history of discovery in science and I shall even venture to mention some, but mainly for the purpose of emphasizing variety over uniformity.
>
> Abraham Pais, 1986[1]

Useful as the tripartite structure of detection, interpretation, and understanding is in appreciating the extended structure of discovery, we are far from showing that this is the only structure of discovery in astronomy, much less in science. In fact, we should be skeptical of such a simple view that a single structure is universal to all types of discovery. After all, we have seen that some classes (gas giant planets, giant and dwarf stars) have been inferred over very long periods rather than detected, while others (dwarf planets) have been suddenly declared. And as we saw in the last chapter, the discovery of new classes of astronomical objects is only one of many categories of discovery that historians and philosophers of science, ranging from Norwood Russell Hanson to Theodore Arabatzis, have distinguished.

Scientists themselves have cautioned against simplifying what they as practitioners recognize as a complex process. Referring to Thomas Kuhn's attempts to find structure in the strikingly numerous discoveries from 1895 to 1905, including the contributions of Roentgen, Bohr, Planck, Einstein, Becquerel, Thomson, and Rutherford, the theoretical physicist Abraham Pais wrote that "Such facile and sweeping generalizations only serve to create serious pitfalls of simplicity. I do believe that one can discern general themes in the history of discovery in science and I shall even venture to mention some, but mainly for the purpose of emphasizing variety over uniformity. After decades spent in the midst of the fray I am more than ever convinced, however, that a search for all-embracing principles of discovery makes about as much sense as looking for the crystal structure of muddied waters." In short, Pais confessed to "very strong reservations concerning the usefulness, let alone the necessity, of a search for general patterns or laws of history, specifically the history of discovery."

We should say at once that we are not searching for any such laws of history, still less for a prescriptive formula for successful discovery, though, for obvious reasons, that has been a subject of considerable discussion. But we do take to heart Pais's plea to seek variety in the structure of discovery. While it is beyond the scope of this volume to discuss in detail the structure of discovery in all the categories that Hanson, Arabatzis, and others have laid out, in this chapter we examine – in no way exhaustively – several other classes of discovery in order to gauge the varieties of scientific discovery. We ask to what extent the telescopic discoveries differ from those made before the telescope, from discoveries of the few classes that can only be inferred indirectly by their effects, and from discoveries of new members of already established classes. We then broaden our examination of the varieties of scientific discovery by drawing comparisons with the discovery of some representative astronomical *phenomena* rather than astronomical *objects*.

7.1 Before the Telescope: Demarcating Pre-Discovery

While the vast majority of astronomical discoveries have been made with the aid of the telescope in one form or another, there are fortunately a few pre-telescopic "discoveries" of new astronomical classes that can be examined as "controlled experiments," with an eye toward any differences they might exhibit compared to discoveries aided by instruments. Long before the telescope, and long before discovery was a prized goal in science and "discoverer" a coveted title for the natural philosopher or scientist, several astronomical phenomena had been apparent to the human eye. Among these were the Sun, Moon, planets, and fixed stars, both as objects in themselves and generators of phenomena such as lunar and solar eclipses that have been part of the human landscape from time immemorial. We argue, however, that these objects were neither discovered nor classified in the modern sense, except perhaps as the most rudimentary categories of a human mind that still lacked the concept of *kosmos* (order). Even by the simplest definition of discovery as the detection of something new, they do not pass the test as true discoveries in prehistory. Because planets, stars, and satellites were later discovered as members of a particular class (e.g., giant and dwarf stars), we might call their simple, obvious, and at times spectacular appearance in the day and night sky (as in the case of eclipses) another example of a "pre-discovery phase."

We have already mentioned the idea of a pre-discovery phase, especially in connection with the few nebulae or "nebulous stars" that could be seen with the naked eye. Their case – as well as rarer naked-eye objects such as comets, meteors, and novae ("new" stars) – is rather different from Sun, Moon, planets, and fixed stars. Although comets, meteors, and novae were also undoubtedly seen from prehistoric times, in their case we can begin to discern a rudimentary form of discovery, if only amounting to the detection of these objects as irregular rather than regular phenomena of the sky. Though there was interpretation in abundance of transient phenomena such as comets, having to do with the meaning of such appearances and their characterization as omens, a true understanding of their nature was long in coming.[2] Indeed, in the Western tradition, comets and meteors were believed to be meteorological phenomena, in accordance with Aristotle's dictum that the celestial regions were unchangeable and incorruptible. Novae were harder to explain but much rarer, and so more easily explained away. Comets, meteors, and novae therefore have a true and recoverable history of discovery as astronomical objects.

Comets, also known to antiquity as hairy stars or broom stars (*hui xing*), were among the most spectacular and frightening of phenomena in the night sky, and occasionally the day sky. Though they were undoubtedly seen in pre-recorded history, the first recorded comet was around 1059 BC in China. In subsequent centuries they were often associated with famous historical events, both in the East and West.[3] For example, the daylight comet of 44 BC appeared just after the assassination of Julius Caesar, including the seven days during which Augustus Caesar held games in honor of his predecessor and adopted father. Interpreted as a sign of Julius Caesar's apotheosis, the same comet had most likely been seen first by the Chinese.[4] In 1066, just before William the Conquerer's invasion of England, a comet appeared, as described in the *Anglo-Saxon Chronicle*: "Easter was then on the sixteenth day before the calends of May. Then was over all England such a token seen as no man ever saw before. Some men said that it was the comet-star, which others denominate the long-hair'd star. It appeared first on the eve called 'Litania major', that is, on the eighth before the calends of May; and so shone all the week."[5] This turned out to be the famous Halley's comet, making one of its periodic seventy-six-year appearances.

Through all of this history, comets were believed to be atmospheric phenomena, at least in the Western world dominated by Aristotelian thinking. Though the Chinese apparently regarded them as astronomical by their very designation as broom "stars," it is impossible to say who in ancient China might have first detected and interpreted comets as astronomical phenomena. By contrast, it was only in the late sixteenth century that the Danish astronomer Tycho Brahe first identified comets as a class of astronomical objects. In this

Figure 7.1. From pre-discovery to discovery: the Great Comet of 1577, as seen in Europe, November 12, 1577. Comets had been recorded for thousands of years, but such casual observations do not constitute discovery. Only when Tycho Brahe demonstrated it was an astronomical, rather than a meteorological, phenomenon, was it detected as a new class of astronomical object – the first phase of discovery. Not everyone, including Galileo, accepted it as a new class. This contemporary woodcut is by by Jiri Daschitzsky, and the heading is translated "Concerning the fearful and wonderful comet that appeared in the sky on the Tuesday after Martinmass [November 12] of this year 1577 (Petrus Codicillus a Tulechova, 1577). Courtesy of the Department of Prints and Drawings of the Zentralbibliothek, Zurich.

the comet of 1577 played the crucial role (Figure 7.1). On the afternoon of November 11, 1577 the thirty-one-year-old Tycho, living under royal patronage at his Uraniborg Observatory on the Danish (now Swedish) island of Hven, was catching fish for his evening meal when he noticed a bright star in the western sky. Ruling out Venus and Saturn, Tycho watched as darkness fell and a long tail appeared. We are fortunate to have his description: "In the year of Christ our Savior's birth 1577, on 11 November in the evening soon after sunset, this new birth in the heavens revealed itself, namely a comet with a very long tail, and the body of the star was whitish, though not with the bright gleam of the fixed stars but somewhat darkish, almost like the star Saturn in appearance, which indeed at that time stood not far away from it. Its tail was great and long, curved over itself somewhat in the middle, of a burning reddish dark color like a flame penetrating through smoke."[6] The sight must have been spectacular, for Tycho tells us that the tail stretched over 22 degrees. Tycho knew immediately it was a comet; though it was the first he had seen, he was well aware of the phenomenon from scientific literature, public pamphlets about their meaning, and astronomical hearsay. Cometary pre-discovery history was therefore an immense help, for Tycho recognized a well-known, if totally misunderstood, phenomenon not even believed to be astronomical.

By our analysis, these casual observations do not constitute the "detection phase" of discovery, defined as the detection of an object as a new astronomical class.[7] So far, Tycho's observations were no different from that of many others who had observed comets through history. But, unlike other comet viewers who saw only omens in the appearance of comets, Tycho immediately went to his well-equipped Uraniborg observatory, trained his radius and later his quadrant instruments on the intruding object, and began to record his observations. For the next two-and-a-half months, whenever clear, Tycho continued making and recording observations, even as the cometary tail grew shorter and shorter, until it was scarcely visible when he last saw the comet on January 26. Most important, using these observations to measure parallax, Tycho was able to conclude that the comet was located far beyond the Earth's atmosphere, in fact well beyond the Moon, at a minimum distance some 230 Earth radii, corresponding to a parallax that could not have exceeded 15 minutes of arc. Because the lower bound of the Moon's orbit was at that time estimated at 52 Earth radii, the comet was clearly shown to be a celestial body, contrary to Aristotle's teaching. Or, at least it was shown to Tycho's satisfaction; given the cosmological implications not everyone accepted the result, and even Galileo in his *Assayer* (1623) still argued comets were optical phenomena that could not have parallaxes.[8]

Tycho's distance determination of the comet of 1577 thus constituted the first phase of discovery – detection of a new class of astronomical objects, bringing by inference all those cometary appearances in the past into the realm of astronomy. The new view of cometary distances in turn now required totally different interpretations for the nature of comets, having nothing to do with the "exhalations" of the atmosphere. Although these physical interpretations marked a rudimentary beginning in the understanding of comets, and although properties such as brightness, position, motion, and morphology could be observed without the telescope, even a basic understanding of these objects depended on knowledge of their physical nature. Such an understanding would be a long time coming, despite advances in the telescope. Even at the beginning of the twentieth century, after spectroscopy had revealed some of their constituents, comets were believed to be dusty sandbanks that released gases as they approached the Sun. Only in 1950 did Harvard astronomer Fred Whipple revolutionize comet science when he suggested that comet nuclei were "a conglomerate of ices" including water, ammonia, methane, carbon dioxide, or carbon monoxide combined with meteoritic materials.[9] About the same time, the Dutch astronomer Jan Oort suggested that comets originated from a vast reservoir of objects at the outer edges of the solar system, now known as the Oort Cloud.[10]

All of this, one could argue, was part and parcel of the interpretation and basic understanding stages, mature understanding dawning only when these conjectures were proven by spacecraft observations. From the time of their proof as astronomical objects, therefore, comets follow a clear pattern of detection, interpretation, and increasing understanding over a period of more than four centuries. Unlike some objects not visible to the naked eye, however, comets have a pre-discovery phase prior to their determination as astronomical objects in 1577, a phase not irrelevant to the subsequent discovery process. Moreover, even in this pre-telescopic era, we see the effect that worldview can have on discovery. Most comet observers in the Western world, including the public and "scientific" observers of the comet of 1577, were content enough with Aristotle's view of these objects as meteorological phenomena that they failed to conceive of the astronomical option, even had they possessed the means to measure parallax. On the other hand, as his biographers have shown, Tycho was skeptical of Aristotle's terrestrial-celestial dichotomy, more inclined to the view of Paracelsus that the universe was a dynamic cosmos subject to change.[11]

Yet another important factor figured in Tycho's identification of the celestial nature of comets, involving what turned out to be another new class of object, for five years earlier he had observed a "new star" (*stella nova*) in the heavens. Like comets, new stars had undoubtedly also been noticed in prehistory, though this would have required more astute observational powers than bright comets. And they had occasionally been seen in recorded history, as described in David Clark and Richard Stephenson's classic volume *The Historical Supernovae*.[12] Such a phenomenon was very dramatic at a time when the heavens were still believed to be immutable. When Tycho first observed the object on November 11, 1572, it was as bright as the planet Jupiter. It soon equaled the brightness of Venus in the sky, at -4.5 magnitude, and was observed in daylight for two weeks. It began to fade at the end of the month, but the star remained visible for some time after that, vanishing completely by March 1574. Meanwhile, Tycho, detaching himself from any Aristotelian preconceptions, made observations for parallax of the object. Finding none, the only conclusion was that the new star was in the celestial regions; since no parallax was detected at all, no upper bound could be set to its distance.[13]

Five years before his determination that the comet of 1577 was astronomical in nature, therefore, Tycho had proven that the phenomenon of new stars was as well. More than that, on that occasion he also implicitly predicted that comets were astronomical, so Tycho was predisposed both by worldview and previous observation of the new star to make the parallax measurement on comets that no one else could have because of the accuracy of Tycho's instrumentation.

Like comets, the interpretation and understanding of the new stars themselves was also long in coming. Only three decades later, observers first spotted another *stella nova* on October 9, 1604, three years after Tycho's death. When the clouds parted in Prague, where Johannes Kepler had gone to work with Tycho and now took his place as imperial mathematician to Rudolf II, Kepler himself saw the object on October 17, studied it in detail, and wrote about it two years later in his volume *De Stella Nova*. There we find five observations made either by Kepler himself or by others in his presence using Tycho's instruments, as well as a summary of the many observations he had gathered from others. As for Tycho's new star, Kepler concluded from parallax measurements that the star of 1604 was celestial, far beyond the sphere of the Moon.[14] Although the lack of parallax proved the new stars were astronomical objects, the physical nature of these objects could not be determined until their true distances were known. Because they were in the realm of the stars rather than in the solar system, distance determinations took much longer than comets; the first stellar parallaxes were not determined until 1838, and others were a long time coming. Lacking distance information, the immense power of the phenomenon could not be realized, much less its nature and cause.

Even when their immense distances did become known, an understanding of their physical nature remained elusive. And not until the 1920s and 1930s were absolute magnitudes and luminosities well enough known to distinguish what we today call novae from an even more astounding phenomenon: supernovae. As these luminosities became known, already in 1920 the Swedish astronomer Knut Lundmark spoke of "giant novae," then "much more luminous novae" (1923), and "upper class novae" (1927); in 1929, Walter Baade called them "Hauptnovae" [chief novae]; and in 1929 Hubble called them "exceptional novae." Walter Baade and Fritz Zwicky coined the term "supernovae" in print in 1934. Two years after James Chadwick's discovery of the neutron in 1932, they theorized that "supernovae represent the transitions from ordinary stars into neutron stars."[15] From 1934 to 1939, they found many more supernovae with telescopes at Mt. Wilson, and Zwicky theorized in more detail that a supernova represented a massive star explosion that resulted in a high-density star only 12 miles in diameter, 100 million times the density in a white dwarf.[16] Only in the 1960s did these prove to be yet another new class of objects – the neutron stars observed as pulsars. And only in the 1960s were the lesser novae themselves understood in physical terms, as members of close binary systems, in which a white dwarf, for example, draws material from a cooler star, resulting in brightness increases of up to ten magnitudes.

By 1941, enough supernovae had been observed that Rudolph Minkowski could separate them into two types based on their spectra. "Spectroscopic observations indicate at least two types of supernovae. Nine objects . . . form an extremely homogeneous group provisionally called 'type I.' The remaining five objects . . . are distinctly different; they are provisionally designated as 'type II'," he wrote. Further, "Supernovae of type II differ from those of type I in the presence of a continuous spectrum at maximum and in the subsequent transformation to an emission spectrum whose main constituents can be readily identified. This suggests that the supernovae of type I have still higher surface temperature and higher level of excitation than either ordinary novae or supernovae of type II."[17] These designations were soon adopted by astronomers such as Zwicky and Baade, and the latter found that supernovae light variations could also be classified by the same two types. What we now know as white dwarf supernovae (Minkowski and Baade's Type I, our Type Ia) exhibit a rapid rise to maximum brightness over two weeks, then a steady decline over a few weeks, with a "half life" of about fifty days during which they successively fade to half their previous brightness. Type II supernovae remain at maximum brightness for several weeks and dim more slowly than Type I.

Both Tycho's new star of 1572 and Kepler's of 1604 are now believed to have been Type Ia white dwarf supernovae. Astronomers continue to learn more about supernovae every day, thus illustrating the complex nature of discovery.

The story of the new stars therefore illustrates how lengthy and complex the process of discovery can be when it comes to a mature, and even a basic understanding of an object. The same can be said of many of the other classes of stars, ranging from dwarfs to giants and supergiants. Though they were seen in the sky as points of light, no one would claim that dwarfs and giants were detected or discovered *as new classes* before Hertzsprung and Russell had enough distance and luminosity data to separate them into classes. The same is true of the other classes; even as their spectral phenomenology was being classified, the identity of stars as dwarfs, or giants, or something in between or beyond, awaited accurate distance determinations. As we shall see in the next chapter, the parsing of stars provides a prime example of classification as both a pre-discovery and a post-discovery activity with respect to their physical nature.

* * *

Meteors present a somewhat different case of pre-telescopic objects than do comets and new stars. Once again the phenomenon had been widely observed throughout history (Figure 7.2), and much more frequently than comets or new stars; catalogues of meteors record observations dating back at least to the seventh century BC, and they were, of course, observed long before that. The Roman poet Virgil wrote, "Oft you shall see the stars, when wind is near, shoot headlong from the sky and through the night leave in their wake long whitening seas of flame."[18] The question was, where did they originate? Like comets, from the time of Aristotle's *Meteorology* about 340 BC, meteors were believed to be atmospheric phenomena for more than two millennia (thus their name). Author Mark

Figure 7.2. Pre-discovery: Depiction of a meteor storm that took place in the early morning hours of November 13, 1833. Although suspected to be of cosmic origin, only in 1863 did H. A. Newton prove this to be true. Observations prior to this time can be considered pre-discovery observations. Today the event is known as the Leonid meteor storm, which occurs periodically about every thirty-three years. This depiction first appeared in April 1888. It was painted by the Swiss artist Karl Jauslin and engraved by Adolf Vollmy, based on a description by Joseph Harvey Waggoner, who saw the event when he was thirteen. It first appeared in a religious publication of the Seventh-Day Adventists, who invested the event with religious meaning. See David W. Hughes, "The World's Most Famous Meteor Shower Picture," *Earth, Moon and Planets* 68 (1995): 311–22. Courtesy Seventh Day Adventist Church.

Littmann called this "perhaps the longest run of any scientific error," because unlike comets, the celestial nature of meteors took much longer to uncover.[19] This is because Aristotle was actually half correct: meteors are an atmospheric phenomenon, but with a celestial cause: a meteoroid from outer space incinerating in the Earth's atmosphere. Thus, meteors are in fact an atmospheric phenomenon, but astronomically induced, even if by the admittedly small objects technically known as "meteoroids." Mostly the size of a grain of sand, they can nevertheless produce spectacular meteor showers, and occasionally, meteor storms.

Prior to the Space Age, knowledge of the existence of meteoroids in outer space depended entirely on the observation and interpretation of meteor showers in the Earth's atmosphere, the only way they became visible. And, unlike new stars and comets, strictly speaking the discovery of meteoroids by way of the meteor phenomenon did not occur in the pre-telescopic era. Although the suspicion that meteors were of cosmic origin dates at least to Edmond Halley in the eighteenth century, and was given credence in the early nineteenth century by claims that meteoritic rocks on Earth were of extraterrestrial origin, it was not until 1863 that meteors themselves were definitely proven to be astronomical. The proof was based largely on observations of what we now call the Leonid meteors, so called because their radiant is in the constellation Leo. The Perseid "August meteors" also played a role, but the Leonid "November meteors" tended to storm every thirty-three years, and thus demanded a more immediate explanation. The Leonid storm of 1833 peaked in the Eastern part of North America and gave birth to modern meteor studies. Yale astronomer Denison Olmstead wrote that "Probably no celestial phenomenon has ever occurred in this country, since its first settlement, which was viewed with so much admiration and delight by one class of spectators, or with so much astonishment and fear by another class. For some time after the occurrence, the 'Meteoric Phenomenon' was the principal topic of conversation in every circle, and the descriptions that were published by different observers were rapidly circulated by the newspapers, through all parts of the United States."[20] From these many observations Olmsted collated data relating to weather, time and duration, number, variety, sound, and apparent origin, and concluded that the meteors of 1833 originated beyond the Earth's atmosphere.

The clinching argument for the cosmic origin of meteors, however, came only in 1863, when Yale astronomer Hubert A. Newton determined from past cycles that the November meteors repeated in intervals of sidereal years, not tropical years, a sure sign of a celestial phenomenon.[21] Meteors were therefore "discovered" as an astronomically induced phenomenon only 150 years ago; their interpretation as due to meteoroids entering the Earth's atmosphere was proven only later, and a full understanding of them in terms of the origins of meteoroids is still ongoing. In this scheme, all previous observations of meteor showers, and even Halley's suspicions, were part of the pre-discovery phase.

What do we learn about the nature of discovery from pre-telescopic objects? Like the other naked-eye objects (nebulae as eventually resolved into their various classes, or the Sun, Moon, and stars before they became the object of scientific study), *all pre-telescopic objects have one thing in common: a pre-discovery phase.* In their case, however, it is often a phase even before they are recognized as astronomical objects, unlike the nebulae, for example, and indeed unlike a few telescopic objects such as Uranus and Neptune, both of which were "seen" before they were "discovered" as planets. We learn further that a determination of distance is essential to even a basic understanding of these new classes of astronomical objects, and this indeed often proves to be the case with telescopic objects also. Although the basic pattern of detection, interpretation, and understanding holds as it does with telescopic objects, ironically pre-telescopic objects took longer to understand.

The concept of pre-discovery, however, is not limited to naked-eye objects. In fact, as hinted at in the last chapter and as can be seen from Appendix 2, most new classes of objects have some history of observation before they are announced or declared as new classes. This includes all planets, observed as "wanderers" in the sky before they were separated as terrestrial, gas-giant, and (still controversially) ice giants; all classes of stars, ranging from dwarfs to giants and supergiants, unrecognized for decades from their "phenomenology" of spectral lines as different classes of "things in themselves"; and all classes of galaxies, observed as nebulae long before they were separated out first as one new class, and then many classes of objects. We come to the following important conclusion: *casual sightings of astronomical objects with the naked eye, or telescopic observations that go unreported, unrecognized, or undistinguished as new classes of objects, constitute what we shall call a pre-discovery phase.* Moreover, the idea of a pre-discovery phase may be expanded to those objects that were predicted from theory, including neutron stars and black holes, even if the theory played no role in leading to the actual discovery. Finally, as we saw in the last section, the discovery phase itself is bracketed not only by the pre-discovery phase, but also the post-discovery phase, during which (except for hidden objects such as black holes) properties of the object, rather than the object itself, are discovered. Recognition of these bracketing phases helps us to focus on the anatomy of discovery itself.

7.2 Indirect Discovery: Evidence of Things Unseen

In contrast to the instrument-aided direct discovery of astronomical objects discussed in the last chapter, there is another category that needs examination: instrument-aided indirect discovery, or, to put it in Arabatzis's terms, "the discovery of entities that are unobservable in principle." Indeed, a few such "objects" exist in astronomy, namely in the form of energetic particles known as cosmic rays and solar wind. Just as Arabatzis and others have written on the discovery of the electron, Hanson on the positron, and many others on the discovery of other elementary particles, so one cannot observe cosmic rays and the solar wind except by their indirect effects. Yet they are indeed objects, if extremely small objects that make the meteoroids discussed in the last section

Figure 7.3. Indirect discovery: Victor F. Hess detected cosmic rays in 1912 after a series of daring high-altitude balloon flights. The detection was not made directly with a telescope, but inferred indirectly from ionization effects of the high-energy particles.

look large by comparison. At the other extreme of mass for indirect detection are black holes, objects that by definition cannot be directly detected since they are so dense that light cannot escape. How does the necessity of indirect detection affect the structure of discovery in these three very different cases?

The first thing to notice about cosmic rays is that they were not discovered with the telescope. Rather, the twenty-nine-year-old Austrian physicist Victor F. Hess first detected them in 1912 during the ninth of a series of daring balloon flights (Figure 7.3).[22] These dangerous hydrogen-gas balloon flights carried Hess and three ionization chambers to great altitudes, culminating at 17,500 feet (5350 meters) on August 7 of that year. And why was Hess motivated to undertake these dangerous flights? Because on March 30, 1910, the Dutch Jesuit high school instructor Theodore Wulf had hauled an electroscope to the top of the Eiffel Tower for four days and shown that it lost its charge, an effect that could only be caused by ionization of the air. The radioactivity that Henri Becquerel had found fourteen years earlier was a source of ionizing radiation, as the alpha and beta rays of radioactive material create ions that conduct electricity. But that radioactivity from the Earth, it was calculated, would be absorbed by the intervening 1000 feet of air at the top of the Eiffel Tower. Wulf concluded there must be another source of radioactive emission in the upper part of the atmosphere, and Hess postulated more generally that "a hitherto unknown source of ionization may have been in evidence in all these experiments."[23]

The obvious next step was to go higher in an attempt to determine how ionizing radiation changed as a function of distance from the surface of the Earth, and it was for this reason that Hess found himself in the dangerous situation of making high-altitude balloon flights. The ionization chambers Hess carried on his flights were greatly improved over those invented by Wulf. Hess found that ionizing radiation fell off at an altitude of about 6000 feet (2000 meters), which made sense for radiation emanating from Earth due to radioactivity. But surprisingly he found that radiation levels increased above that. This he ascribed to "radiation of very great penetrating power [which] impinges on our atmosphere from above and still evokes in the lowest layers a part of the ionization observed in closed vessels."[24]

In this "detection" phase of discovery, for which Hess received the Nobel Prize in Physics in 1936, there was little doubt that an ionizing agent had been detected emanating from outer space. Just before the outbreak of World War I in 1914, physicist Werner Kolhörster ascended to 9300 meters in the last of his five balloon ascents, and measured ionization six times greater than at ground level, thus confirming Hess's result.[25] But the nature of the ionizing agent was more difficult to deduce, and even its extraterrestrial nature was sometimes called into question. As late as 1924, at a meeting of the American Physical Society, Caltech physicist Robert Millikan asserted that "the whole of the penetrating radiation is of local origin."[26] After further experiments, by 1926 he agreed that Hess's phenomenon originated beyond Earth, and in one of the great misnomers of science, he coined the term "cosmic rays."[27] A great debate took place in 1932 between Millikan and the physicist Arthur Holly Compton at the Christmas meetings of the American Association for the Advancement of Science in Atlanta, with Millikan arguing the observed phenomenon consisted of rays, photons similar to light created when hydrogen or other elements fused in interstellar space. Compton, on the other hand, argued they were charged particles. In the previous months he had mounted a worldwide campaign to observe cosmic rays, and himself traveled 50,000 miles in search of a latitude effect, which he and others found and interpreted as an uneven distribution of charged particles due to the influence of the Earth's magnetic field. Millikan, on the other hand, saw no such effect. The argument between the two Nobelists was closely followed by the press, notably the *New York Times*.[28]

We now believe that Compton was right: cosmic "rays" are not light rays but atomic or subatomic particles. We also believe that because primary cosmic rays with energies below 1 MeV do not reach the Earth, and because very high-energy cosmic rays are rare, Hess, Millikan, and Compton were observing showers of secondary particles produced when cosmic rays originating outside the solar system, but within the galaxy, hit molecules in the atmosphere. The different categories of cosmic rays based on energy, composition, and source were only gradually separated out over time; they range from low-energy anomalous cosmic rays from the local interstellar medium to extremely energetic extragalactic cosmic rays. Ironically, cosmic ray physics gave rise to high-energy physics, which in the 1950s began to accelerate its own charged particles in increasingly large accelerators, whose results then built high-energy physics into one of the most robust fields of "big science."

Although cosmic rays could not be seen with a telescope, composed as they were of atomic and subatomic particles, from the astronomical point of view, and in the Three Kingdom system of Appendix 1, they constitute a new Subfamily of astronomical objects known as energetic particles. Nor were they the only class of this Subfamily. It was also in the 1950s that the landmark discoveries were made of energetic particles emanating from the Sun, which, surprisingly, was now shown to emit matter as well as light into space. The realization of this "solar wind" as a reality came only slowly. British astronomer Richard Carrington first suggested the concept of particles from the Sun in 1859, after observing a solar flare.[29] By the 1930s, scientists realized the Sun's corona must be about a million degrees Celsius (1.8 million degrees Fahrenheit). In 1955, Sydney Chapman calculated the properties of such a hot gas and concluded that it must extend beyond Earth's orbit. Indirect observational evidence of particle emission from the Sun came in 1951 from the German scientist Ludwig Biermann, who postulated that this steady stream of particles caused a comet's tail to always point away from the Sun. James van Allen added another piece of the puzzle with his discovery of Earth's radiation belts using his detector on the first American satellite, Explorer 1. Van Allen attributed these particles to the solar plasma, but proof of the source remained elusive.

The landmark event in recognizing the solar wind as a class of energetic particles came in 1958 when astronomer Eugene Parker wrote a seminal paper on the subject of interplanetary gas in the *Astrophysical Journal*, and coined the term "solar wind."[30] Parker's theory of the solar wind was not widely accepted at first; in fact his *Astrophysical Journal* paper was initially rejected and only published later following the intervention of the journal's editor, Subrahmanyan Chandrasekhar. But satellites confirmed the solar wind in the 1960s, at first tentatively with the brief measurements of Luna 3 in 1959 and Explorer 10 in 1961, and then more persuasively with the electrostatic analyzer aboard Mariner 2 on its way to Venus. It was Mariner 2 that established the continuous flow of the solar wind, its composition, velocity, temperature, and density; the Pioneer spacecraft and many others have since refined our knowledge. The solar wind was even studied from the ground using the newly discovered technique of interplanetary scintillation. Anthony Hewish, who received the 1974 Nobel Prize for the discovery of pulsars using the same technique, recalled of his work in the early 1960s, "Since interplanetary scintillation, as we called this new effect, could be used in any direction in space I used it to study the solar wind, which had by then been discovered by space probes launched into orbits far beyond the magnetosphere. It was interesting to track the interplanetary diffraction patterns as they raced across England at speeds in excess of 300 km/second, and to sample the behaviour of the solar wind far outside the plane of the ecliptic where spacecraft have yet to venture."[31]

We now know this "solar wind" consists of a stream of charged particles, primarily electrons and protons, ejected from the Sun at speeds of 200–600 miles per second, about a million miles per hour. This corresponds to energies of 10–100 electron volts, where air molecules at room temperature have energies of about 1/40 of an electron volt. The acceleration mechanism is believed to be the magnetic field associated with sunspots. These solar energetic particles are also sometimes termed "solar cosmic rays," but the term "cosmic rays" refers more commonly to much higher energy particles originating beyond the solar system that Hess discovered. By contrast, the sources of solar wind include solar flares and corona mass ejections from the Sun.[32] The solar wind also creates the solar system's heliosphere.

Because the solar wind is unseen even through a telescope and detected only indirectly, it may seem to be of little importance. But it is arguably of more direct importance to human activity than the macro-objects more commonly known in astronomy. In fact, the discovery of the solar wind as a new phenomenon, a new class of objects, has been the subject of much research precisely because it is a component of space weather, and its effects include geomagnetic storms that affect power grids on Earth, as well as aurora and other solar-terrestrial phenomena. The Solar Wind Composition Experiment was one of the first experiments employed during *Apollo 11*, the first manned lunar landing in 1969. It was repeated on *Apollo 12, 14, 15*, and *16*. It consisted simply of an aluminum foil sheet, 1.4 by 0.3 meters, deployed on a pole facing the Sun. This was exposed to the Sun for periods ranging from 77 minutes on *Apollo 11* to 45 hours on *Apollo 16*. The embedded particles were returned to Earth for analysis, and among the gases detected were helium-3, helium-4, neon-20, 21, and 22, and argon 36. But the Solar Wind Spectrometer on *Apollo 12* and *15* confirmed that more than 95 percent of the solar wind is in the form of equal numbers of electrons and protons.[33] It has been estimated that the solar wind removes about 2 billion pounds of matter from the Sun every second; even so, during its 5 billion year history, only 0.01 percent of the Sun's mass has been lost through the solar wind. And

Figure 7.4. Indirect discovery: Hubble Space Telescope astronomers described this image as "seemingly conclusive evidence" for massive black holes, in the form of a gas disk spiraling into the supposed object, which is itself unseen by definition. The spiraling was determined by Doppler velocity measurements of different parts of the spiral disk, and the mass of the black hole could be inferred from the velocities. The inferred black hole is at the center of the active galaxy M87, 50 million light-years from Earth. The highly collimated relativistic jet characteristic of accretion disks is also evident. The Space Telescope Science Institute news release, dated May 25, 1994, was titled "Hubble Confirms Existence of Massive Black Hole at Heart of Active Galaxy." NASA, the Space Telescope Science Institute, Johns Hopkins University, Applied Research Corp., and the University of Washington.

what is true of the Sun is true of the stars: in fact, some stars have much stronger stellar winds, including the massive Wolf-Rayet stars. And stellar winds are now believed to greatly influence galactic evolution.

There is one final astronomical object that is by definition only indirectly detected – black holes. As we saw in Chapter 5, black holes were long a theoretical entity before an observational one. The speculations of John Michell and Pierre-Simon de Laplace constitute a pre-discovery phase. In 1916 Martin Schwarzschild, on the other hand, was the first to deduce black holes based on a robust theory – the solutions of Einstein's equations of general relativity. Another landmark was passed in 1939, when Oppenheimer and his graduate student George Volkoff speculated on what would happen if a star of very large mass, greater than that which produced a neutron star, would collapse. Later in the year, with another student, Hartland Snyder, Oppenheimer concluded it would collapse into a singularity, what we today call a black hole. Princeton physicist John Wheeler christened them "black holes" in 1967.

Because very few people believed they actually existed, least of all astronomers, their search was delayed. But following the discovery of pulsars in 1968, interest increased. By the mid-1970s, Roger Blandford and his colleagues showed a spinning accretion disk would produce jets, and they had been observed in quasars. Still, this was circumstantial evidence, possibly caused by a supermassive star rather than a collapsed object. In 1994, the Hubble Space Telescope found an unusual number of stars around M87 (attributed to the attraction of the immense gravity of its black hole), and in the same year astronomers announced "seemingly conclusive evidence" in the form of velocity measurements of orbiting gas at its center (Figure 7.4). By 1997, Hubble astronomers were suggesting that most galaxies, including our own, harbored black holes. Clearly, then, black hole theory preceded black hole observation by many decades.

These three cases of indirect discovery demonstrate that, like direct discovery, there is no getting around the interpretation and understanding phases following detection (and that, especially in these cases, must essentially include interpretation even in their detection). Even following Kolhörster's confirmation in 1914 of Hess's 1912 observation of the effects of a penetrating radiation at high altitudes, some doubt remained whether the source was extraterrestrial. The interpretation phase stretched out via the Millikan-Compton debate in the 1920s and 1930s over particles versus rays, shading into the process of an increased understanding as different classes of cosmic rays were separated, including extragalactic cosmic rays in the 1930s and 1940s and anomalous cosmic rays in the 1970s. Similarly, solar wind had been conjectured in the mid-nineteenth century by Carrington, suspected by Biermann and others, but not detected until the 1950s and 1960s via spacecraft. That detection was strengthened by the theoretical work of Parker, which also aided in the interpretation and understanding of solar wind. In the case of black holes, theory preceded detection by fifty years, and the discovery of different classes of black holes based on mass stretched out over decades in the 1990s and beyond.

Finally, lest we think that indirect discovery is not as important as direct discovery, we recall that the "dark matter" hypothesized by astronomers by definition also cannot be directly detected. Yet, by its indirect effects, it is today not only believed to be as real as planets, stars, and galaxies, but also to make up 23 percent of the matter-energy in the universe, with all the objects we have discussed heretofore, including black holes, comprising less than 5 percent. The mysterious "dark energy" comprises the remaining 72 percent. But have dark matter and dark energy really been "discovered," any more than when nineteenth-century physicists pronounced the ether to be "real" because of its perceived necessity based on its presumed effects? The anomalous rotation rate of galaxies indicates dark matter should exist in order to explain gravitational effects. The accelerating universe indicates dark energy "must exist" to account for the acceleration. But neither has been detected, interpreted, or understood, only inferred, and arguably inferred in a much weaker sense than, say, the standard classes of stars. Indirect discovery likely has a big future, but it needs to be approached with caution.

7.3 Discovery of New Members of a Known Class

We now depart from our usual analysis of new classes of astronomical objects to study the discovery of new members of a class already established. While discoveries of new classes are for the most part unpredicted, they *can* be predicted, as evidenced by neutron stars and black holes. Discoveries of new *members* of a class, on the other hand, are less predicted than expected following the discovery of the first member of a class. Does the structure of discovery change in such cases?

Planetary satellites offer an excellent laboratory for addressing this question. Following Galileo's discovery of the Jovian moons, how was the discovery of subsequent members of the class affected? The most obvious effect was that astronomers could now *search* for new satellites around other planets, something Galileo had not preconceived, or even conceived as a possibility. The terrestrial and Jovian moons were the only satellites known in the solar system until Christiaan Huygens discovered Titan around Saturn in 1655. The discovery was not the result of a conscious search, but a serendipitous outcome of studying the rings of Saturn with a telescope he and his brother Constantijn had built. Only four other moons were discovered in the seventeenth century: Iapteus and Rhea in 1671 and 1672, and Tethys and Dione in 1684, all by Giovanni Domenico Cassini (with his large telescope) and all around Saturn. Thus, nine satellites were known at the end of the seventeenth century. By the end of the nineteenth century, twenty-one satellites were known to exist other than our own Moon. Today, swelled by spacecraft observations, the total is 167, some little more than large rocks but others entire worlds of their own comparable to our Moon in size but extremely varied in physical conditions.

This is not to say that a conscious search for new members of the class of satellites could not have its own drama. To the contrary, each satellite has its own story; the circumstances of each discovery vary, but the drama can be well illustrated in one famous example: Asaph Hall's discovery of the two tiny moons of Mars. Hall had been an astronomer at the U.S. Naval Observatory in Washington, DC, since 1862 when he decided to search for a moon of Mars at its next favorable opposition in 1877. He knew that many had looked for satellites around Mars to no avail; in fact, with few exceptions the search had been abandoned since William Herschel's failed attempt in 1783, and one exception – a search in 1862 by the German astronomer Heinrich d'Arrest – had also failed. The first condition was therefore that Hall had to be motivated. The approaching favorable opposition of Mars, coupled with the availability of the new 26-inch refractor at the Observatory – the largest in the world – might have been all the incentive he needed. But there was one more crucial motivator. In December 1876, Hall had determined that the rotation rate of Saturn was in error by more than a quarter hour; such an error, enshrined in the canonical textbooks for many years, "made me ready to doubt the assertion one reads so often in the books that 'Mars has no moon,'" Hall recalled.[34]

Motivated by the defiance of conventional wisdom, armed with the largest telescope of its type in the world, Hall also had one other tool in his armory to guide his thinking: He knew from D'Arrest's calculations that a search more than 70 arcminutes from the planet would be useless, for beyond that distance the gravity of Mars could not hold any moons. On August 10, 1877, Hall thus began searching very near the planet, with the glare of the planet blocked by being kept just outside the field of view. At first he found nothing, and again nothing early the night of the August 10, but later that night he glimpsed a faint object. Cloudiness intervened and on the 15th he again saw nothing. Finally, on August 16 he saw the object again and found – as Galileo had more than 250 years before in the case of Jupiter – that the object was moving with the planet. On the night of the 17th, Hall found a second moon closer to the planet. He was puzzled by this second moon: it would appear on different sides of the planet on the same night, leading him to believe there might be two or three inner satellites. But he watched it on the nights of the 20th and 21st, "and saw that there was in fact but one inner moon which made its revolution around the primary in less than one third the time of the primary's rotation – a case unique in our solar system." Hall observed the satellites with the 26-inch until October 31, which gave him enough data to determine their orbits. He chose to name the inner satellite Phobos, and the outer one Deimos.

It is tempting to attribute Hall's success to the fact that he had the largest telescope in the world at the time, at least in the form of the long-focus refractor. But Hall himself subsequently wrote that even the 9.6-inch telescope, on the Naval Observatory grounds since 1845, could have discovered the satellites of Mars. Indeed, Hall's colleagues did detect them afterward using this instrument, as did Edward C. Pickering and his assistants using the 15-inch refractor at Harvard, and Henry S. Pritchett using the 12-inch Clark refractor at Morrison Observatory in Glasgow, Missouri. There is no doubt the moons of Mars could have been found much earlier, despite their small size, for photometric observations at Harvard showed that the diameter of the outer moon was about 6 miles and the inner moon 7 miles, only slight underestimates of what we know today. By comparison, the Galilean moons of Jupiter vary in diameter from about 2000 miles for Io and Europa to more than 3000 miles for Ganymede and Callisto.

Thus, in the case of the moons of Mars, detection took place more than 250 years after the discovery of the telescope, and perhaps a century later than it could have with the telescopes available to William Herschel and others. The interpretation stage was much faster than in Galileo's case, for Hall was searching for his quarry and knew when he had reached his predetermined goal. He immediately had a basic understanding of his object as analogous to the many satellites of other planets previously discovered. Mature understanding – the post-discovery stage – took much longer, awaiting spacecraft observations a century later to image the satellites and determine their physical nature, much less their still-disputed origin. In short, after Galileo, never again did astronomers have to make the intellectual leap he struggled with in January 1610: that secondary bodies, today known as moons or satellites, can in principle circle a primary planet. That kind of leap is reserved for the discovery of a new class. Indeed, in the annals of planetary satellite discovery, some made the leap too fast, claiming they had found even more moons of Mars, moons of Venus, and even another moon of Earth! Very shortly after the discovery of Neptune in 1846, William Lassell believed he had discovered rings around the planet with the aid of his 24-inch reflector. Such was the quest for glory through discovery, raising the interesting problem of how to divest oneself of a "failed" discovery.[35]

By contrast with planetary satellites, which had nine members of its class (including the Moon) in the first century after the discovery of the moons of Jupiter, the first planetary rings to be discovered after Saturn's were those of Uranus in 1977, more than 350 years after the first member of the class had been detected. Indeed, so much time had elapsed that the discovery was serendipitous; no one was searching for rings of Uranus or any other planet. On March 10, 1977, James L. Elliot, Edward W. Dunham, and Douglas J. Mink were observing occultations of a star by Uranus from the Kuiper Airborne Observatory, flying over the Indian Ocean. They were intending to use the occultation to study the Uranian atmosphere. But instead of observing a single occultation by the planet, the star blinked five times. Such blinking might have been caused by any number of things, including instrumental error. But, now forced to interpret the data staring them in the face, with the example of Saturn in the background, it did not take long for the observers to conclude that the blinking was caused by a set of rings as each ring in turn occulted the star (Figure 7.5).[36]

Aside from Saturn, the rings of Uranus remains the only other ring system discovered from the ground (an airborne observatory being close enough to

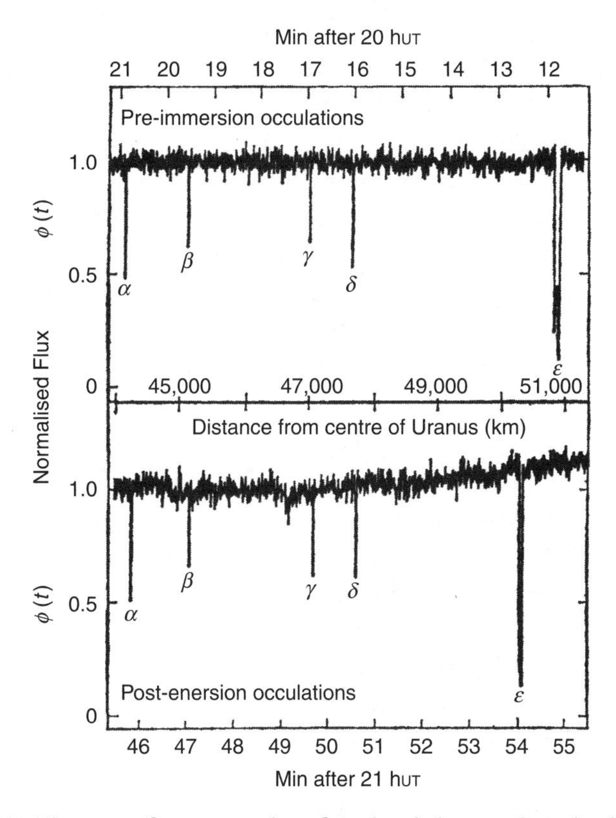

Figure 7.5. Discovery of a new member of an already known class, the rings of Uranus. Photometric measurements made in 1977 of an occultation of the star SAO 158687 by Uranus unexpectedly showed five pre-immersion and five post-immersion occultations, labeled here alpha, beta, gamma, delta, and epsilon. The three-channel photometer was attached to the 91-cm telescope aboard the Kuiper Airborne Observatory. Knowledge of the rings of Saturn allowed a quick interpretation of these observations. Reprinted with permission from Macmillan Publishers, Ltd., from J. L. Elliot, E. Dunham, and D. Mink, "The Rings of Uranus," *Nature* 267: 328–30, copyright, 1977.

qualify as "ground" by comparison with Earth orbit). But in 1979, Voyager 1 unexpectedly detected the solar system's third set of planetary rings, a thin ring system around Jupiter, as reported by Brad Smith and his colleagues in the journal *Science*.[37] Additional rings of Uranus were found in 1986 by Voyager 2, and two outer rings were discovered in 2003–2005 by the Hubble Space Telescope, for a total of thirteen Uranian rings known by 2008. The discovery of the rings of Neptune by Voyager 2 in 1989 completed the discovery of ring systems among the gas giants of the solar system.[38] By this time rings were all the rage. Some claimed (spuriously, it now seems) that the Saturnian satellite Rhea had rings;

and in 2009, the Spitzer Space Telescope actually found a huge ring shining in the infrared around Saturn, far removed from its visible ring system.

The long duration between the discovery of Saturn's rings and those of Uranus raises an interesting question: does the discovery of a single novel object constitute the discovery of a new class of objects? More to the point, the Moon was a unique object until Galileo's discovery of the moons of Jupiter in 1610; it was not a member of a class of objects as we would now conceive it. But, as we saw in Chapter 2, the existence of Earth's Moon allowed Galileo to conceptualize what he was viewing - if only after a matter of weeks - even to guess the physical nature of the objects by analogy. By contrast, although Herschel declared the new class of objects known as asteroids within sixteen months of Piazzi's detection of its first member, it took a half-century for widespread acceptance of this new class, *and the discovery of numerous new members was essential to its acceptance as a new class.*

Pluto offers a reverse case: Clyde Tombaugh and the Lowell Observatory astronomers almost immediately declared it a planet in the wake of a conscious search and upon discovering an object that fit the definition of a planet by its motion analogous to the other planets. Nor did this change when it was realized shortly afterward that Pluto might be quite small, based on the admittedly slim line of reasoning from its brightness and albedo. Seventy-six years passed between its detection in 1930 as a planet and its formal designation as a "dwarf planet," a new class of object that astronomers declared, by vote of the International Astronomical Union, was not a planet at all. This time, the discovery of new objects similar in nature to Pluto was essential to even proposing the existence of a new class - dwarf planets. We will evaluate this phenomenon - criteria for class status - more fully in the next chapter. Meanwhile, the lesson in discovering new members of a class is that the discovery can be more quickly framed by comparison with the previous members, sometimes too quickly due to spurious observations as in the case of the non-existent satellites of Venus and rings of Rhea, sometimes too quickly because of a lack of data as with Pluto, and sometimes not quickly enough as in the case of asteroids.

7.4 "Object" Discovery vs. "Phenomena" Discovery in Astronomy

As we saw in the last chapter, "phenomena" are one of the many types of discovery philosophers of science have identified. Although in general, physical phenomena such as electromagnetism are detected in the laboratory through experiment, astronomical phenomena are most often discovered using a telescope and its many accoutrements. In his book, *Cosmic Discovery*, Martin Harwit lumps such phenomena together with objects as a single category of discovery under the heading of "cosmic phenomena."[39] But because an object such as a meteoroid is distinct from the meteor phenomena it causes, or a planet distinct from a phenomenon such as a planetary magnetic field it causes, and so on, there may well be a fundamental difference in their modes of discovery.

More to the point, some of the landmark discoveries in astronomy have not been of objects, but of phenomena. Distinguished in this fashion, thirty-six of Harwit's forty-three "Cosmic Phenomena" are in fact objects, though not necessarily new classes of objects as we define them as members of a specific taxon in a classification system such as the Three Kingdom system of Appendix 1. (Several, such as magnetic variables, X-ray stars, and flare stars, are one level or more below the Class taxon in this system). The seven remaining in Harwit's list are indeed what we would call phenomena, or perhaps more accurately, the processes, entities, and properties that give rise to phenomena: cosmic expansion, interstellar magnetic fields, the x-ray and microwave background, masers, the gamma-ray background, and gamma ray bursts. Similarly, Marcia Bartusiak's useful anthology of primary astronomy writings in *Archives of the Universe*, billed as "A Treasury of Astronomy's Historic Works of Discovery," employs an even broader definition of discovery. Her seventy-five episodes of discovery range indiscriminately from *objects* such as we have described, including asteroids, nebulae, pulsars, and quasars; to *properties of objects* such as the spherical Earth, sunspots, and the spiral arms of the Milky Way; to *phenomena* such as the expansion of the universe, the acceleration of the universe, cosmic radio waves, and the cosmic microwave background.[40]

We test the hypothesis of a difference in "object" discovery versus "phenomena" discovery by looking at three of these well-known cosmic phenomena: the expansion of the universe discovered in the 1920s, followed by the accelerating universe in the 1990s; radio waves of cosmic origin detected in the 1930s, eventually giving rise to radio astronomy; and the 3 degree microwave background first detected in the 1960s, interpreted as a remnant of the Big Bang.

With the idea of the expanding universe we return once again to Edwin Hubble and his landmark announcement on New Year's Day, 1925, that galaxies exist beyond our own, as described in Chapter 5. As more and more galaxy distances were determined, using Henrietta Leavitt's Period-Luminosity relation for Cepheid variables as well as other methods, Hubble proposed a velocity-distance relationship: the more distant the object as determined by one of his methods, the faster it was moving away from us, as shown by redshifted spectra, a relation that came to be known as "Hubble's law." The relationship is usually shown graphically with velocity on the y-axis and distance on the x-axis, with a straight line running diagonally from the origin, the classic linear relationship between two sets of data. The interpretation of that relationship, however, is another matter entirely, and this is where we come to the expanding universe.

Because of its importance to modern astronomers, the development of the velocity-distance relation has been the subject of detailed research among historians of astronomy, notably by John North, Norris Hetherington, Helge Kragh, and Robert Smith.[41] Their research makes clear that no one pointed a telescope at a single object or set of objects and "discovered" the velocity-distance relation. Moreover, such a discovery required more than the unadorned telescope; just as Leavitt's Period-Luminosity relation required photometric observations to determine the period, and a series of steps to determine the luminosity before its relationship with the period could be established, so the velocity-distance relation required spectroscopic observations to determine spectral redshifts that were then related to radial velocities, which in turn could be related to distances determined by other methods. More than that, the radial velocities of many galaxies had to be observed.

In this effort to determine the essential radial velocities, Lowell Observatory astronomer V. M. Slipher was the pioneer, even if he had no initial expectation of measuring radial velocities. In 1912, after six years of perfecting his technique, Slipher determined the first radial velocity of the brightest of the spiral "nebulae," the Andromeda Nebula (M31), which was nevertheless a very faint object for spectroscopic observation. At this time, we recall, the extragalactic nature of the spiral nebulae was still undetermined, some still arguing they were solar systems in formation or clusters of stars within our Galaxy. When he began observing M31, Slipher himself believed he was observing a solar system in formation. By 1914, Slipher had determined the radial velocities of fifteen nebulae. The results were so alarming as to call into question the interpretation of redshifts as Doppler shifts, the standard interpretation for stellar radial velocities: on such an assumption the Andromeda nebula appeared to be approaching the Earth at about 300 km/second, while eleven of the fifteen spirals exhibited recessional velocities ranging up to 1100 km/second. Such velocities implied they were star systems beyond our Milky Way.[42] Slipher went on to find about forty spiral nebulae redshifts before he exhausted the power of his 24-inch refractor and spectroscope to measure even fainter nebulae.

Meanwhile, as historians of science have pointed out, the Dutch astronomer Willem de Sitter, interested in applying Einstein's relativity theory to astronomy, in 1917 had derived an alternative solution to the equations of Einstein's general theory of relativity, predicting a relationship between redshift and distance. "If continued observation should confirm the fact that the spiral nebulae have systematically positive radial velocities, this would certainly be an indication to adopt the hypothesis B [de Sitter universe] in preference to A [Einstein universe],"

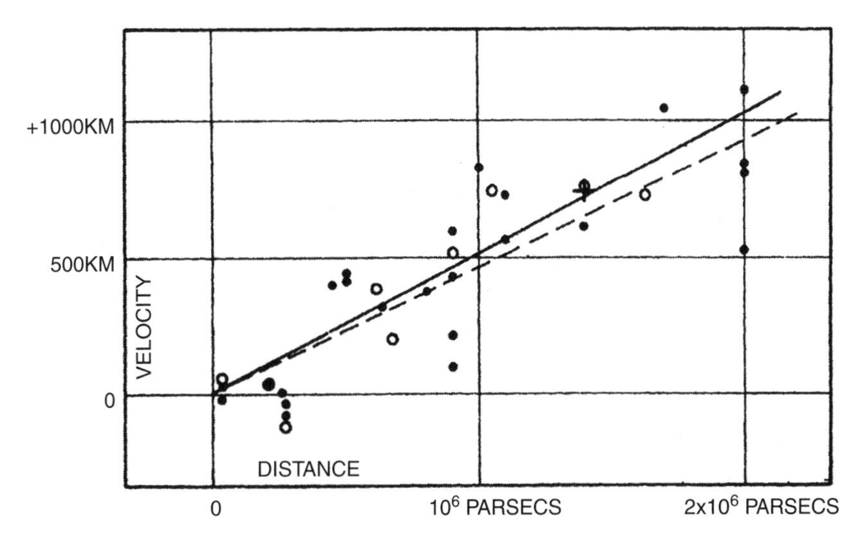

Figure 7.6. Discovery of new phenomena: The velocity-distance relation, a kind of discovery distinct from the discovery of new classes of objects, and in general more complex. The plot shows the apparent velocities of twenty-four galaxies to a distance of 2 million parsecs (about 6 million light-years). Based on this plot it appeared that the velocity of a galaxy increased by 500 kilometers per second for every million parsecs. This factor, later dubbed the "Hubble constant," is now believed to be between 67 and 73 (km/sec) per million parsecs. Hubble and Humason themselves were very cautious, however, preferring to call the measured redshifts "apparent velocities," and never accepting the idea of an expanding universe, which required even more assumptions. From *PNAS* 15 (1929): 168–73.

de Sitter wrote. By the 1920s, this result stemming from the "de Sitter effect," aided by the writings of Arthur Eddington, had come to be widely accepted.[43] It is notable that the de Sitter effect held that a redshift would be observed for more distant galaxies because the frequency of light decreases, rather than because of any recessional velocity; thus, de Sitter's universe (like Einstein's) was still static. But it did produce a redshift in any case. Hubble knew of de Sitter's work by 1926, but it was only in 1928 that he decided to investigate a possible velocity-distance relation. Enjoying a vast advantage over Slipher with the state-of-the-art instruments at the Mt. Wilson Observatory, Hubble consciously proceeded to examine the relationship indicated by de Sitter's theory and hinted at by Slipher's radial velocity observations. Hubble's problem was to determine accurate distances, at least accurate enough that a rough relation with velocity might be established. He did this by using the proven techniques of Cepheids and novae in the nearer galaxies where they could be detected, using the brightest stars in more distant galaxies, and finally just the brightness of the galaxies themselves, on the rather shaky assumption that those fainter were more distant. In this way he obtained distances to twenty-four galaxies. In a famous 1929 paper, Hubble plotted these distances against Slipher and Humason's redshifts (Figure 7.6), and decided that a linear relation best fit the data, a fit borne out by observations at increasingly large distances ever since.[44] The velocity-distance relationship became one of the foundations of modern cosmology.

For Hubble, however, his velocity-distance relation was an "apparent" velocity relation, an empirical relationship whose significance was still subject to interpretation. In fact, as historians have pointed out, throughout his life Hubble himself never identified redshifts with Doppler shifts. In their follow-up 1931 paper which provided even more evidence, Hubble and his colleague Humason (who took most of the spectra) wrote that their result "concerns the correlation of empirical data of observation. The writers are constrained to describe the 'apparent velocity-displacements' without venturing on the interpretation and its cosmologic significance."[45] In other words, Hubble himself never accepted the expanding universe.[46] Nor was Hubble alone in this reticence. The irascible Fritz Zwicky preferred an interpretation where a gravitational "drag" on light quanta at increasing distances might produce the redshift effect.[47] Others, however, were not so cautious, and did not hesitate to assert the expanding universe. Some were impressed by the Russian mathematician Alexander Friedmann's finding that one solution to Einstein's cosmological field equations was that the universe was not static, but evolving. Others pointed to the Belgian astronomer Abbé Georges Lemaître's 1927 finding, also stemming from Einstein's theories, that the universe was actually expanding. Nevertheless, Hubble's point – that a relation between apparent velocity and distance is not the same as the expanding universe – is well taken, even though that is the way most of his colleagues, and we today, interpret it. The expanding universe required another step of interpretation.

Who, then, did discover the expanding universe? In an incisive joint paper Kragh and Smith address precisely this question.[48] They conclude that Friedmann's prediction on theoretical grounds of an evolutionary universe is not equivalent to an expanding universe. Lemaître, on the other hand, showed that one Einsteinian model of the universe did indicate an expanding universe, and, crucially, believed based on the astronomical data that the real universe was expanding.[49] Hubble and Humason showed only that there was a relation between velocity and distance. In the estimation of Kragh and Smith, Hubble discovered an empirical law, not the expanding universe, and Lemaître has the better claim to the discovery of the expanding universe. That Hubble became known as the discoverer of the expanding universe was due to attributions after the fact, beginning with Einstein in 1945. Only in the 1950s, mostly following his death, did the attribution to Hubble of the discovery of the expanding universe, and related expressions such as "Hubble's constant," become more common, especially from his student Allan Sandage, Hubble's heir apparent and one of the great cosmologists of the twentieth century. Sandage's attribution was picked up in textbooks, from which it became canonical - a poignant exemplar of post-discovery activity.

When we examine the anatomy of the discovery of the expanding universe, therefore, we find a quite different and more complex pattern than with the discovery of objects discussed heretofore, whether predicted, merely expected, or neither predicted nor expected. The discovery in fact breaks down into two parts, two separate and distinct discoveries: the first is Hubble's discovery of the apparent "velocity-distance" relation, which, as Hubble himself insisted, was just that: a relation between two sets of empirical data, Slipher's radial velocity/redshift data on the one hand, and Hubble's distance data on the other. The existence of such a relation was predicted by de Sitter, though without the expanding universe interpretation. Hubble clearly believed and claimed the *apparent* velocity distance relation discovery as his own. The second discovery was that this velocity-distance relation proved that the universe was actually expanding. This was most certainly not Hubble's discovery, nor did he himself claim it, though it was later imputed to him for a variety of reasons related as much to social as conceptual considerations. Rather, as Kragh and Smith argued, the discovery of the expanding universe belongs more to Lemaître than anyone else. It was Lemaître who not only showed that it followed from Einsteinian theory, but also connected it to astronomical data, even if not his own.

On the other hand, the discovery of the expanding universe as an empirical rather than a theoretical fact indisputably relied on the velocity-distance relation, under the interpretation that the observed redshifts were really Doppler shifts, thus indicating real velocities. The discovery of the expanding universe was consummated only after numerous steps: the observation of radial velocities; the assumption of observed redshifts as Doppler shifts, though not for Hubble himself; the requirement of many objects rather than one to uncover the phenomenon of the velocity-distance relation; a prediction that led Hubble to look for the velocity-distance relation; and, with Lemaître in the lead, the discovery of the expanding universe itself. In contrast to the discovery of new classes of astronomical objects where there is normally no prediction, and in contrast to new members of a class where there is some expectation, with the expanding universe there is an interplay of empirical findings and prediction based on theory. In addition, there is the curious twist that Hubble - the discoverer of the velocity-distance relation himself - did not embrace the predicted expanding universe, even at his death in 1953!

Seventy years after Hubble announced the velocity-distance relation, the startling announcement was made that the expansion of the universe was accelerating. By this time cosmology had advanced by leaps and bounds, but the principles of discovery were similar in some ways. In this case, rather than Cepheids, white dwarf supernovae were the "standard candles" for distance determination. Because all exploding white dwarfs have exactly the same mass (the Chandrasekhar limit), their explosion always results in the same luminosity, as opposed to other types of supernovae, which result from exploding stars of different masses (high masses to be sure), and thus different luminosities. Using this technique pushed to its limits with the Hubble Space Telescope, Adam Riess and his colleagues, in collaboration with astronomers at ground-based telescopes (altogether comprising the "High-z Supernova Search Team), announced in 1998 that the expansion of the universe was accelerating. An even larger "Supernova Cosmology Project" team of some thirty astronomers, headed by Saul Perlmutter, again using Hubble and ground-based telescopes, independently verified the accelerating expansion they had already suspected, but not proven, in 1997, the year before the Riess team published its results.[50] The discovery was such a surprise because it was counter-indicated, in other words, current theory indicated the expansion would be slowing down due to gravity, not speeding up. But there was a discarded theory that had predicted such a phenomenon, for a repulsive force counteracting gravity was Einstein's idea of a "cosmological constant," an idea he considered his greatest blunder after Hubble showed the universe was expanding. Far from being a blunder, the accelerating universe leads to the idea of "dark energy," now believed to represent more than 70 percent of the mass-energy of the universe. The driving force of this acceleration, and the nature of the dark energy, remain complete mysteries. Nevertheless, they and numerous other data have given rise to the Cold Dark Matter (CDM) theory of the universe.

* * *

Even as Hubble was taking up his serious work on the velocity-distance relation in 1928 at the premier observatory on the west coast of the United States, on the east coast a very different discovery was in the making. In that year Karl Jansky, a physicist with a BA degree fresh out of the University of Wisconsin, had joined the newly formed Bell Telephone Labs in New Jersey. Jansky, who had done his undergraduate thesis on vacuum tubes, was assigned to work on the practical problem of trans-Atlantic radio noise. He built his own 4-meter wavelength receiver for the purpose. By August 1929, he began supervising construction of an antenna 29 meters long that rotated on four wheels from a Model T Ford, driven by a ¼ horsepower motor. This antenna, designed for a wavelength of 14.5 meters, was erected at a new site a few miles away in Holmdel, New Jersey, where Jansky began shortwave observations in the second half of 1930.

In the summer of 1931, Jansky began to be intrigued by a night-time weak static moving across the sky. Woodruff T. Sullivan III, a radio astronomer himself and the premier historian of early radio astronomy, has shown that Jansky's first real recognition of an anomaly was in January 1932, when he reported "a very steady continuous interference – the term 'static' doesn't quite fit it. It goes around the compass in 24 hours. During December this varying direction followed the sun."[51] Janksy suspected the static had something to do with the Sun, but his first published article on the subject in 1932 spoke only of a "very steady hiss-type static, the origin of which is not yet known." By the end of the year Jansky was convinced the origin was astronomical, because the static "always lies in a plane fixed in space, at a right ascension of 18 hours and a declination of -4 degrees." He reported his astronomical interpretation in a second publication in 1933, where he refined the declination to -10 degrees, and noted two possibilities for the origin of the waves: the direction toward which the Sun moves with respect to the stars, and the direction in Sagittarius toward the center of the Milky Way Galaxy. In a paper two years later, Jansky bolstered his case that the actual origin coincided with the center of the Milky Way.[52]

Advancing from the recognition of the anomaly to the suggestion of extraterrestrial origin took less than a year. Janksy would write only one more paper on the subject of radio astronomy, in 1935, before returning to more routine and practical work. He died in 1950 at the age of forty-four. Recognition of Jansky's achievement and its potential – no less than an entirely new field in astronomy – was slow in coming, especially in the midst of the Great Depression. Observatories could not afford to hire radio engineers, and so it was left to an avid amateur, Grote Reber, to follow up on Jansky's work ten years later with a radio telescope of his own design and construction (Figure 9.8). During World War II practical work would finally spur the field onward, leading to the discovery of the first radio galaxies and other objects discussed in Chapter 5. Thus, the discovery of a phenomenon (cosmic radio waves) spurred on the discovery of numerous astronomical objects, as would the opening of the electromagnetic spectrum at other wavelengths in the future (see Section 9.2).[53]

* * *

Thirty years after Jansky's last paper on the subject of radio astronomy, and from the same Holmdel station of Bell Laboratories in New Jersey, two astronomers would make a discovery even more fundamental than Jansky's. The discovery was also related to the expanding universe, for it turned out to be evidence for the Big Bang, the prevailing theory for the origin of the universe and its subsequent expansion. In 1964, Arno Penzias and Robert W. Wilson were in the process of converting a horn antenna used for Echo-1 satellite communications into a radio telescope for galactic studies. Like Jansky, they were pestered by a persistent noise in the system; unlike Jansky they wanted to get rid of the noise rather than study it. But they could not get rid of the noise, nor could they identify any atmospheric or ground-based source for it. A few miles away, Princeton physicist Robert Dicke and his colleagues were in the process of setting up an antenna to search for just such radiation, predicted by Ralph Alpher, Robert Herman, and George Gamov in the 1940s, but long forgotten until Dicke and James Peebles reasoned anew that such a relic radiation should exist from the Big Bang, perhaps at a temperature as low as 3.5 degrees Kelvin. When Penzias and Wilson found out about this, they invited the Princeton group to Holmdel, and after discussions, the two groups surmised that Penzias and Wilson had serendipitously found what the Princeton group was just beginning to consciously search for. The Princeton group had been scooped, but the two groups published adjacent papers in the *Astrophysical Journal* for July 1, 1965. Penzias and Wilson's, a mere eight paragraphs long, reported their "noise temperature" observations, while Dicke and his colleagues explained the cosmological consequences. As with Janksy's discovery, the story was front-page news in the *New York Times* for May 21, 1965. Unlike Jansky, Penzias and Wilson shared the 1978 Nobel Prize in Physics "for their discovery of cosmic microwave background radiation." Dicke, Peebles, and their colleagues pointedly did not. Such are the rewards for discovery, clearly usually defined in terms of observation rather than theory.[54]

The discovery of cosmic background radiation is notable both for what came before and what came after the discovery. Not only had the 1948 prediction of Alpher, Herman, and Gamov been forgotten, but the Canadian astronomer Andrew McKellar, whom we last saw in Chapter 3 in connection with the discovery of the first interstellar molecules, had recognized in 1941 that one of them, the cyanogen molecule, was being energized by a thermal background of about 2.3 degrees Kelvin. Neither McKellar nor others realized the significance of this find, in some cases believing it to be a nagging error in their instrumentation. One is tempted to conclude that the correct interpretation is necessary to claim credit for discovery. Clearly detection is not enough. However, Penzias and Wilson only realized the significance of their detection when clued in by the Princeton astronomers, once again a case where "collective discovery" would seem to be the best description of what actually happened.

As with all good scientific discoveries, the detection of the 3 degree (more accurately, 2.73 degree) isotropic blackbody radiation raised still more questions. In particular, if it was isotropic – coming equally from all directions of the sky – how did objects like galaxies form? Shouldn't there be some variation in the temperature of this radiation, however slight, that would provide the "seeds" for these future objects? After hints of such variation from the ground, NASA decided to find out. At Goddard Space Flight Center astronomers began to develop a satellite

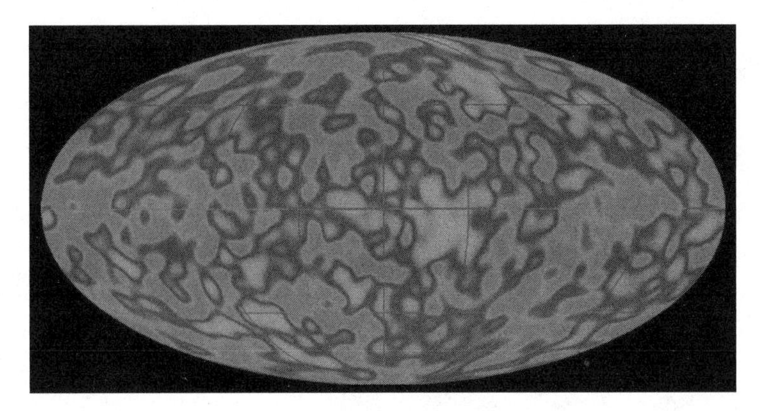

Figure 7.7. Discovery of a new phenomenon: the cosmic background radiation, interpreted as a remnant of the Big Bang. These data, from the Cosmic Background Explorer (COBE), show cosmic microwave background fluctuations to one part in 100,000 in the 2.73 K average temperature of the radiation field. The fluctuations are the imprint of density contrast in the early universe, believed to have given rise to the structures that populate the universe today: clusters of galaxies and vast regions devoid of galaxies. John Mather and George Smoot were awarded the 2006 Nobel Prize in Physics for this work, which has since been refined by the WMAP satellite, which mapped the variations to one part in 1 million, and by the Planck spacecraft. NASA/DMR/COBE Science Team.

that could not only measure variations in the background radiation, but also prove that it was blackbody radiation – required if this was the real remnant of the Big Bang. After numerous delays for various reasons, including the Space Shuttle *Challenger* disaster in 1986, the Cosmic Background Explorer (COBE) was launched in 1989 by a Delta rocket. Although it would return observations for four years, within hours it had demonstrated that the radiation was indeed blackbody.

The first COBE results were announced at the January 13, 1990, meeting of American Astronomical Society. By that time the results from the Far Infrared Absolute Spectrophotometer (FIRAS) instrument, based on nine minutes of data, were in. NASA's John Mather, Principal Investigator for the instrument, displayed the data showing the perfect blackbody curve, as expected if this were indeed related to the Big Bang. He received a standing ovation, a rare event for a scientific meeting. Further results were announced as the climax to the annual meeting of the American Physical Society on April 23, 1992. In a news conference immediately following their presentation, George Smoot, Principal Investigator for the Differential Microwave Radiometer (DMR) instrument, announced COBE had observed "the oldest and largest structures ever seen in the early universe; the primordial seeds of modern-day structures such as galaxies, clusters of galaxies, and so on; huge ripples in the fabric of space-time left over from the creation period" (Figure 7.7). And he uttered what became an instantly famous quote: "If you're religious, it's like seeing God." Mather and Smoot received the 2006 Nobel Prize in physics for "for their discovery of the blackbody form and anisotropy of the cosmic microwave background radiation." Their separate accounts in Smoot's *Wrinkles in Time* and Mather's *The Very First Light* are an insight into what Stephen Hawking called "the scientific discovery of the century, if not all time."[55]

Despite its spectacular results, COBE had returned only a crude map of the variations in the background radiation. Its work was continued by high-altitude balloon experiments in Antarctica, but scientists wanted more. Thus was born the Wilkinson Microwave Anisotropy Probe (WMAP), named after Princeton's David Wilkinson, the same Wilkinson who had co-authored the Dicke paper and who died in 2002. Launched in 2001, the satellite achieved its prime mission by 2003. Whereas COBE had measured temperature differences to one part in 100,000, WMAP measured even more accurate temperature differences, down to one part in 1 million. The satellite also provided more evidence of the rapid "inflation" of the universe at its beginning, verifying and refining the leading theory of the origin of the universe. WMAP also achieved much more. It pinned down the age of the universe, within 100 million years, to 13.7 billion years. It yielded information on the dark matter content of the universe. And it provided unprecedented detail on the origin of the universe and the evolution of the first stars and galaxies. In 2009, it was joined in orbit by the European Space Agency's Planck satellite, which by 2013 had refined our knowledge of the microwave background and the early universe even further.

As with the velocity-distance relation and the expanding universe, there were some dissenters about the cosmological significance of the data. But most saw it as confirmation of the Big Bang theory of the origin of the universe. As Kragh puts it in *Cosmology and Controversy*, his history of the Big Bang and Steady State theories, "the discovery of the cosmic microwave background radiation constituted in effect the final blow to an already dying theory, Fred Hoyle's Steady State theory."[56] The case of the cosmic microwave background demonstrates many aspects of discovery, from serendipity to its collective nature. Above all, it demonstrates how science builds upon itself. Fifty years after the first discovery in 1965, COBE, WMAP, and Planck will have greatly refined the initial discovery, and in the process stimulated numerous other discoveries. However, when it comes to credit and rewards such as Nobel Prizes, there is nothing like the first discovery or definitive confirmation of a new phenomenon, no matter how much it is later refined.

8

Discovery and Classification

> The first step in wisdom is to know the things themselves. This notion consists in having a true idea of the objects; objects are distinguished and known by classifying them methodologically and giving them appropriate names. Therefore, classification and name-giving will be the foundation of our science.
>
> Linnaeus, 1735[1]

> The initial classifier of any unknown subject necessarily begins with no idea of how to choose the key parameters. How then to produce a classification imbued with any fundamental physical significance or predictive power?
>
> Allan Sandage, 2004[2]

> The MK system [for stellar classification] has no authority whatever; it has never been adopted as an official system by the International Astronomical Union - or by any other astronomical organization. Its only authority lies in its usefulness; if it is not useful, it should be abandoned.
>
> W. W. Morgan, 1979[3]

In contrast to the discovery of new laws, processes, or properties, one of the hallmarks of the discovery of localizable natural objects such as we have been discussing in this volume is an almost irresistible temptation to classify them. A new object will barely have been discovered before the human mind tries to determine where it "fits" in the order of things already known. This is undoubtedly an ancient instinct, and one of the basic characteristics of a developing and even a mature science, as basic to astronomy as it is to biology, geology, chemistry, and physics.[4] To put it another way, the recognition of new classes and their classification is an integral part of natural history, and a particularly challenging activity when the discovery involves a new class of objects. The origin of the categories in which we think is a question rarely raised except in the rarefied reaches of philosophy, but it goes to the core of discovery. In discovery it may flare briefly and sporadically, as in the case of Pluto, or at more sustained length as in the case of stellar luminosity types. But once routine, the origin of the categories is often forgotten, even as they determine the way we think of things.

If, as we have argued in Chapter 6, the macrostructure of discovery consists of detection, interpretation, and understanding, and the microstructure of conceptual, technological, and social elements, where does this process of classification fit in? While it might seem a priori that classification marks the end of discovery, in fact it often marks the beginning, even a pre-discovery stage. But in another guise it is also part and parcel of understanding, and so a part of discovery; indeed, as Linnaeus intimated in the statement above, it may be the basis for understanding "the things themselves." Finally, because classification systems are often continually refined, classification may also be a post-discovery activity. In all these roles it is important to understand classification if we are fully to understand discovery in its empirical aspects.

Classification is no simple activity, but a problem of humans "ordering nature," sometimes a reflection of Nature's own order – though even that claim is a philosophical problem of the highest order. Two hundred fifty years after Linnaeus, evolutionist Stephen Jay Gould reflected on both the importance and the problematic nature of classification when he wrote about the biological world "taxonomies are reflections of human thought; they express our most fundamental concepts about the objects of our universe. Each taxonomy is a theory of the creatures that it classifies."[5] This is no less true in astronomy, but unlike biology, classification has not received the systematic attention it deserves from astronomers or its historians, despite well-developed classification systems for particular branches such as stellar and galactic astronomy. This lapse is undoubtedly responsible in part for the confusion surrounding the Great Pluto Debate, and similar lesser-known debates that have engaged, and occasionally enraged, astronomers over the centuries.

The early stages of classification rarely occur with full understanding of "the things themselves," or their relationship to each other in an evolutionary sequence, often held up as the ultimate goal of discovery and classification. When classification begins too early, it is potentially a recipe for trouble. Yet, since classification is the basis for understanding, and because the urge to classify is a property of the human mind, premature classification often occurs. Both classes of objects and entire classification systems may eventually be shuffled or dismissed in a kind of natural selection based on scientific utility and changes in our understanding of reality. Switching an object from one class to another is relatively common as more information becomes available: the controversy over the status of Pluto as a planet, the nature of numerous trans-Neptunian objects, and the problematic classification of objects beyond our solar system as planets or brown dwarfs, are exemplars of this problem. They represent only the tip of the iceberg of the more general problem of the classification of astronomical objects. It is a problem long known in science in general and especially in the biological world, where natural history, taxonomy, and systematics form a significant part of the history of biology. Although classification may seem to some a boring subject, as Linnaeus realized in the eighteenth century, it stands in many ways at the foundation of science. And because classification systems represent theories of the objects classified, they are windows on the discovery process.

In this chapter we discuss the relationship of classification to discovery, in particular the concept of class, the problems of classification, the development of classification systems in astronomy, the acceptance of these systems by way of negotiation and utility, and the relative role of conceptual and social elements. While we are primarily concerned with astronomy, the problem cannot be considered without some context from other sciences, where similar issues have been addressed. We shall find that, in contrast to the microstructure of discovery, the details of which are inevitably socially *influenced*, the concept of class and the development of classification systems are socially *determined*, even while grounded in nature.

8.1 On Class and Classification

Because this is a book about the nature of discovery as it applies to new classes of astronomical objects, it is high time we define what we mean by "class." More precisely, and historically more interestingly, we need to ask what astronomers themselves have meant by class, how classes are negotiated, and whether the meaning of class has changed over the last 400 years of telescopic discovery. This, the philosophical problem of class, or whatever term is used in its place, is a necessary prelude to the problem of classification systems discussed below. The two concepts need to be distinguished, for while it is possible to have a class without a classification system, it is not possible to have a classification system without a class. In a relative sense scientists may declare a new class of object – whether an element, elementary particle, or species – in the sense of a new "kind" or "type" of object. In an absolute sense, where "class" is taken to be a category in a classification system, a class cannot be defined in the absence of the system that determines the taxonomic levels; otherwise it could not be distinguished from other levels. Even then it is often uncertain whether an object should be placed at a certain taxonomic level, or in a higher or lower one. The great Pluto debate was about class, not a classification system, for no classification "system" existed for the few planets of our solar system, even though they were divided into terrestrial, gas giant, and (much later) ice giant planets.

For proper context, and for possible insights and lessons, we begin by examining the problem in other sciences, where objects may constitute a "class" by another name. The oldest sophisticated classification systems (excluding very general ones such as earth, air, fire, and water) are those of biology, stemming from Linnaeus or even earlier from Aristotle, followed by chemistry with its periodic table, and physics with its Standard Model of elementary particles. Specifically, we will examine to what extent the problems of class and classification in astronomy may be seen as analogous or dis-analogous to the problem of species in biology, element in chemistry, elementary particle in physics, and by extension, to the units of any science that employs a classification system – which is to say, most of them that have a tradition of natural history.

No science has developed a more sophisticated understanding of the history, practicalities, and problems of classification than biology, and those parts of its history that have been written reveal the complexity and nuance of the story. The work of historian Paul Farber, for example, codified in his book *Finding Order in Nature*, describes this complexity of the naturalist tradition in biology from Linnaeus to E. O. Wilson, raising questions about "natural order" versus order imposed by humans. At a deeper and more detailed level, in his classic work *The Growth of Biological Thought*, the evolutionary biologist and systematist Ernst Mayr addresses, from the point of view of an historically sensitive practitioner, the history of classification, the epistemology of classification, its philosophical principles, its relation to evolutionary biology, and how all of this fit into various scientific worldviews throughout history. And in his book, *The New Foundations of Evolution on the Tree of Life*, Jan Sapp has explored how genomics as applied to the microbial world heralded a profound shift not only in biological classification, but also in biology itself and the very ideas of its evolution. All these studies provide a framework for approaching classification and discovery in astronomy.[6]

The fundamental problem in any classification system is the unit of classification. In biology this unit is a species, and it is no surprise that the very concept has been the subject of considerable controversy through history. For Aristotle, species were "eternal, immutable, and conceptually distinct as triangle and gold. They are part of the fundamental makeup of the universe."[7] Linnaeus held this Aristotelian view, except, as one witty historian of biology has remarked, in the Christian era he placed "the origin of species in the Garden of Eden and their demise at the Second Coming": God created the primary species, and they had hybridized since to produce additional species. Darwin famously explicated the origin of species by natural selection, but he did not establish "a determinate relationship between phylogeny and classification," that is to say, he did not understand the evolutionary connections among species, certainly not at any deep level. Such connections awaited the evolutionary synthesis in the twentieth century.[8]

Over the years, practitioners, historians, and philosophers developed a multitude of definitions of species depending on the focus of their own work. In an attempt to bring order to these definitions, Mayr divided all species concepts into types: (1) the typological species concept of Linnaeus, simply meaning "kind," or "an entity that differs from other species by constant diagnostic characteristics," in the same way a mineralogist speaks of species of crystals classed by symmetry, or a physicist of nuclear species; (2) the nominalist species concept, in which species are arbitrary mental constructs with no basis in reality; (3) the biological species concept stemming from the naturalist tradition that different populations do not interbreed; and (4) the evolutionary species concept favored by some evolutionists such as George Gaylord Simpson, construed as a lineage evolving separately from others.[9]

Mayr analyzes all these interpretations in considerable detail, making it clear that he and most biologists favor the biological species concept. Mayr's own classic definition according to that concept, stemming from his book *Systematics and the Origin of Species* in 1942, is that species are "groups of actually or potentially interbreeding natural populations, which are reproductively isolated from other such groups," a definition still widely used today.[10] In his view, it was this concept – centering on interbreeding rather than degree of difference – that raised the species concept above the mere typological classes of inanimate objects, "one of the earliest manifestations of the emancipation of biology from an inappropriate philosophy based on the phenomena of inanimate nature." This biological species concept, Mayr emphasizes, "utilizes criteria that are meaningless as far as the inanimate world is concerned."[11] Finally, Mayr insisted that this species concept is not invented by taxonomists or philosophers, but actually exists in nature.

Mayr's work thus provides our first lesson in clarity for astronomy, for, by contrast to biology, in astronomy we are clearly dealing with his first definition of species as a class, namely, typological. In this definition "membership in a class is determined strictly on the basis of similarity, that is, on the possession of certain characteristics shared by all and only members of that class. In order to be included in a given class, items must share certain features which are the criteria of membership or, as they are usually called, the 'defining properties.' Members of a class can have more in common than the defining properties, but they need not. These other properties may be variable – an important point in connection with the problem of whether or not classes may have a history."[12] While this definition of class has been largely abandoned in biology as too restrictive, it serves quite well for astronomy, with a few exceptions we will address later.[13]

Mayr also insists on the importance of precision in nomenclature, for example, in understanding the difference between a *category* and a *taxon*, where "category" designates a level or rank in a hierarchic classification, and a "taxon" is a taxonomic group of specific organisms. Categories are an abstraction, while taxa are groups of concrete zoological or botanic objects to be placed in the categories. Thus, Linnaean terms such as (in descending order) kingdom, phylum, class, order, family, genus, and species are categories, while any given species such as a robin or *Homo sapiens* is a taxon. It may seem a subtle point, but when we speak of a "class" of objects in astronomy as a category, this is different from the comets, asteroids, spiral galaxies, and so on that we place in the class. Moreover, in such a hierarchical classification system, the very term "class" takes on a technical meaning not to be confused with other categories of the system, nor with other concepts such as "natural kind." These distinctions are important to avoid confusion in astronomy no less than in biology.[14]

These problems are far from academic, as generations of biologists can attest. The practical problems of species with real-world implications are strikingly brought home in the case of prehistoric human ancestry, where debates still rage about the origin of our own species, *Homo sapiens*. To take only one aspect, the latest ancient DNA research shows that human ancestors interbred with at least two different kinds of archaic humans, including Neandertals, producing surviving children. But how can Neandertals retain their usual designation as members of a different species if we now know they interbred? By Mayr's definition they must be members of the same species. Paleoanthropologists, however, consider Neandertals a distinct species because the anatomical differences are "an order of magnitude higher than anything we can observe between extant human populations." Moreover, they point out, about 330 closely related species of mammals interbreed and a third of them produce hybrids. In other words, the nature of a species depends on which characteristics one chooses to classify, and there is as yet no agreed upon measure for how much morphologic or genetic difference is necessary to define a new species.[15]

The stakes are raised when evolution enters the picture, for species are not only considered the basic taxonomic units in the Linnaean hierarchy, they are also often viewed as the evolutionary units of the organic world.[16] This, of course, is a major difference from astronomy. While evolution in the sense of development takes place in astronomical objects, and indeed cosmic evolution based in part on the relationships of objects such as galaxies, stars, and planets is now seen as central to astronomy (Chapter 10), there is no "descent" by modification in any strong sense, and no genetic mechanism such as natural selection acting on variation. Though variation exists in abundance in the astronomical world, it is subject to the laws of gravity, thermodynamics, and perhaps the emergence of complex systems. Thus in this sense the "species" of astronomy, which in this volume we term simply "classes" following more common usage in the field, are not fundamental units as "element" and "elementary particle" are in chemistry and physics. With the exception of nuclear physics, in chemistry and physics there are no transitions between different chemical elements or different atomic particles - no intermediate states - as there are in biology and astronomy. Yet astronomy does have a temporal variability of the objects classified, as a solar-type main sequence star, for example, transitions to a red giant and eventually a white dwarf. Thus, astronomical classes retain some of the ambiguities of biological species, but not their evolutionary mechanisms, introducing both practical and theoretical problems for classification.

Temporal variability aside, the problem of species is essential to the larger problem of classification, which Mayr defines as "the ordering of organisms into taxa on the basis of their similarity and relationship as determined by or inferred from their taxonomic characters." By this definition, taxonomic characters are decisive, but, as Mayr points out (and as Allan Sandage points out for astronomy in one of the opening quotations of this chapter), the problem is what characteristics are most useful and indeed legitimate in taxonomic analysis.[17] This dilemma, made more acute because the purpose at hand may change over time, often leads to competing classification systems. For example, the most commonly used classification system in biology in the latter half of the twentieth century was the Five Kingdom system, which was developed by R. H. Whitaker in 1959, gained increasing support over the following decades, and became embodied in Lynn Margulis's compilation by that name.[18] The Five Kingdoms of this system are the Prokaryotes (bacteria), the Prototista such as algae and protozoa, Fungi such as mushrooms, molds, and yeasts, Animalia, and Plantae. But as Jan Sapp has documented, in the late twentieth century another competing system arose, the Three Domain system of Archaea, Eucarya, and Bacteria, developed during the late 1970s and 1980s by Carl Woese and colleagues, based on comparing the evolutionary sequences of 16S ribosomal RNA, giving rise to the new field of molecular phylogenetics (Figure 8.1). This system, adopted by many molecular biologists because it is arrived at through

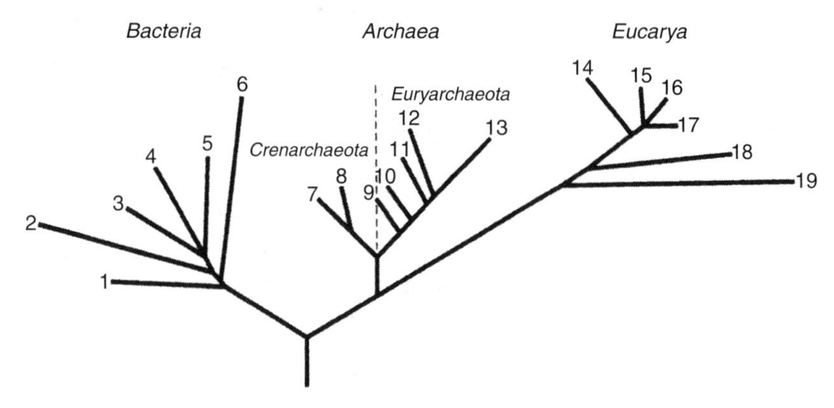

Figure 8.1. The "Three Domain" system of molecular biologists, delineating the domains of Archaea, Eucarya, and Bacteria, put forth by Carl Woese and colleagues in 1990, was based on evolutionary sequences of 16S ribosomal RNA. The numbers correspond to different groups of organisms. The system represents an entirely different method of classification than the "Five Kingdom" system of the macrobiologists. From Carl Woese, Otto Kandler, and Mark Wheelis, "Towards a Natural System of Organisms: Proposal for the Domains Archaea, Bacteria, and Eucarya," *PNAS* 87 (1990): 4576–79: 4576. With permission of Carl Woese.

the techniques of molecular biology, is seen by its founders as a more "natural" system than its Five Kingdom rival.[19] Both rival classification systems, it should be noted, are hierarchical in nature, a series of nested categories based on increasingly refined characteristics, until one reaches species, "the unit of evolution." But their implications for evolution are very different.

While the Three Domain system was seen as a signal advance by some, it was severely criticized by more traditional leading macro- and microbiologists including Mayr, Theodosius Dobzhansky, George Gaylord Simpson, and Margulis. To this day many non-molecular biologists still use the Five-Kingdom system; in Margulis's latest compilation, titled *Kingdoms and Domains*, she made use of the molecular sequence data but found it necessary to "reject the bacteriocentric three-domain scheme on biological, evolutionary and pedagogical grounds." Specifically, she found that Woese's Three Domain system did not do justice to symbiogenesis (also known as endosymbiosis), whereby former bacteria fuse to generate new individuals, a process that she saw as a major source of innovation in the evolution of eukaryotes. She also saw the Five Kingdom system as much more robust in the sense that it used, in its latest form, not only Woese's molecular data, but also morphological, developmental, metabolic, and other criteria. Pedagogically, she thought the proliferation of kingdoms in Woese's system "defeats the purpose of manageable classification." And it is not irrelevant that Margulis was the pioneer in the theory of endosymbiosis, and that traditional biologists saw molecular biologists encroaching on their territory of evolution, most galling because of the shifts in funding toward the newer methods. A drawn-out and still ongoing process of negotiation among methods and classification systems ensued. All of this will prove illuminating as we examine class and classification systems in astronomy.[20]

* * *

That chemistry has undergone a similar classification process in its attempt to understand order in nature is evident in two recent books, Michael Gordin's *A Well-Ordered Thing: Dmitrii Mendeleev and the Shadow of the Periodic Table* (2004), and Eric Scerri's *The Periodic Table: Its Story and Significance* (2006). Both show how Mendeleev's periodic table emerged out of the famous "Karlsruhe Conference," a three-day meeting in southern Germany convened in September 1860 at the behest of organic chemist August Kekulé, with the purpose of standardizing some of the most basic concepts in chemistry, including "atom," "molecule," and the calculation of atomic weights. As Gordin points out, this was "the first time that chemists from across Europe gathered in one place to resolve central scientific issues," and thus was an important stage in the professionalization of chemistry as an international science.[21] The meeting, especially at the instigation of the Italian chemist Stanislao Cannizzaro, revived Amedeo Avogadro's rather obscure 1811 hypothesis to standardize atomic weights, a standardization that would prove essential to Mendeleev, who was only twenty-six when he attended. Moreover, even as astronomers were beginning to apply the new technique of spectroscopy to heavenly bodies in the wake of the work of Kirchhoff and Bunsen, in the 1860s an array of new elements was being discovered in the laboratory via the same technique, adding to the number of elements in which some pattern might be detected. Mendeleev not only formulated his periodic table (Figure 8.2) and the periodic law based on standardized atomic weights, he also predicted where other elements should exist based on gaps in the periodic table.

Chemical elements thus conform to a typological definition of class, wherein members of the class are distinguished by their unique property, the atomic structure of each particular element, which largely explains the chemical behavior of each element.[22] More specifically, we now know, thanks to the work of the English physicist H. G. J. Moseley shortly before his death at age twenty-seven in the infamous Gallipoli Campaign of 1915, that each element is defined by its atomic number, equal to the number of protons in the nucleus, uniting chemistry and atomic physics. Whereas this definition of class is impractical for biology, where species evolve, it serves chemistry well.

But, as in biology, finding order in chemistry was elusive, as is evident in the fact that Mendeleev's periodic table emerged as only one of six

			Ti = 50	Zr = 90	? = 180.
			V = 51	Nb = 94	Ta = 182.
			Cr = 52	Mo = 96	W = 186
			Mn = 55	Rh = 104,4	Pt = 197,4
			Fe = 56	Ru = 104,4	Ir = 198.
			Ni = Co = 59	Pl = 106,6	Os = 199.
H = 1			Cu = 63,4	Ag = 108	Hg = 200.
	Be = 9,4	Mg = 24	Zn = 65,2	Cd = 112	
	B = 11	Al = 27,4	? = 68	Ur = 116	Au = 197?
	C = 12	Si = 28	? = 70	Sn = 118	
	N = 14	P = 31	As = 75	Sb = 122	Bi = 210?
	O = 16	S = 32	Se = 79,4	Te = 128?	
	F = 19	Cl = 35,5	Br = 80	J = 127	
Li = 7	Na = 23	K = 39	Rb = 85,4	Cs = 133	Tl = 204.
		Ca = 40	Sr = 87,6	Ba = 137	Pb = 207.
		? = 45	Ce = 92		
		? Er = 56	La = 94		
		? Yt = 60	Di = 95		
		? In = 75,6	Th = 118?		

Figure 8.2. The first published form of Mendeleev's periodic table, based on standardized atomic weights, one of six competing systems for ordering the chemical elements. This and further iterations of the Periodic Table emerged out of a particular culture and won out over competing forms due to simplicity, utility, and advocacy. This sheet is entitled "An Attempt at a System of Elements, based on Their Atomic Weight and Chemical Affinity," dated February 17, 1869. From Mendeleev, *Periodicheskii zakon. Klassiki nauki*, ed. B. M. Kedrov (Moscow: Ixd. An SSSR, 1958), p. 9.

competing systems. As Eric Scerri puts it, "The discovery was made, essentially independently, by six diverse scientists who differed greatly in their fields of expertise and in their approaches."[23] In France, the geologist Alexandre De Chancourtois was the first to discover periodicity in the data, but his three-dimensional system was too complicated and not well-rendered by his publisher. The British chemist John Newlands recognized "the law-like status of chemical periodicity," but was largely ignored because he compared periodicity to musical octaves. His colleague William Olding also designed periodic systems but denied the lawfulness of chemical periodicity. In the United States, Gustavus Hinrichs developed a spiral periodic system but seemed to descend into mysticism by comparing the dimensions of the solar system to the dimensions of the atom. Most importantly, Lothar Meyer produced a mature periodic system, but did not make any predictions. Only Mendeleev brought together all the properties of simplicity, accuracy, and the prediction of undiscovered elements in his system. So who discovered the periodic table and the periodic law? Scerri couches his conclusions in terms of six "independent discoverers," and the "leading discoverer" Mendeleev – whose form was actually adopted, in no small part also because of Mendeleev's relentless public relations effort. The bottom line is that his system – even today fundamentally non-hierarchical in structure by comparison to biology – "created the biggest impact on the scientific community," the *sine qua non* for the spread and adoption of a scientific discovery.[24]

In short, much more than empirical science is at work in the emergence of the periodic table out of the many forms independently proposed. Moreover, as Gordin's book demonstrates, the periodic table that "won out" can only be understood in the context of Mendeleev's personal history and the broader cultural and political history of Russia.[25] How Mendeleev arrived at his first form of the periodic table in 1869, and how the periodic table evolved over the years, is both a fascinating cultural story and an expression of a truth grounded in nature – that there is order in the way elements are structured, and in how they react chemically. But at the same time, classification is the product of an individual or groups of people whose backgrounds inevitably affect the ultimate form of the ordering.

One finds similar lessons in the story of the discovery and classification of elementary particles, a story especially associated with Murray Gell-Mann, who received the 1969 Nobel Prize in physics for his work. This history is well-known and need not be repeated here except to emphasize its classification aspects. While the discovery of protons, neutrons, and electrons did not inspire a classification system, the development of particle accelerators resulted in such a profusion of particles that classification soon became necessary. By 1954, when a dozen particles were known, Enrico Fermi is reported to have remarked when asked by a student to name them, "Young man, if I could remember the names of these particles, I would have been a botanist."[26] By the end of the decade, thirty subatomic particles were known, and another sixty or so were discovered by the mid-1960s. Gell-Mann's contribution was the concept of the Standard Model, according to which all matter (fermions) may be classified as leptons (electrons or neutrinos) or hadrons. Leptons are elementary, but hadrons may be further classified as baryons or mesons, where baryons consist of elementary quarks, which in turn form protons and neutrons, then atoms and molecules. The unified theory known as the Standard Model not only classified quarks and leptons, but also the force carrier particles such as the photon, gluon, and Z and W bosons. And, like the periodic table, it made predictions for particles, all of which have now been found, most recently the Higgs boson at the Large Hadron Collider at CERN in Geneva, Switzerland, announced in 2012. As with biology and chemistry, there were social, institutional, and cultural aspects of these discoveries; their role in the construction of the Standard Model has been analyzed in Andrew Pickering's book *Constructing Quarks*.[27]

* * *

Our brief excursions into the role of the species concept in biological taxonomy, the periodic law arising from chemistry's periodic table, and the Standard Model in physics all raise basic questions for the very idea of classification in astronomy.

It is first of all important to be aware of the differences. In his work Mayr insists that the classification of animate objects is not the same as the classification of inanimate objects because living things have an evolutionary relationship, while inanimate ones do not. "When one is dealing with evolving biological populations – and that is what species of organisms are – one cannot expect the simplicity and unambiguousness that one encounters among parameters in the physical sciences."[28] This, of course, is true for many inanimate objects like books and rocks. In astronomy, however, matters are not quite as simple as Mayr might have thought. As we emphasized earlier, astronomical objects do have evolutionary relationships, even if they were unknown at the time the objects were classified. But unlike biology astronomical classes, while they may in some cases be related in an evolutionary sense, they have no genetic relationship, much less descent from a common ancestor, even if they have descent from a common event – the Big Bang 13.7 billion years ago.[29] Moreover, unlike chemistry and physics, classes of objects in astronomy are not the building blocks on which matter is constructed. In this case, there is no "unit of classification" except the object itself, and unlike biological species there are no isolated reproductive groups with protected genotypes. It is of interest, therefore, to see how astronomers have historically defined "class."

Despite these differences, there are lessons to be learned, beginning with the definition of "class." Surely the typological definition of class promises to be useful, whereby "membership in a class is determined strictly on the basis of similarity, that is, on the possession of certain characteristics shared by all and only members of that class. In order to be included in a given class, items must share certain features which are the criteria of membership or, as they are usually called, the "defining properties." A second lesson is the importance of precision in nomenclature, distinguishing the abstract categories in a classification system from the concrete taxa placed into the categories, using "class" to have only a very specific meaning, and so on. Third, although no evolution by natural selection exists in astronomy, cosmic evolution does exist, and we need to be aware of its implications for classification. Fourth, we should not be surprised to find competing classification systems in astronomy, and should beware of proclaiming an epistemologically favored system; the importance of utility and pedagogy, however, should not be underestimated. Fifth, we will be on the lookout for the meaning of astronomy's tendency to have hierarchical systems as in biology, rather than non-hierarchical ones as in chemistry's periodic table. Sixth, we need to consider the role of community in classification, already evidenced in Chapter 1 in the Pluto/planet debate, where different astronomical communities proposed different classification criteria. Finally, we will keep an eye out for the role of simplicity, accuracy, prediction, and utility in astronomical classification.

The inextricable connections between the concept of "species" and the Linnaean classification, between "element" and the periodic table, and between "elementary particle" and the Standard Model demonstrate how closely discovery and classification are intertwined, sometimes with an underlying theoretical basis. Armed with these questions and insights from other sciences, we now explore the connection between classification and discovery in astronomy.

8.2 Discovery and Classification in Astronomy: In Search of "The Thing Itself"

In order to understand the relationship between discovery and classification in astronomy, we need to distinguish between class, the process of classification, and the construction of classification systems. Clearly, the idea of "class" is raised early on in the discovery process, as astronomers detect a new object or phenomenon and try to fit it in the body of canonical knowledge. Just as elements were discovered before the periodic table, and as species were discovered before Darwin suggested their evolutionary relationships, so in astronomy, implicitly or explicitly, Galileo declared class status (or at least novelty) for planetary satellites and what turned out to be rings, Herschel for planetary nebulae and globular clusters, and Lord Rosse for spiral nebulae – all before their true nature was known, much less any evolutionary relationships. Although they were new classes, at the time of their discovery there was no question of classifying them as part of any coherent system with so few members of the class known. The first requisite for a classification system is to have a large enough representative sample of objects to classify.

Despite astronomy's status as the oldest of the sciences, this prerequisite for classification was not fulfilled for any group of objects until the last half of the nineteenth century. Only then did new techniques begin to reveal physical characteristics of objects that had been perceived, for all of history, as mere pinpoints of light in the night sky. Astronomy's first classification efforts centered on stars and developed with advances in stellar spectroscopy, a technique that proved surprisingly suited to unraveling a mystery that the philosopher

August Comte once held up as the prototypical problem that would never be solved – the composition of the stars.[30] Once spectroscopy was invented, there was no stopping the classifiers and the construction of numerous classification systems. But spectroscopy was not the only novel method to inspire classification, as natural history in astronomy moved into high gear. Spurred on by photography, another technique that not only enabled ever fainter points of light to be seen but also revealed structure in some cases, the classification of galaxies became possible, again spawning a variety of rival systems. Ironically, the classification of the nearest objects – the planets in our own solar system – came last, both because of the difficulty in determining their composition and because the discovery of planets beyond our solar system did not occur until the last decade of the twentieth century. In short, classification depends on data, usually in large amounts, and only when such data are gathered during a science's "natural history" phase can a classification system be constructed, even if classes have been declared with each new discovery. Surprisingly, depending on the definition of "discovery," classification may flourish well before any discovery is made.

Phenomenology: Classification as a Pre-Discovery Activity

The commonsense view of science holds that objects are discovered (we would say "detected"), then distinguished from those objects already known, then classified, as in biology and the early chemical elements. The historical record shows, however, that just the opposite is often the case: *objects are classified according to some observable characteristic (such as a spectrum), an exercise in "phenomenology;" then discovered during the extended triple process of detection, interpretation, and understanding; and finally classified again in the sense that physically meaningful classes are delineated.* Nowhere is this ironic and rather more complex role of classification in relation to the discovery of new classes of astronomical objects better illustrated than in the realm of the stars. There the sheer number of objects demanded classification, but the complexity of the physical mechanisms underlying the sorting activity assured that the initial classification, the phenomenology of spectral lines, did not necessarily equate with understanding the objects themselves. To be sure, many discoveries of a different kind were necessary before the discovery of what we today consider the standard stellar classes. But whether considering what are today the well-known Harvard spectral classes or the luminosity classes of stars as discussed in Chapter 4, a great deal of classification of spectral lines took place long before the true nature of the objects themselves was discovered.

Consider the stars. Spectroscopy was the key to classification and eventually to understanding, but first chemists and physicists had to discover what spectral lines meant. While the English chemist and mineralogist William Wollaston detected dark lines in the solar spectrum already in the early nineteenth century, he completely failed to recognize their importance, concluding they were boundaries between the continuous color spectrum of sunlight already seen by Newton two centuries before. It was left to the German optician Joseph von Fraunhofer to rediscover the lines a few years later, label the brightest according to an alphabetic scheme, measure the positions of some 350 of them, and conclude they were indeed an intrinsic property of sunlight. To this day they are called the "Fraunhofer lines," after the pioneer who died at the young age of thirty-nine, but not before laying the foundations for a new science.

Fraunhofer's interaction with the solar spectrum, however, was a mapping, not a classifying activity. Had the spectra of the other stars been identical to that of the Sun, there would have been no need for classification. But it turns out that although the Sun is a star, it belongs only to a particular class of star; solar spectral lines represented only the rudimentary beginning of the variety of stellar spectra we know today. And even when that classifying activity took on a furious pace in the late nineteenth century and developed into increasingly elaborate classification systems, the physical meaning of the lines remained obscure. The discovery of giants and dwarfs, subgiants and supergiants, white dwarfs, supernovae, and other stellar classes all lay in the future, well after the empirical classification of spectral lines became the main activity of the nascent astrophysics. Yet the physical meaning of those classes lay hidden in those lines, only awaiting their Rosetta Stone, or a number of Rosetta Stones, for decipherment.

The classifying activity did not take long to begin, but the decipherment was a long, drawn-out process marked by numerous false starts. The unraveling of the mystery of stellar spectra began with Fraunhofer himself. In 1817 he reported lines in the spectrum of Venus similar to those in the Sun, indicating reflected sunlight, and found "three broad bands which appear to have no connection with those of sunlight" in the spectrum of Sirius, the brightest star in the sky. By 1823, he had observed the Moon, Mars, Venus, Sirius, Castor, Pollux, Capella, Betelgeuse, and Procyon. He reported that the spectrum of Betelgeuse in Orion contained countless fixed lines, a few in common with the Sun, but others completely unknown. Thus Fraunhofer is the founder of stellar spectroscopy, but after his death in 1826, no one followed up his work for decades.[31] Not only was the sample still too small to detect any patterns and begin the classification process, but more fundamentally, no one yet realized these spectra held the key to the nature of the stars. An entire discipline lay fallow for more than three decades.

Even Fraunhofer did not attempt to explain the lines in the solar and stellar spectra. His role was the purely empirical work of mapping the lines, as a cartographer maps the Earth without knowing the underlying geology and mineralogy. Only gradually, most notably through the work of Gustav Kirchhoff and Robert Bunsen around 1860, did physicists unlock the key to the Fraunhofer lines on the basis of laboratory experiments with different elements. Laboratory spectra of lithium, sodium, potassium, and calcium, for example, would display the same spectral lines if they were present in the Sun or other stars. Even as this led to the discovery of new elements on Earth and played into the many forms of the periodic table, including Mendeleev's, it also led to the identification of the stellar spectral lines with specific elements.

Kirchhoff and Bunsen's Rosetta Stone decipherment of solar and stellar spectral lines revealed the chemistry of the stars and opened the door to a flood of data that for the first time could yield rudimentary classes of stars based on that phenomenology of spectral lines. As in other sciences, an intense period of natural history activity began, as budding astrophysicists gathered stellar spectra at an increasing pace, as Darwin would gather beetles, barnacles, and butterflies.[32] The Italian astronomer Giovanni Battista Donati published the spectra of fifteen stars in 1862, enough for him to classify them into three categories as white (Sirius), yellow-orange (Capella, Arcturus), and red (Betelgeuse, Antares). The following year Lewis M. Rutherfurd in the United States and James Carpenter at Greenwich published observations of a few dozen stars, and both divided them into two or three groups according to their spectral lines. All of this was a prelude to the work of the Italian astronomer Angelo Secchi, often considered the father of stellar spectral classification. From 1863 to 1868, Secchi devised several classification systems, ending in 1868 with four "Types," and a fifth Type for those that did not fit any of the others. By 1878, Secchi had published a catalogue of 366 objects classified in this way, and had obtained spectra of almost 4000 more stars yet to be examined. In contrast to biology, all of these stellar classification systems were linear rather than hierarchical, and constituted the beginnings of spectral types, without knowing exactly what that meant.

But physical interpretation was not long in coming, indeed, had already begun. Hermann Vogel in Germany, and William Huggins and Norman Lockyer in England, among others, were devising their own classification schemes, and already there was a search for physical meaning and talk of an evolutionary sequence. "The only rational classification of the stars by their spectra might be achieved, if one starts from the point of view, that in general the evolutionary state of the respective heavenly bodies is mirrored in their spectra," Vogel wrote in 1864.[33] Based on the strength of the absorption lines, Vogel believed this sequence was white stars, yellow stars such as our Sun, and red stars. A quarter century later Lockyer proposed his own evolutionary scheme known as the "meteoritic hypothesis," whereby vaporized colliding meteors produced nebulae and comets, which then condensed to form stars. The stars' contraction resulted in higher and higher temperatures, ending in cooling and extinction phases. In this scheme, spectral lines were indications of temperature, the higher temperatures being young stars, and cooler temperatures representing older stars. From its origin in 1887 to its most developed form in 1900 in a book titled *Inorganic Evolution*, Lockyer presented a rational classification system that he believed to be imbued with physical meaning. Moreover, it was based on laboratory observations of spectra, whereby spectral lines were stronger at higher temperatures, according to a "dissociation" of metals theory that Lockyer also elaborated. Despite the fact that Lockyer anticipated Henry Norris Russell's division of stars into giants and dwarfs, and even Meghnad Saha's ionization theory in a crude way, his work was never widely accepted and was subject to a great deal of criticism.[34] Lockyer, however, is not credited with the discovery of classes of dwarfs and giants, nor even with the discovery of the temperature sequence as now understood. His classification system, therefore, is a pre-discovery activity with respect to true knowledge of the existence of those physically meaningful classes.

Matters are a bit more complex with respect to the discovery claims for the spectral classification work that took place at Harvard in the four decades following 1885, for that work as a whole *can* claim to have discovered the spectral types and the true stellar temperature sequence as we know it today.[35] As we have seen in Chapter 4, by far the most robust stellar classification system was developed at Harvard under the supervision of E. C. Pickering (Figure 8.3). Here Williamina Fleming classified the spectra of 10,351 stars and produced the Draper Memorial Catalogue in 1890; Antonia Maury undertook a more detailed study of the brighter stars, resulting in a catalogue of spectra for 681 stars published in 1897; and Annie Jump Cannon produced a series of catalogues culminating with the Henry Draper Catalogue including 240,325 stars, published in nine volumes from 1918 to 1924 (Figure 8.4). There followed the Henry Draper Extension between 1925 and 1936, yielding a total of more than 359,000 stars classified by Cannon between 1896 and 1941. This surely represents not only unparalleled individual efforts (with the help of an even larger corps of women assistants), but also an unparalleled natural history operation for any science over a similar period. Cannon, especially, deserves to be known as the Linnaeus of stellar astronomy; she classified so many stars that her system became the standard – no one wanted to repeat that massive effort. The question of what it all meant, however, only gradually became clear with the discovery of more physically meaningful stellar classes.[36]

Figure 8.3. Part of the so-called Pickering's harem, women who worked at Harvard College Observatory as human "computers." Only a few worked on spectral classification of stars under Observatory Director Edward C. Pickering's supervision. The image, taken on May 19, 1913, includes Pickering (center top), and Annie Jump Cannon (darker dress at top center); Maury is not present, and Fleming had died in 1911. At far left is Margaret Harwood, who took the image of Russell seen in Figure 4.1. Complete identifications are available in Barbara Welther, "Pickering's Harem," *Isis* 73 (1982): 94. Courtesy Harvard College Observatory.

The classification systems devised for these catalogues evolved with time and reflected increasing detail in the spectra. Fleming's system was based on the four Secchi spectral types, Type I being divided into A, B, C, and D; Type II divided into E, F, G, H, I, K, and L; III was M and IV was N. O was used for special spectra with bright lines (now known to be "Wolf-Rayet stars"), P for planetary nebulae, and Q for all others not classified A through P. Maury, however, used a more detailed classification notation for her higher resolution spectra, consisting of twenty-two groups from I to XXII, supplemented by "collateral divisions" labeled a, b, and c; as we have seen in Chapter 4, hints of the giant/dwarf dichotomy lay within these divisions, in particular the "c" division with its broadened spectral lines, which soon caught the attention of Ejnar Hertzsprung. Cannon adopted Fleming's system, added decimal subdivisions of spectral types from B to M, and rearranged the letters to reflect what she believed was both a temperature sequence and an evolutionary sequence, the now famous sequence O, B, A, F, G, K, M (recalled by thousands of students through the mnemonic "O Be A Fine Girl Kiss Me," to which was later added "Right, Now, Smack" (and just recently L and T classes, for which one shudders to think of the potential mnemonics).[37]

The idea that classification is not the same as discovery is indicated by the fact that Fleming, Maury, Cannon, and even Pickering are rarely credited with

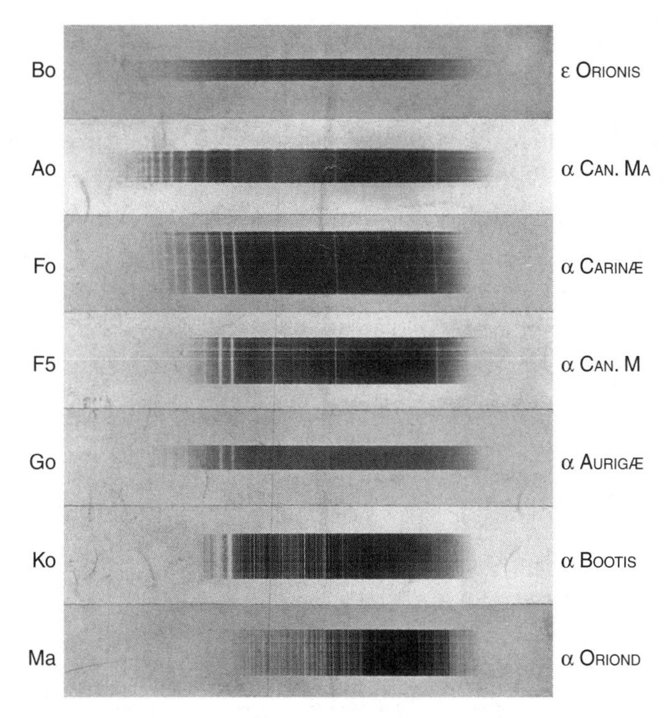

Figure 8.4. Classification as a phenomenological pre-discovery activity: spectral lines used to construct the Harvard system, as found in the frontispiece to the Henry Draper Catalogue. These representative spectral types from B through M, representing colors blue through red, were later found to represent a temperature sequence. *Annals of the Astronomical Observatory of Harvard College Observatory,* 91 (Cambridge: The Observatory, 1918).

"discovering" star types such as O, B, A, F, G, K, and M. Theirs was primarily a classification process. But when a large and representative number of objects are involved, classification may be important to the discovery process, normally associated with physically meaningful discovery rather than any phenomenology such as spectral lines. As a group the Harvard classifiers may well be credited with the discovery of what was widely perceived (and is now known) to be a true temperature sequence after Cannon's reordering, even though it ended up not being the evolutionary sequence they suspected. Most important, along with the second Rosetta Stone inherent in the distance and luminosity determinations gathered by others, their spectral types would prove crucial to Hertzsprung and Henry Norris Russell and the construction of the H-R diagram, which clearly delineated the classes of giant and dwarf stars. And even the discovery of those physical classes was not the end of the story, with more stellar classes and the physical basis for their spectra based on ionization theory (yet another Rosetta Stone) still to be discovered. Thus, while University of Michigan astronomer Ralph Curtiss declared the empirical classification of astrophysical spectral lines substantially complete by 1929, he emphasized this achievement was not the same as understanding the star itself: "It remains to develop a rational theory of astrophysical spectra and of stellar spectra in particular. Such a theory should connect the spectrum of a star with fundamental stellar characteristics including effective temperatures, surface gravity and atmospheric composition. But we are still far from being able to do this."[38]

Thanks to more and more distance and luminosity determinations, the giant and dwarf classes of stars would eventually be followed by others, as we saw in Chapter 4. Moreover, in his dying days in 1929, Curtiss could pay tribute to Saha's temperature-ionization explanation for spectra as a partial explanation for the spectral lines classified in the Draper system. Most important, he could state that "the outcome of all investigations to date has been to confirm the practical value of the empirical system of the Draper Catalogue." In other words, the originators of the Draper system had chosen their spectral characteristics wisely, in a way that eventually aligned with the physical properties of stars. It was not at all obvious that this should turn out to be the case. But even this was not enough, and in the wake of the discovery of dwarfs and giants and indications of other physically meaningful classes, another stellar classification system sprang up that would not replace, but supplement, the Harvard-Draper system, adding another dimension and demonstrating that there is also a role for classification after discovery.

The Thing Itself: Classification as a Post-Discovery Activity

As Ralph Curtiss had hinted in 1929, even after the discovery at Harvard of a temperature sequence, and the Hertzsprung-Russell discovery of the giants and dwarfs during the period between 1905 and 1913, some astronomers were not satisfied. They foresaw the need for more refined classification at a new level, resulting in an even more robust classification system: "The classification system of the future will undoubtedly be based on physical principles in addition to temperature-ionization and will be expressed numerically in terms of definite parameters ... at least three parameters will be necessary. They include a quantity based on temperature, another depending on the abundance of neutral atoms, and a third depending on the abundance of ionized atoms."[39] A few years later Yerkes astronomer Otto Struve emphasized that two stars with equal atmospheric temperature and pressure do not necessarily produce identical spectra; there would be differences due to other physical parameters such as the abundance of elements.

Not all astronomers agreed; Russell at Princeton and Cecilia Payne-Gaposchkin and Donald Menzel at Harvard did allow that two stars with equal atmospheric pressure and temperature would produce identical spectra only if their surface gravities, relative atomic abundances, and other unidentified factors were all identical. But they felt that the criteria for the Harvard classification "express the most conspicuous features from type to type. It is doubtful whether more outstanding bases for classification could be selected." While spectral line intensities and profiles are necessary for a complete *description* of a spectrum, they emphasized, description is different from classification: "We cannot admit that a mere listing of line intensities *constitutes* sufficient classification; classified material differs from tabulated data as a rogues' gallery differs from an undiscriminating photographic album." As John Hearnshaw has pointed out, the Harvard astronomers had gone from defending the Harvard system from being too complex as compared with some of the early classification systems, to defending it from being too simple compared with what might be done in light of ever-increasing amounts of new and more sophisticated data.[40] The debate illustrates one of the primary characteristics of classification systems: they must have a kind of "Goldilocks" quality – not too simple and not too complex, but "just right" in the sense of being useful to the practitioner.

Struve, however, was not to be mollified, perhaps because, as Hearnshaw has again pointed out, he was an analyzer rather than a classifier. Struve was the director of Yerkes Observatory, and it is therefore not surprising that it was at Yerkes that such a system was born. The agent of change was his young employee W. W. Morgan, who arrived at Yerkes in 1926 and would remain there for more than sixty years (Figure 8.5). Following his thesis under Struve on peculiar stellar spectra, Morgan received his PhD in 1931, and shortly thereafter set about creating "a two-dimensional spectral classification that would be determined completely from the appearance of spectral lines, bands, and blends." Morgan recognized that Walter S. Adams and his associates at Mt. Wilson had created a two-dimensional temperature-absolute magnitude system already in the years 1914–1917, but in his view it was subject to serious errors in calibration of the luminosities from the absolute magnitudes, calibrations based on trigonometric parallaxes. In other words, as calibrations of absolute magnitudes were changed, luminosities were also changed. "The unsatisfactory nature of this situation became obvious to me in the mid 1930s," Morgan recalled many years later. His system "was to be autonomous; that is, it would be defined completely by the spectra of standard stars, without having to appeal to any theoretical picture. The classification process would then be a judgment of 'like' or 'not like' for an unclassified spectrum in comparison with the spectra in the array of standard stars."[41]

Based on the mass-luminosity relationship, color temperatures (T) and trigonometric parallaxes, by 1937 Morgan had estimated surface gravities (g) for a sample of bright stars, including supergiants, giants, dwarfs, and one white

Figure 8.5. William W. Morgan, the driving force behind the Morgan-Keenan-Kellman system for spectral classification. Unlike the phenomenological spectral line classification, Morgan sought to classify "the thing itself," as in Figure 8.6. Photograph by David DeVorkin, courtesy AIP Emilio Segrè Visual Archives.

dwarf. When he plotted log g versus log T, he recognized the resulting plot as a form of the H-R diagram, in which "the main sequence is shown to have the characteristic of approximately constant surface gravity over the whole range from A0 to M . . . Each successive luminosity group is separated in g from the others."[42] Morgan saw that this could function as the basis for a new "two-dimensional" classification system, and the following year he introduced five luminosity classes ranging from supergiants (I) to bright giants (II), normal giants (III), subgiants (IV), and main sequence stars (V).[43] Under this system the criteria for luminosity classification were the ratios of line intensities, rather than the line intensities themselves that the Harvard astronomers had ridiculed. This system, assembled with the help of Philip C. Keenan and Edith Kellman and therefore known as the MKK system, was published in 1943 as the *Atlas of Stellar Spectra, with an Outline of Spectral Classification.* It became one of the canonical publications in the history of astronomy. The *Atlas* consisted of fifty-five prints of Yerkes' 40-inch spectra; future classifiers depended heavily on the spectra of its "standard stars," which sometimes changed along with the system (Figure 8.6).[44]

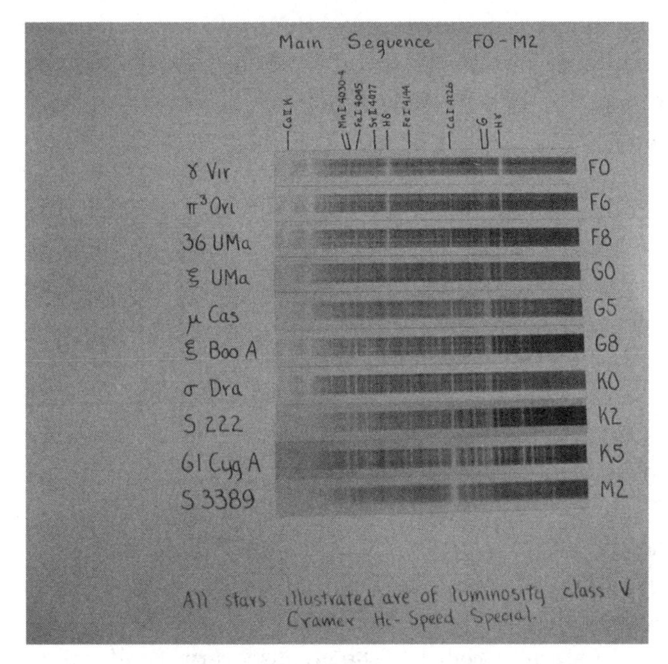

Figure 8.6. Classification as a post-discovery activity: the MKK system, or Morgan-Keenan-Kellman system, defined standards for luminosity classes, which represent real physical classes such as giants and dwarfs (see Figure 4.3). Seen here are the standard stars for main sequence dwarfs (luminosity class V) with Harvard spectral types F0 through M2. The full spectral designation would be, for example, M2V. One of the fifty-five prints from W. W. Morgan, P. C. Keenan, and E. Kellman, *Atlas of Stellar Spectra with an Outline of Spectral Classification* (Chicago: University of Chicago Press, 1943).

The driving force behind the MKK system is notable, and has been revealed by Morgan himself. "Since youth, I have been entranced with what I later discovered Husserl has called the 'thing itself,'" he wrote in 1988. "In my field of astronomy it could be a star (that is a particular star), a star cluster, a galaxy, or a cluster of galaxies." In biology, he noted, it could be a tree or a sparrow. In either case, he asserted "*the 'thing itself,' whether breathing or not, growing or not, has a fundamental property or existence far deeper than any conceptual construction.* And how does this bear on the classification of stars, galaxies, and clusters of galaxies? It induces a very deep feeling of respect for those things *for themselves*, as independent structures of Nature" (italics in original). Morgan's conclusion is worth quoting in full, since it was the animating feature behind his career and the MKK system: "It is of supreme importance to the creation of our operational morphologies that we use the 'things themselves' as our picture-language for stellar spectra, and direct photographs for galaxies and clusters of galaxies. The derived morphologies must be 'true' to the information revealed in the 'things' (or specimens), *not* from superposition of formal constructions on the 'things.' After the specimens have been arranged in the order indicated by the *observed* forms, we have brought into existence morphological types of lasting value and richness."[45] Ironically, Husserl is considered the founder of the "phenomenology" movement, which concentrates on phenomena rather than the thing itself, a thing that Kant said could never be reached by the human mind, but that Morgan considered of "supreme importance" in classification: For Morgan the discrimination of the thing itself, the specimen, from its spectra or optical forms, was everything.

Philosophical underpinnings aside, the MKK system (later known as the MK after it was expanded in the absence of Miss Kellman) stands as an example of post-discovery classification by reference to the discovery of physical classes of stars, including supergiants, bright giants, giants, subgiants, and main sequence stars. Those classes had for the most part been discovered by the time the system was constructed; Morgan, Keenan, and Kellman are therefore not the discoverers of these classes, but the originators of a more refined post-discovery classification system. The canonical luminosity classes now accepted in astronomy are therefore preceded by the pre-discovery classification systems ranging from Secchi to the Harvard/Draper catalogues, a phenomenological exercise undertaken without a knowledge of the physical nature of the stellar classes, and followed by the MKK classification, the best attempt to represent "the thing itself." The necessity and utility of the latter was (and occasionally still is) questioned, but it has survived the test of time. Today stars are typically completely represented by both systems: Sirius is an A1V, a main sequence star of A1 spectral type; Capella is a G8III, a bright giant of G8 spectral type; and Betelgeuse in Orion is an M2I, a supergiant of spectral type M2.

Stellar classification yields a final lesson: the case of the hypergiants. In Morgan's 1938 paper, classes Ia and Ib had been designated as the brightest supergiants. In 1956, Michael Feast and Andrew Thackeray suggested that even brighter supergiants (greater than absolute magnitude -7) be called super-supergiants.[46] In 1971, Philip Keenan suggested a new class, designated 0, to define hypergiants, and in the next revision of MK in 1973, Morgan and Keenan wrote, "In the new and tighter network of supergiants the highest luminosity class, 0, is defined by the four reddest of the brightest stars in the Large Magellanic Cloud. These are the stars originally called 'super supergiants' by Feast & Thackeray."[47] Class 0 hypergiants are distinct from what are called Ia-0, supergiants/hypergiants, the latter being an example of the common practice of classifying intermediate luminosity classes, as in II–III.

The subdivision of class I into subclasses Ia and Ib on the one hand, and the declaration of a new class 0 of hypergiants on the other, seems somewhat arbitrary. But Keenan justified the new class 0 by saying the situation was confused for stars of luminosity class Ia and brighter, and that the super-supergiants discovered in the Large Magellanic Cloud "provide the means of anchoring the upper end of the luminosity sequence for these types," for which he placed the absolute magnitudes at around −9, 500,000 times solar luminosity.[48] The class was only sporadically and inconsistently used for several decades, but is now coming into more common use with the discovery of more such objects. One can argue whether hypergiants should be given separate class status; in the end it is an arbitrary designation. Like the other MK classes, the class of hypergiants was never officially adopted. But common usage now makes these stupendous objects – ranging to about 100 solar masses and on the verge of instability – part of the system. A classification system can, therefore, evolve. But it must not evolve too much, or it will become useless over long time periods. Utility is the watchword.

* * *

Even as early efforts at star classification were ongoing, a second effort at astronomical classification began in connection with the variety of nebular shapes revealed with ever-larger telescopes. As we saw in Chapter 5, the primary problem with these nebulae was the lack of known distances, many of which turned out to be much greater than those of the stars whose classification we have discussed so far. The discovery that some of them were objects outside of our own galaxy, in other words island universes or galaxies of their own, was definitively proven only in 1924 with Hubble's use of Cepheid variables as distance indicators, a result formally announced on the first day of 1925. And as we saw in Chapter 5, galaxies in the form of nebulae had long been detected prior to that time, having been listed in the catalogues of Messier and the Herschels, among others, in increasing numbers based on visual observations during the century from the 1770s to 1860. Following this natural history phase, John Herschel published his *General Catalogue of Nebulae and Clusters of Stars* in 1864, containing 5079 objects. By the 1880s, the *New General Catalogue* compiled by J. L. E. Dreyer held 7840 objects, and the addition of two *Index Catalogues* raised the number to more than 13,000, altogether a reasonable database constituting a formidable task for the classifier.[49]

The task of galaxy classification, however, was very different from that of stars. As Morgan pointed out, "The ordered appearance of the [stellar] spectrograms contrasts sharply with the disorder of the forms of many galaxies; and even in the case of galaxies possessing a degree of order, we find them in a fantastic variety. This clearly requires a differing point of view toward the classification problem . . . we can never expect to obtain the clean, detailed matching of specimen form to standards that we take for granted in the case of stellar spectra."[50] Small wonder, then, that galaxy classification would turn out to be a rather more convoluted process than star classification.

These problems notwithstanding, like the stars, classification of nebulae took place as a pre-discovery activity long before the extragalactic nature of galaxies was discovered. As discussed in Chapter 3, William Herschel himself even attempted a rough classification of nebulae, most of which proved to be galaxies, but only the photographic surveys beginning about 1890 revealed the faint galactic structures that proved to be crucial to their classification. The German astronomer Max Wolf constructed one such classification system in 1908 (Figure 5.1), used extensively until the 1940s.[51] Meanwhile, in his 1917 dissertation, Edwin Hubble had measured the positions and described the morphologies of faint nebulae in a few selected fields he himself obtained using the 24-inch reflector at Yerkes Observatory. Hubble placed his nebulae in one of the twenty-two categories of Wolf's classification system, which he considered no more than a temporary filing system "until a significant system shall be constructed." This is precisely what Hubble undertook in his early classification system in 1922, elaborated in 1926 (Figure 5.3), and extended post-discovery in his classic book *The Realm of the Nebulae* (1936), the system still used in modified form today.

As we have emphasized, classes are the basis for classification systems, and galaxies once again illustrate the problem of criteria for class status. As Sandage, one of the great observers and classifiers of galaxies, has noted, the primary problem is to decide what characteristics have physical significance: "The first step of any new science has always been to classify the objects being studied. If a proposed classification scheme is 'good,' progress is achieved. This leads to deeper understandings, and the subject advances. A 'bad' classification scheme, by contrast, usually impedes progress." The problem is not simple: "the initial classifier of any unknown subject necessarily begins with no idea of how to choose the key parameters. How then to produce a classification imbued with any fundamental physical significance or predictive power?"[52]

This was the problem facing early classifiers, long before the distances and physical nature of the galaxies was known. Description was not enough: "Simple description, although not sufficient as a final system, is often an important first step... But as a classification develops, a next step is often to group the objects of a set into classes according to some continuously varying parameter. If the parameter proves to be physically important, then the classification itself becomes fundamental, and often leads quite directly to the theoretical concepts." Wolf's system, which did not and could not distinguish nebulae inside our galaxy from real galaxies, used a linear sequence based on morphology: it proceeded from amorphous forms with no spiral features to fully developed spirals and everything in between. It thus distinguished only two of what we now consider the canonical Hubble classes, ellipticals and spirals, and many believed that ellipticals would eventually be resolved into spirals. In neither case did Wolf's classification correspond to what Morgan would have called "the thing itself" - not necessarily our current description of it, but what it "really" is.

Hubble's original system, by contrast, was considerably simpler, not taking into account the great variety of spiral forms. As shown in Chapter 5, its classification criteria were based only on morphology, not only distinguishing spirals, elliptical, and irregulars by 1926, but also subclasses such as E0 (spherical) to E7 (elongated or "lenticular") elliptical galaxies, all arranged in a "tuning fork" diagram that Hubble also believed represented an evolutionary sequence (Figure 5.6). While some astronomers preferred the more detailed Wolf classes, Hubble's system won out, no doubt in part due to its simplicity, in part to its usefulness, and in part to Hubble's subsequent status as the great authority on galaxies following his proof of their extragalactic nature. Hubble's system also illustrates the problem of shifting classes: what he called lenticular galaxies and were for him the most elongated class of ellipticals equivalent to the E7 class, are today what Hubble hypothesized in 1936 as a hypothetical S0 class intermediate between ellipticals and spirals.

For galaxies, as for stars, there were competing classification systems: not only those of Wolf and Hubble, but also those developed independently by Knut Lundmark (ellipticals, spirals, and Magellanic Cloud–type irregulars) in 1926 and 1927, and by Harlow Shapley based on concentration and other properties.[53] Lundmark's system was a particular point of contention with Hubble. In his 1926 article Hubble noted acerbically that Lundmark, who was present at the Cambridge meeting and appointed a member of the IAU Commission on Nebulae, published a classification system in 1926 which "except for nomenclature, is practically identical with that submitted by me. Dr. Lundmark makes no acknowledgments or references to the discussions of the Commission other than those for the use of the term 'galactic'." The two systems were not, in fact, "practically identical," but the claim shows how possessive scientists can be of their own work. Nor was Hubble enamored of Shapley's system, which Sandage later characterized as "individual descriptions with no underlying physical basis. There was no physics behind it, hidden or otherwise. Shapley's classification system was useless." A good indication of the truth of Sandage's statement is that Shapley himself abandoned his own system four years later.[54]

The Hubble-Lundmark dust-up is somewhat ironic in light of recent evidence brought to light regarding the influence of John H. Reynolds on one significant part of Hubble's classification system - the spirals. Reynolds was not an astronomer, but a highly successful and wealthy British industrialist, who devoted much time, energy, and funding to astronomy, founding his own observatory and constructing with his own hands its 28-inch reflector. Amateur status notwithstanding, he has recently been called "England's foremost observational expert on the morphology of the nebulae" during Hubble's time.[55] Reynolds was well enough known in astronomy that Hubble was in correspondence with him – indeed, although Reynolds had no formal training in astronomy, he was a member of the IAU Commission on Nebulae and Star Clusters, which included Hubble and V. M. Slipher, among others. In 1920, at Hubble's urging and six years before Hubble's own published scheme, Reynolds published a scheme of seven classes of spiral galaxies, designated I through VII, some of which are clearly identical to Hubble's Sa, Sb, and Sc designations in his 1926 paper (Figure 5.3). Although Hubble acknowledged Reynolds in private correspondence, he does not refer to Reynolds at all in his definitive 1926 classification paper, and only mentions him in passing in connection with spirals in his 1936 *Realm of the Nebulae*.[56] True, spiral nebulae were only a part of Hubble's system, but an important part.

It is a matter of history that Hubble's system triumphed, at least it has so far. Classification was, and remains, a robust post-discovery activity, as astronomers proposed modifications, extensions, and variants to the Hubble system, and even totally new systems. In 1959, the French American astronomer Gérard de Vaucouleurs argued that Hubble's ordering of spiral galaxies, based only on how tightly the spirals were wound and whether they were barred or unbarred, was not enough. He developed another system that includes more refined criteria for bars and spiral arms, and adds rings to the classification. And in a parallel to what happened in stellar classification, the following year the Canadian astronomer Sidney van den Bergh introduced the concept of luminosity classes for galaxies.[57]

Meanwhile, a few years after Hubble's death in 1953, none other than W. W. Morgan, the originator of the MKK system for stars, developed his own galaxy classification system based on the correlation between morphology and stellar population in the central regions of galaxies. Until this time, unlike stars, morphology rather than spectroscopy had played the primary role in galaxy classification. With Morgan's background as a pioneer in stellar classification, it was only natural that he should develop a system based not only on morphology, but also on spectral type of the composite stellar population; he was still in pursuit of "the thing itself." Since its publication in 1958, however, the Morgan classification has rarely been applied, proof that a classifier in one area of astronomy may not be so successful in another.[58]

Given the difficulties in correlating galaxy morphology with physical significance, it is remarkable that galaxy classification systems proved as good as they did. In a significant way, the jury is still out on Hubble's galaxy classification system, in its original or extended forms: whether the characteristics chosen as the basis for the classification system are fundamental to the galaxies themselves, and particularly their evolution, is still an open question. Forty years ago Sandage himself judged it too early to tell; surprisingly the same holds true today despite advances.

* * *

Ironically the planets, those objects seen since antiquity and closest to us, have the least developed classification system in all of astronomy. While this may seem odd, this circumstance is due almost entirely to one fact: the small sample size. At most three classes of planets exist in our solar system: 4 terrestrial planets, 2 gas giants, and (provisionally and controversially) 2 ice giants. And then there is Pluto, the small object that precipitated a large problem, one of class, not of a classification system.

Only a little more need be said about Pluto in the context of this chapter. We recall from Chapter 1 that based solely on its motion, upon its discovery Pluto was placed into the well-established class of 'planets," one of the "wanderers" among the background fixed stars. Doubts arose about planet status based on its low mass, especially as determined after the discovery of Charon in 1978. Only when numerous other Pluto-like objects were found beyond Neptune, some larger than Pluto itself, did the clamor for its membership in a new class of objects become so insistent as to instigate official action at the International Astronomical Union in 2006, with what effect all the world knows.

In hindsight, from one point of view the problem of Pluto was that it was classified with too little information, before "the thing itself" was known. Had its true characteristics been known in 1930, it might never have been classified as a planet. From this point of view, Pluto (like the asteroids) was assigned to a class before it was known that it was only one member of a large number of similar objects in what is now known as the Kuiper Belt; even when that did become known, it was unclear whether it should be declared a new class of "dwarf planets." On the other hand, Pluto illustrates a deeper problem that we have occasionally seen in the context of stars and galaxies: the fuzziness of the class concept in the case of boundary objects that seem to fall somewhere between distinct established classes. This is certainly a problem in biology, and it has occasionally reared its head even in astronomy. Although seen in rudimentary form as early as attempts to distinguish between comets and planets based on incomplete observations, some objects defy classification even with much more complete information. This is clearly seen in the discovery in 1977 of the object known as Chiron, orbiting between Saturn and Uranus and labeled by the IAU as minor planet 2060 Chiron. But in 1988, cometary activity was discovered on Chiron, and a cometary coma was detected a few years later. Was it a minor planet or a comet? In the end Chiron was designated both minor planet 2060 Chiron and comet 95P/Chiron and placed in a class of objects dubbed "Centaurs," not half man, half horse, but half comet, half asteroid.[59] This subclass of minor planets and comets, of which several are known and thousands suspected, is reserved for objects that show the characteristics of both.

Viewing Pluto as such a boundary object between a terrestrial planet and a Trans-Neptunian object, astronomer Michael F. A'Hearn, one of the leaders in the Pluto debate, called for a dual classification based on what seemed strong arguments. In an article written several years before the Pluto debate reached its climax in Prague, A'Hearn drew on the case of Archaeopteryx, a classic example from the world of biology. Discovered in 1861 shortly after Darwin's *Origin* was published, the fossil immediately became enmeshed in the controversy over the evolutionary relationship between dinosaurs and birds. It had the feathers and wishbone characteristic of birds, but teeth and a pelvis resembling dinosaurs. A'Hearn pointed out that there was strong disagreement for more than fifty years about how to classify Archaeopteryx, and even after a century some were still arguing. Pluto, he emphasized, also had dual characteristics: a spherical shape and thin atmosphere characteristic of planets, and dynamical features including moderate eccentricity, high inclination and libration with Neptune characteristic of TNOs. The lesson for A'Hearn: "If we spend the next several decades disputing the classification we will be distracted from the entire point of classification. The important point is that Pluto should be looked at from both sides since, like archaeopteryx, it is on or near the boundary between two classes and thus can shed unique light on the relationship between the groups . . . the key lesson from archaeopteryx is to look at Pluto from both sides and not spend decades, or even centuries, disputing the classification." While this seemed an eminently practical conclusion, it is a matter of record that it was not the solution adopted by the IAU, perhaps demonstrating astronomers' predilection for clarity, not duality. Nevertheless, in practice astronomers will undoubtedly learn more about both TNOs and planets by considering its dual characteristics.[60]

Despite its present notoriety, the problem of Pluto will undoubtedly be seen as a passing fad in the sweep of history – albeit one with deep cultural and scientific implications. Much more important is that the golden age of planetary classification has now begun, in the wake of the discovery of thousands of planets beyond our solar system, and in the context of the discovery of more objects in the Kuiper Belt and Oort Cloud (which also undoubtedly exist in other planetary systems). Although it is likely that most of the more than 2000 extrasolar planet candidates found so far fall in the gas giant class, and a few in the terrestrial planet class, one of the Kepler planets (Kepler 7b), as well as the planet known as WASP-17b, appear to have the density of styrofoam, different enough from the others to perhaps constitute a new class. The latter has 1.6 Saturn masses, but 1.5 to 2 Jupiter radii, yielding a density of 6–14 percent that of Jupiter.[61] Only further observations will determine whether these objects truly constitute a new class of planets.

Meanwhile, attempts already are being made to classify extrasolar planets based on their physical characteristics. One study finds that the extrasolar planets discovered thus far across five orders of magnitude in mass can be accommodated into the three classes already known in our own solar system: terrestrial planets, gas giants, and ice giants. Another study by the Italian astronomer S. Marchi distinguishes five classes based on planetary mass, semimajor axis, eccentricity, stellar mass, and stellar metallicity.[62] Whether the latter are true new classes of objects, or types of a particular class, remains to be seen and is dependent on the definition of "class." Thus, in the end and despite being grounded in nature, the declaration of a new class of astronomical objects is a socially determined exercise, and the construction of any classification system doubly so, as class is piled upon class in the attempt to order nature.

* * *

Aside from stars and galaxies, and the coming plethora of planets, numerous smaller classification systems have been developed over the decades of particular classes of astronomical objects, ranging from asteroids and comets, to nebulae and star clusters, to binary stars and supernovae. Each system has been formulated reflecting the knowledge of its time and according to diverse methods that have proven variously useful to the communities that study these objects. Each could be analyzed along similar lines to our discussion of planets, stars, and galaxies in terms of pre- and post-discovery classification. But having seen how astronomy's main classification systems developed in relation to the discovery of their major classes, the question now arises about the possibility of a more comprehensive classification system for all of astronomy's objects, analogous to the hierarchical classification system of biology, or the more linear sequences of physics and chemistry – or something entirely different. By way of closing this section, we briefly depart the realm of history to consider that epistemological problem, and propose one possible comprehensive system as an exercise in classification. The construction of such a system illustrates all the challenges of class and classification systems we have discussed in this chapter.

Astronomy's Three Kingdoms: A Comprehensive Classification System for Astronomy

Astonishingly, in stark contrast to biology, physics, and chemistry, no comprehensive classification system has ever been constructed for astronomy. What would such a system look like, and based on what principles? With the idea that such an exercise can shed light on our subject, in Appendix 1 we present one such system devised over the last fifteen years for pedagogic use, and presented here in full for the first time.[63] This so-called Three Kingdom system begins with the three Kingdoms of planets, stars, and galaxies, stipulates six Families for each Kingdom, and distinguishes eighty-two distinct classes of astronomical objects. Like biology, it is hierarchical, extending from Kingdom to Family to Class, with the possible extension to further categories lower in the hierarchy such as Type and Subtype. As in biological classification it occasionally adds an intermediate Subfamily level wherever useful. With the benefit of hindsight (and with utility in mind), the system incorporates some classes as they have historically been defined, and adds others as they might be defined in a more coherent and consistent system.

In constructing such a system, one immediately runs into the problem of how to define the categories of Kingdom, Family, and Class. The three Kingdoms adopted here (planets, stars, galaxies) are the three canonical divisions adopted in textbooks for almost a century, since it became clear that galaxies were indeed a separate "realm of the nebulae," as Hubble put it. For each Kingdom, six astronomical Families are delineated, based on the object's origin (Proto-), location (Circum- and Inter-), subsidiary status (Sub-), and tendency to form systems (Systems), in addition to the "Central" Family (planet, star, or galaxy) with respect to which the other Families are defined. These considerations give rise to astronomy's eighteen Families, and the symmetry of the six Families of each Kingdom reflects their physical basis in gravity's action in all three Kingdoms.

Then we come to the all-important category of Class, the focus of this volume. While classification systems in stellar and galactic astronomy have proven very successful in the pre- and post-discovery process where the idea of class evolved with those systems, attempts to delineate astronomical classes in a more general way have not proven so successful. Thirty years ago, in his pioneering book *Cosmic Discovery*, astronomer Martin Harwit pointed out that most cosmic phenomena (objects and otherwise) differ from each other by properties such as scale, energy output at various wavelengths, pulsations, and so on. "Most astronomers would agree that whenever we detect some strikingly new pattern of events in our observations, we sense that we are dealing with an entirely new process – a new phenomenon," he wrote. "What we mean by *striking* in this context must of course, be settled; but most individuals would agree that differences spanning many orders of magnitude – differences amounting perhaps to at least a factor of 1,000 in some trait – would label such a newly observed object or set of events as strikingly different from anything previously discovered." In short, the definition of a new class would center on "the strikingly disparate appearance of different phenomena." Harwit went on to make a heroic attempt to define such classes in terms of a set of observational filters, but – in an echo of early forms of the periodic table, and discarded systems of stellar and galaxy classification – the result was so complex and impractical that astronomers have largely ignored it.[64]

Another way of approaching the question of the definition of class is by looking at history, where (partial exceptions like stars and galaxies notwithstanding) classification has often been ad hoc, haphazard, and historically contingent on circumstance. If astronomical history demonstrates anything, it is that the classification of astronomical objects has been based on many characteristics, depending on the state of knowledge and the needs of the particular community. For example, planets could be divided according to their physical nature (terrestrial, gas giant, and ice giants) or, as the recent discovery of planetary systems has taught us, by orbital characteristics (highly elliptical or circular), proximity to their parent star ("hot Jupiters"), and so on. Historically, binary stars have often been classified by the method of observation as visual, spectroscopic, eclipsing, and astrometric, or (after more information became known) by the configuration or contents of the system, such as a white dwarf binary, or (after even more information) by the dominant wavelength of its electromagnetic radiation, as in an X-ray binary. While these overlapping systems have served astronomers well, and illustrate how the same object may be classified in many ways, such designations are the source of much confusion among students, not to mention indecipherable to the public.

History also demonstrates that at the time of discovery, by the very nature of the problem, it is sometimes difficult to decide if a new class of object has been discovered. Perhaps by analogy with the Earth's Moon, Galileo decided relatively quickly that the four objects he first saw circling Jupiter were satellites, proof that the Moon was not unique, but a member of a class of circumplanetary objects (even if he did not speak in terms of "class"). But the object he first saw surrounding Saturn was not at all obviously a ring, and awaited the interpretation of Christiaan Huygens more than forty years later. Even in the late twentieth century it was not immediately evident that pulsars were neutron stars, or that quasars were active galactic nuclei, both qualifying in the end for new class status.

Inconsistency notwithstanding, the criterion that astronomers have most often used in the astronomical literature for determining class status – and the one we adopt for the Three Kingdom system – is the physical nature of the object. In the planetary Kingdom, for example, rather than orbital characteristics, the definition of planetary classes in our own solar system has been based on their physical characteristics as rocky, gaseous, or icy in composition; pulsar planets have also been distinguished by being inferred as physically very different again due to the extreme nature of their environment and probable different origin. As we have noted, new classes of planets will undoubtedly be uncovered as observations of extrasolar planets progress, but thus far not enough is known about their physical nature to do so. Many of the extrasolar planets discovered so far are believed to be gas giants; many are close to their stars and thus called "hot Jupiters." The first terrestrial extrasolar planets have also been claimed, in the form of "Super-Earths" and the first rocky transiting system, CoRoT-7b.

This history indicates that a comprehensive classification system for astronomy can perhaps do no better than to use the typological definition of class largely discarded by biologists, as discussed in Section 8.1: "membership in a class is determined strictly on the basis of similarity, that is, on the possession of certain characteristics shared by all and only members of that class. In order to be included in a given class, items must share certain features which are the criteria of membership or, as they are usually called, the 'defining properties.' Members of a class can have more in common than the defining properties, but they need not. These other properties may be variable – an important point in connection with the problem of whether or not classes may have a history."

But what is the unit of classification for astronomy? As we have seen, for physics it is elementary particles. For chemistry it is the elements, defined by atomic number in the periodic table. For biology it is species at the macro level, giving rise to biology's "Five Kingdoms," still favored by some macrobiologists, and genetic sequences of 16S ribosomal RNA at the molecular level, giving rise to Carl Woese's "Three Domains" of Archaea, Bacteria, and Eucarya – favored by most molecular biologists.[65] For astronomy, the unit of classification adopted here is the astronomical object itself, and with some theoretical justification. For as strong and weak forces are dominant in particle physics, and as the electromagnetic force is dominant in chemistry (except for nuclear chemistry), so in astronomy is it the weakest but most far-reaching force of gravity that predominantly acts on and shapes these astronomical objects. Though other considerations such as hydrostatics and gas and radiation pressure come into play, gravity is the determining factor for the structure and organization of planets, stars, and galaxies, their families and classes of objects. To put it another way, the strong interaction holds protons and neutrons together and allows atoms to exist; the electromagnetic interaction holds atoms and molecules together and allows the Earth to exist; and the gravitational interaction holds astronomical bodies together and allows the solar system, stellar systems, and galactic systems to exist.[66] Gravity is thus a prime candidate – the one adopted here – to serve as the chief organizing principle for a comprehensive classification system for all astronomical objects.

Where does such a definition of class lead in the construction of a classification system? In the stellar Kingdom we have seen that stellar spectra were first classified on what turned out to be a temperature sequence (the rearranged Harvard system with its familiar O and B stars and so on), and later on a luminosity scale, the Yerkes/MKK system with its dwarfs, giants, and supergiants. Which to choose to delineate "classes" for stars in a more comprehensive system for astronomical objects? We argue that the Yerkes/MKK system, as a more evolved two-dimensional system based on spectral lines sensitive not only to temperature, but also to surface gravity (g) and luminosity, should be favored. As astronomers Richard Gray and Christopher Corbally recently put it in their magisterial volume *Stellar Spectral Classification* in connection with the luminosity classes, "Stars readily wanted to be grouped according to gravity as well as according to temperature, and this grouping could be done by criteria in their spectra."[67] The resulting luminosity classes (main sequence, subgiant, giant, bright giant, and supergiant labeled from Roman numeral V to I respectively), together with the stellar endpoint classes (supernova, white dwarf, neutron star, and black hole) not only have significance in the evolutionary sequence, but also have a real history of discovery that can be uncovered. W. W. Morgan delineated these luminosity classes to begin with because he realized each grouping of stars formed a sequence of near constant log g (surface gravity).[68] Thus, gravity as a sculpting force for stars was recognized already by the founders of the MKK system as the dominating force for the luminosity classes.

The choice of luminosity for stellar classes does not subordinate the Harvard system of spectral types. As the originators of the Yerkes/MKK system argued, it is simply the case that their system contains more information and better represents the physical nature of stars, as astronomers gradually separated them (during the thirty years from 1910 to 1940) into supergiants, bright giants, giants, subgiants, and main sequence dwarfs. In other words, since 1943 with the Yerkes/MKK system, modern astronomy has a formal two-dimensional temperature-luminosity system with distinct classes, building on the HR diagram, which was literally a two-dimensional plot beginning in 1914, and building on the Mt. Wilson system developed immediately thereafter that Morgan found unsatisfactory because of calibration problems with the dimension of absolute magnitude. Both the Harvard and the Yerkes systems are represented in the full designation of a star, as in Sirius (A1V) as a main sequence star with spectral type A1.

Thus, choices for class status become more clear-cut once there is a guiding principle such as physical meaning, which goes to the heart of Morgan's quest for "the thing itself." Again in the stellar Kingdom, for the interstellar medium instead of "diffuse nebulae" (a morphological classification), classes in the Three

Kingdom system are distinguished according to physical constitution of the nebulae: gas (cool atomic neutral hydrogen, hot ionized hydrogen, and molecular), and dust (reflection nebulae). These categories are used in astronomy and subsume classifications based on morphology that are historically contingent. In the galactic kingdom, galaxy morphology (elliptical, lenticular, spiral, barred spiral, and irregular) also reflects compositional differences (as Morgan's galaxy classification system showed), so the principle of physical meaningfulness still holds.

As in any system, there will be ambiguities. These can be mitigated by a system of classification principles. For the Three Kingdom system these include the following when it comes to the determination of classes and the placement of objects in classes:

(1) Classes are delineated based on the physical nature of the object, defined as physical composition wherever possible.
(2) An object should always be placed in its most specific Class.
(3) To the extent possible, Classes already in use are retained, as in the luminosity Classes of the MK system and the Hubble Classes for galaxies, supplemented by new knowledge.
(4) The recommendations of the International Astronomical Union are followed; e.g., a dwarf planet is not a Class of planet.
(5) Potential, but unverified, Classes are not included.

Appendix 1 is the result of applying these principles to astronomical objects.[69] For those who do not recognize their favorite objects, it is likely because they exist at a taxonomic level below that of Class. The plethora of variable stars, for example, are not classes of objects in this system, on the same level as giants, dwarfs, and so on. Rather, they are Types of these stars that could be elaborated in a more complete system.

As we have emphasized, classification in astronomy has similarities and differences with classification in biology, chemistry, and physics. The most obvious difference between the classes (species) in biology and the classes in astronomy, at least as depicted in our putative Three Kingdom system, is the sheer number of species. E. O. Wilson, the Harvard naturalist who is one of the chroniclers of the diversity of life, has estimated that by 2009, 150 years after Darwin's *Origin of Species*, some 1.8 million species had been discovered and described, out of perhaps tens of millions that now exist. And this does not include what Wilson (in a rare astronomical analogy employed in the domain of biology) calls the "dark matter" of the microscopic universe, which could be tens or hundreds of millions of species of sub-visible organisms.[70]

The number of species or classes in astronomy is obviously put to shame by the effusive and creative diversity of biology, no matter how one defines class or what classification system one uses. In terms of number, astronomy's classes, at least as defined in the Three Kingdom system, are more comparable to elements in chemistry (93 natural and 15 artificial), or to the phyla (32) and classes (90) in just one of Lynn Margulis's Five Kingdoms (*Animalia*) of biology, which contains almost a million species by itself. Any such comparison depends not only on how one defines a class of astronomical objects, but also whether the classes as defined here in the 3K system are really analogous to species in the biological hierarchy of classification, or to elements in the linear classification. That is also a matter of definition, and in part a subjective matter based on relation to higher and lower categories in the system. One can argue whether a giant star of Luminosity Class III in the MK system should be called a class or a type, but one cannot argue that a particular member of the class, a type of giant star such as an RR Lyrae, for example, should be placed at a higher level in the system than the class of which it is a member.

This classification exercise also illustrates a problem that astronomical taxonomy has in common with biological taxonomy: classification characteristics do not necessarily conform to evolutionary relationships. The class of giants as defined by the MKK system definition was not precisely the same as the class of giants that Henry Norris Russell declared about 1910, nor is it entirely coextensive with the evolutionary states of the giant stars as known today. Russell's definition (and the Mt. Wilson system) was based on size and luminosity, as determined by their distances and apparent magnitudes, which could be converted to luminosity. The MKK definition was based on spectroscopy, in particular line ratios defined by standard stars. If an unclassified star matched the standard in a spectroscopic sense, it became a member of that class, such as a giant, without regard to its internal structure or evolutionary status. While luminosities and MK definitions are still used, today astrophysicists often think of giant stars and other stellar classes in terms of their evolutionary state, which for a giant is normally undergoing core helium fusion, but varies depending on the star's mass and where it stands in the spectral temperature sequence. Moreover, a particular class may be adjusted based on new data; in the early 1990s the Hipparcos satellite determined distances ten times more accurate than ground-based parallaxes, and correspondingly more accurate luminosities. The data showed that many of the luminosities were in error, and in the post-Hipparcos era, the modern concept of a giant (core helium fusion with shell hydrogen burning via the CNO cycle) is by no means co-extensive with MK class III defined by spectral line ratios. Nevertheless, the general classes of stars remain, but with a broader definition than determined by the MK system.

In short, astronomical classes have evolved in a way analogous to biology, where "the way it looks" (the phenotype) was primary in the Five Kingdom classification embraced by zoologists, as opposed to the deeper structure based on genetic makeup (the genotype). But whereas in biology Woese's Three Domain system caused an uproar in biology with its finding of a completely new Domain and different relationships for parts of the classification system, the classification of stars by how they physically operate rather than by how they appear has thus far led to broader thinking with only minor adjustments.[71]

Because all classes and classification systems are socially constructed, the Three Kingdom system for astronomy is not the only system that could be proposed. But in the end, like the other classification systems discussed in this chapter, its *raison d'etre* and its staying power are dependent on its accuracy, simplicity, and utility, both in scientific and pedagogical terms. Such features are an asset for astronomical classes and classification systems in general, discussed in the last section.

8.3 Negotiation and Utility in Discovery and Classification

As we have emphasized, the problem of Pluto was one of class, not placement in a classification system for planets, since such a system did not exist at the time of the IAU intervention, and will not exist until more is known about a large enough sample of extrasolar planets. Like Archaeopteryx, Pluto possessed some of the characteristics of two established classes, planets and Trans-Neptunian objects. Rather than dual classification, the IAU opted to define a new class. Clearly, the disputed outcome for Pluto was based on negotiation, and, unusually for science, an official vote. Diverse communities had a stake in the outcome: planetary scientists interested in planetary composition and structure versus dynamicists more interested in their orbital motions; astronomers trying to separate planets from smaller bodies at the lower end of the mass spectrum versus those trying to separate planets from brown dwarfs at the upper end; astronomers with individual and nationalistic agendas, and so on. In an insightful article in *Social Studies of Science*, historian of science Lisa Messeri put these negotiations in the context of what Peter Galison has called "trading zones," an idea taken from anthropological observations of interactions among different cultures. She points out that prior to the Pluto controversy the term "planet" functioned well (was "smoothly traded") among different subdisciplines of astronomy such as dynamicists and planetary scientists, as well as with the public "cultural cosmology." Afterwards, however, the first official definition of planet excluded some of the ideas of the dynamical, planetary science, and cultural views of Pluto, so the trading zones broke down.[72]

Whether or not one adopts this unusual terminology for astronomy, clearly Pluto was not the first to undergo this process of negotiation in the field of astronomy. Nowhere have we seen it more starkly displayed than in the private correspondence between Herschel and Piazzi in the process of negotiating class status for the new object he discovered on the first day of the nineteenth century. A few months after the discovery, with some knowledge of how small the object really is, Herschel writes:

> Moreover, if we were to call it [Ceres] a planet, it would not fill the intermediate space between Mars and Jupiter with proper dignity required for that station. Whereas, in the rank of Asteroids it stands first, and on account of the novelty of the discovery reflects double honour on the present age as well as on Mr. Piazzi who discovered it. I hope you will see the above classification in its proper light, as so far from undervaluing your eminent discovery it places it, in my opinion, in a more exalted station. To be the first who made us acquainted with a new species of primary heavenly bodies is certainly more meritorious than merely to add what, if it were called planet, must stand in a very inferior situation of smallness.[73]

Piazzi fervently wishes his newly discovered object to be a planet; Herschel argues that it would be much better to be a "new species of primary heavenly bodies," and suggests it be identified as a new class of object he dubbed "asteroid." This suggestion stirred considerable controversy. A few weeks after Herschel's suggestion, Heinrich Olbers, the discoverer of Pallas (the second such object), wrote, "I agree with you, honored Sir, in your sagacious suggestion that Ceres and Pallas differ from the true planets in several respects, and the name *asteroid* seems to me to fit these bodies very well." But Carl Friedrich Gauss, who had calculated the orbit of Ceres, disagreed, writing in reply to Olbers that Herschel (1) "doesn't announce it as being a modest proposal, but rather says simply 'I call them,' and (2) that his reason in Ceres' case consists in that it now 'is out of the zodiac.' That shows a very biased and, it seems to me, unphilosophical outlook." Others, including Laplace and Piazzi himself, weighed in on what was both a class and a nomenclature controversy, but in the long run "asteroids" won out on both counts.[74] Small wonder that Pluto, in the much more complex setting of international institutions, became the object of so much controversy two centuries later. And Ceres is now officially a dwarf planet also.

Classes are still created today, not always with such controversy as Pluto. As we saw in Section 2.3, although the separation of the classes of terrestrial and giant planets had already occurred in the mid-seventeenth century in terms of size, the redesignation of giant planets as gas giants in the middle of the twentieth century was ushered in without much ado. And the declaration of a class of ice giant planets came only at the century's very end, tentatively and still without any certainty it will be accepted by the astronomical community. Unlike Pluto, declaring a new class of ice giant planets caused no stir among the public, perhaps because it was too esoteric, perhaps because whereas even schoolchildren knew there were nine planets in the solar system, with Pluto in the exalted position of the most distant, few cared about the much larger Uranus and Neptune as long as they were not "demoted" from planet status, a demotion that would have required changing textbooks.

More recently and at another level, astronomer C. R. O'Dell has stated that it took considerable courage for him to declare, among the first images of Orion from the Hubble Space Telescope in 1992, that he had detected a new class of objects which he believed to be protoplanetary disks. He dubbed them "proplyds," a designation that "the competition" has not yet fully accepted in their interpretation or the nomenclature. Examples could be multiplied, but all these events raise the broad question of how classes and classification systems are negotiated in astronomy. To explore this negotiation, we turn once again to astronomy's major classification systems, those for stars and galaxies.

* * *

How is it that, of all the one-dimensional stellar classification systems devised, the Harvard/Draper system came to be accepted? The classic example of how "classes" were transformed into a classification system, and of negotiating a classification system universally adopted by consensus, is embodied in the events of the first fifty years of stellar spectroscopy, resulting in what is still the most robust and complex classification system in all of astronomy. In particular, David DeVorkin has shown how the "Draper," or "Harvard," system of spectral classification of stars now widely used was accepted through a process of negotiation among astronomers, headed by E. C. Pickering.[75] Within ten years of the beginning of stellar spectroscopy, he demonstrates, there were eight competing systems of stellar classification attempted by four astronomers, including five from Angelo Secchi and those of Hermann Vogel and Norman Lockyer, as well as Edward Pickering's Harvard system. By 1901, there were twenty-three systems. It took fifty years for a negotiated consensus to appear at the meeting of the International Solar Union in 1910, and DeVorkin concludes that by 1921, "mainly through inertia, the Harvard Classification did become the universal standard" – quite different from the declaration of Pluto by official vote. To this day the Henry Draper catalogue of stellar spectra is well known to astronomers and their students, as is its association with Pickering, Annie Jump Cannon, and Wilhelmina Fleming.[76]

It might have been otherwise. In his summary of stellar classification systems up to 1929, the expert stellar spectroscopist Ralph Curtiss remarked that Lockyer's system had been judged too harshly, probably because of its "strange nomenclature and untenable theories." But, he pointed out, the overall sequence in Lockyer's system had much in common with the Cannon system, and moreover had been reached through his search for a temperature sequence based on solar spectra. Like Secchi, Harvard classifiers came to their sequence by classifying spectra according to color and line characteristics. "It may, therefore, be considered accidental that the Harvard order proved to be a temperature sequence of valuable physical significance. If it had not, some modification of Lockyer's classification would probably have been in use today. To Lockyer and his associates we owe a distinct debt for their development and application of temperature criteria for the classification of stellar spectra."[77]

A different scenario played out when the Harvard system was supplemented with its second (luminosity) dimension in the MK system, sometimes simply dubbed the "Yerkes system." Therein lies a tale. For, while Yerkes astronomers, including Director Otto Struve, gave the MKK system a glowing review – pointing to its generality, empirical basis, and its two-dimensional properties in which data were arrayed "not in a linear sequence, but in a rectangular pattern of pigeonholes, like the mailboxes at a post office," the response was much colder from rivals at Mt. Wilson, where two-dimensional classification based on spectroscopic method for determining luminosities had originated with Walter Adams and Alfred Joy in the years around 1917. Commenting on the Morgan, Keenan, and Kellman *Atlas* published in 1943, Joy himself wrote that "in the opinion of the reviewer this nomenclature [for the luminosity classes] adds little in convenience and detracts attention from the physical picture represented by the Russell diagram . . . The new system for designating luminosity . . . merely adds new names for groups previously recognized."[78] From Harvard there was silence. And much later Allan Sandage, Hubble's successor and by then the guardian of Mt. Wilson's reputation, wrote that Morgan and Keenan, noting certain calibration problems with the Mt. Wilson data, took the "drastic step" of abandoning the assignment of numerical values to absolute magnitudes, instead assigning luminosity classes. "These authors simply re-sorted the Mount Wilson main-sequence, subgiant and giant-type stars into discrete classification bins, which they labeled as luminosity classes I to V without calibration (at first) in actual absolute magnitudes. In this way MKK did away with the continuum of the absolute-magnitude variations assigned by Mount Wilson and Victoria, replacing them with two-dimensional adjacent luminosity boxes." In doing so, the MKK system in his view had "taken a step backward and "abandoned the analog continuum of traditional stellar classifications for the digital intervals of luminosity classes."[79]

Sandage's harangue did not end there. "The MKK authors acknowledged their predecessors only minimally, especially concerning the discovery of the spectroscopic method itself for determining luminosities," he wrote. For good measure, he added that the MKK authors had incorrectly attributed proof that the spectral sequence was a temperature sequence to Maury, Cannon, and Lockyer rather than Adams and Hale, his Mt. Wilson predecessors who, in his view, had used sunspot spectra and laboratory data to do so.[80]

And yet it was the MKK system that triumphed. Why? By contrast to the Harvard system, the MKK system was not negotiated but gradually accepted due to its usefulness. As Sandage admitted, "MKK published their summary of the classification problems using a set of photographs. It proved to have a decisive impact because it simplified spectral classification for beginning students. Now, anyone could classify spectra! A visual atlas of star types was far easier to understand than a text description and a table of line ratios." Still, referring to the MK system as "the modified Mount Wilson/Victoria system of luminosity classification," sixty years after the first publication of the MKK system, Sandage regretted that an illustrated approach versus "dry-as-dust academic language" had "gradually blurred the California origins of the two-dimensional classification."[81]

Thus, it was not through negotiation or vote that the MK system triumphed, but through simplicity and utility. As Morgan wrote in 1979, and as remains true today, "The MK system has no authority whatever; it has never been adopted as an official system by the International Astronomical Union – or by any other astronomical organization. Its only authority lies in its usefulness; if it is not useful, it should be abandoned."[82] And more than three decades after the original MKK system was published, he concluded, "there is no doubt that the MK classification system should be described as a preliminary system; by doing so, we increase the possibility of its having a useful life in future decades of astronomical research."[83]

* * *

Negotiation, simplicity, and utility may also be seen at work in the case of galaxy classification, where once again Morgan – ever the classifier – took on a controversial role, this time with very different results. As we have seen, despite rival classification systems from Lundmark, Shapley, and others, it was the Hubble sequence that was accepted, due in part to its simplicity, accuracy, and the authority of Hubble. In the opinion of Sandage there was no question of the early Wolf system becoming the permanent system. Even though it was used into the 1940s, it is now forgotten except by historians, a victim of its lack of physical meaningfulness. The same may be said of Shapley's system, which he himself abandoned. Lundmark's system, on the other hand, seems to have been more a victim of Hubble's priority and authority.[84]

At the same time, as Sandage has emphasized, the Hubble sequence was not immediately accepted without controversy. Some astronomers, including J. H. Reynolds who (as we saw above) had proposed yet another classification system in 1920 prior to even Hubble's earliest, thought the Hubble sequence too simple, despite the fact that Hubble had borrowed some of Reynolds's spiral classes without attribution. While Reynolds praised Hubble for recognizing that not all galaxies were spirals (a still contentious point that Reynolds had broached in 1920), he criticized Hubble's system of early, intermediate, and advanced spirals and barred spirals, later designated as Sa, Sb, Sc and SBa, SBb, and SBc. "This classification of spirals seems to me to be altogether too simple for the great range of types to be found in the many hundreds of examples now known through photographs with large reflectors . . . The problem I have always found in attempting a general classification of spiral nebulae is that one meets case after case where a special class is required for the individual object. Spectral classification of stars is a simple and straightforward matter compared with this."[85] Reynolds and others called for a more complex classification system that included many more galactic features. Hubble replied that any early general classification should be as simple as possible, though indeed a few years after Hubble's death Erik Holmberg and then Sidney van den Bergh refined the Hubble system, the former providing ten spiral types rather than three, the latter adding a galactic luminosity class, not to mention the refined system of Gérard de Vaucouleurs.

The Hubble classification system based on morphology had its limitations, as Morgan, who devised his own rival system, was quick to point out: "There is an important point concerning the usefulness of morphological systems in general: such systems cannot be definitive in nature if the field of specimens is incompletely explored. A good example of this is furnished by attempts to use the classical Hubble form-classification system for the classification of galaxies which were found later to be strong radio sources."[86] Given this truth, it is remarkable that Hubble's system was the one finally adopted, and elaborated in ever more subtle form even today.

With regard to the discovery process, galaxy classification and star classification thus bring home similar lessons. Although one might have thought that classification comes at the end of the discovery process, following detection, interpretation, and at least basic understanding, the historical record is considerably more complex. Classification systems for stars were developed long before there was even a basic understanding of their physical nature, much less an evolutionary sequence. Secchi, Lockyer, Fleming, Cannon, and Maury are not normally considered discoverers but outstanding classifiers. They were classifying spectra whose meaning was not yet known. Yet, their classifications held physical meaning that had to be discovered, as indeed it was beginning with the discovery of giant and dwarf stars of the H-R diagram. These discoveries were then followed by further classification, now of absolute magnitudes and luminosities rather than spectral types. The classification of galaxies and planets follows a similar pattern, displaying both pre-discovery and post-discovery roles.

In all of their classification work astronomers appear to have a clear goal in mind, a driving force behind their classification mania. "The goal of any classification system, either in biology or astronomy, is to reveal underlying physical properties," one astronomer wrote, and furthermore, in the case of galaxies to highlight those properties "which may lead to understanding the formation and evolution of galaxies." In a similar vein, the purpose of classification, Sandage wrote, "is to look for patterns from which hypotheses that connect things and events can be formulated."[87] Thus, whether preceding or following discovery, the Holy Grail of classification is to find evolutionary patterns. And, as we shall see in Chapter 10, in the end discovery and classification in astronomy are synthesized into the ultimate master narrative of the universe – cosmic evolution.

Part IV Drivers of Discovery

9

Technology and Theory as Drivers of Discovery

The history of astronomy has been defined by discoveries made with instruments of ever-increasing power (and cost). The exponential growth in capability of the telescopes has been crucial for maintaining the stream of discoveries in astronomy.

R. D. Ekers and K. I. Kellermann, 2011[1]

It would be a mistake . . . to opt for a technological determinism, according to which, in explaining the "successes" or "failures" of individual astronomers, the only issues in play were the sizes of telescopes available to them.

Robert Smith, 2009[2]

Because instruments determine what can be done, they also determine to some extent what can be thought. Often the instrument provides a possibility; it is an initiator of investigation.

Albert van Helden and Thomas Hankins, 1994[3]

Throughout our discussion of the discovery of new classes of astronomical objects we have seen both technology and theory at play, as well as other factors involving personality, social, and institutional context. The question now arises whether telescopes or theories are the primary drivers of discovery, or, indeed, if either of them are primary drivers given the other factors also in the mix. The question is of more than historical interest, for if discovery is the essence of science, and either technology or theory is the primary driver of discovery, then funding one or the other is a practical policy issue, especially in the era of big science and scarce resources. Indeed, astronomers routinely argue to funding agencies that bigger and more sophisticated telescopes will generate new discoveries, with all the subsequent rewards. But if this is true, does large telescope aperture accompanied by better detectors generate discoveries, or do other factors predominate?

The importance that contemporary astronomers attach to technology is well illustrated in a document compiled by a group of radio astronomers at the turn of the millennium listing more than a dozen discoveries enabled by radio telescopes, resulting in eight Nobel Prize winners. "It is important that those who are in a position to filter research ideas, either as grant or observing time referees, as managers of facilities, or as mentors to young scientists, not dismiss as 'butterfly collecting,' investigations which explore new areas of phase space without having predefined the results they are looking for," they wrote. "Progress must also allow for new discoveries, as well as for the explanation of old discoveries." And, they were quick to add, "New telescopes need to be designed with the flexibility to make new discoveries which will invariably raise new questions and new problems."[4] The astronomer Martin Harwit, in his path-breaking book *Cosmic Discovery*, discussed similar issues about the relation of technology to discovery, concluding that technology was indeed the primary driver of discovery. These claims raise a host of questions having to do not only with theory and technology, but also with funding, social and institutional factors, and the importance of natural history. In short, they highlight the broader context of the drivers of discovery.[5]

In this chapter we first explore the extent to which telescopes and their increasingly sophisticated detectors have been "engines of discovery" of new classes of astronomical objects. We then measure our findings against the degree to which theory and prediction have led to the discovery of new classes. Technology and theory are, of course, not mutually exclusive as drivers of discovery. In the never-ending search for funding, radio astronomers have indeed claimed that discoveries "were the result of the right people being in the right place at the right time using powerful new instruments, which in many cases they had designed and built. They were not as the result of trying to test any particular theoretical model or trying to answer previously posed questions, but they opened whole new areas of exploration and discovery. Rather many important discoveries came from military or communications research; others while looking for something else; and yet others from just looking."[6] Although this perspective is shared by scientists in many other fields, we shall find this is not always so.

Such distinctions between telescopes, techniques, and detectors, and between technology and theory, raise even more questions: if theory does not explicitly lead to discoveries, what role does theory play in discovery as an extended process? To what extent did the opening of the electromagnetic spectrum for astronomy, in other words, radio, infrared, ultraviolet, X-ray and gamma-ray telescopes, spur the discovery of new classes of objects? What was the role of the Space Age in the discovery of such classes? And what was the role of social and institutional factors? While the discovery of new classes of astronomical objects is admittedly only a part of discovery in astronomy, it is an important part, and our analysis will perhaps shed light on the drivers of discovery in other areas of astronomy and even science in general. As with the other parts of this book, by focusing on the discovery of new classes of objects, we utilize a finite database of historical experience from which empirical conclusions can be drawn for this class of discoveries with some assurance, and for other classes of discovery only with cautious extrapolation subject to further research.

9.1 Telescopes as Engines of Discovery

Throughout our discussion of narratives and patterns of astronomical discovery, the importance of technology has been apparent. The improvement of telescopes, the use of new techniques including photography, spectroscopy, and photometry, and the invention of new electronic detectors such as charge-coupled devices (CCDs) has had a demonstrable effect on the discovery of new classes of objects in astronomy. Stars could not begin to be classified before spectroscopy, galaxy classification depended on the new photography, and the definition of a planet was changed when CCD detectors found objects so small that it forced astronomers toward new definitions of the small bodies of the solar system. The idea of technology as an engine of discovery is not new in astronomy – in fact it seems commonsensical – and that role has been recognized in several contexts in the history of science, including elementary particle physics, where particle accelerators are the *sine qua non*.[7] But that does not mean technology is the only, or even the primary, driver of discovery.

Two approaches offer themselves in analyzing the role of technology in astronomical discovery. First, one can look at the development of astronomical technology over the last four centuries and gauge to what extent expectations were met as measured by new discoveries. Put in a more focused way, to a first approximation we can ask to what extent did the largest telescopes in the world discover new classes of objects? Second, one can look at the new classes of astronomical objects and see what role technology played in their discovery. These approaches are complementary; though the latter is more specific to our subject, we first take a look at some technological landmarks in astronomy, the largest telescopes of their era, to test the common-sense hypothesis that the largest telescopes discovered most of the new classes of objects.

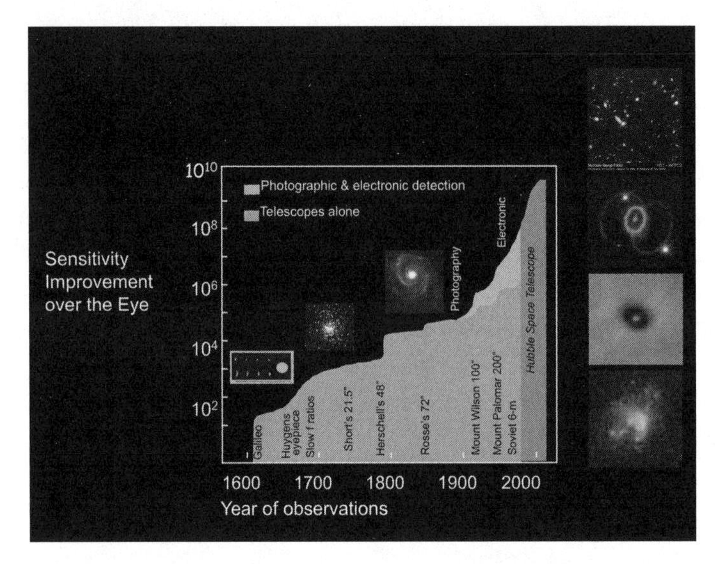

Figure 9.1. Sensitivity improvement of telescopes and detectors over the eye during 400 years. In the first three centuries of the telescope the sensitivity improved about 100,000-fold due to increasingly larger-aperture instruments, with a notable jump at William Herschel's telescopes. Over the next century photography and then electronic devices improved sensitivity another 1000 times. And with the Hubble Space Telescope above the Earth's atmosphere, sensitivity improved yet again by a factor of one hundred. Over the 400 years of the telescope, astronomers and craftsmen have thus improved our ability to see faint objects by a factor of 10 billion. Courtesy NASA, Space Telescope Science Institute. An early version of this plot appeared in Harwit, *Cosmic Discovery*, p. 175.

Figure 9.1, adapted from Martin Harwit's 1980 *Cosmic Discovery* volume and often used by Hubble Space Telescope astronomers to illustrate Hubble's pride of place among telescopes, shows the improvement in sensitivity over the eye that telescopes have enabled, in other words, our ability to detect faint objects over the last four centuries of telescopic astronomy. It shows that in the first three centuries of the telescope the sensitivity improved about 100,000-fold due to increasingly larger-aperture instruments. Over the next century, photography and then electronic devices improved sensitivity another 1000 times. And with the Hubble Space Telescope above the Earth's atmosphere, sensitivity improved yet again by a factor of one hundred. Over the 400 years of the telescope, astronomers (and the craftsmen who built them!) have thus improved our ability to see faint objects by a factor of 10 billion.

Galileo's example would seem to be a promising start to the thesis that technology was paramount to discovery, for as we saw in Chapter 2, within a few months of turning his telescope heavenward in 1609 (Figure 9.2), the nascent

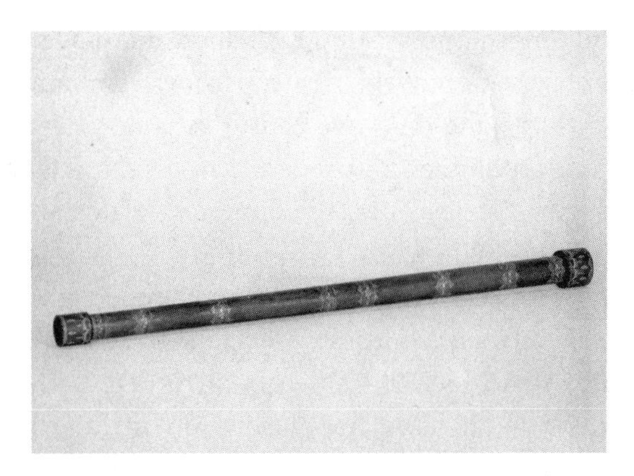

Figure 9.2. Telescopes became engines of discovery in astronomy, beginning with Galileo. The early models, built in 1609 and 1610, magnified several tens of times and had very small fields – about one-third of the Moon. This replica of an early Galilean telescope is held by the Smithsonian Institution; the original is in the Museum of History of Science in Florence, Italy. Smithsonian Institution, courtesy AIP Emilio Segrè Visual Archives.

astronomer had discovered the satellites of Jupiter, planetary rings, and star clusters, unaccountably leaving Nicolas Peiresc (or Johann Cysat) to discover a short time later the first of yet another class of object: the diffuse cloud now known as the Orion Nebula. Less known, as we saw in Chapter 4, is that Galileo and his colleague Benedetto Castelli observed the first double star in 1617, as well as what Galileo thought was a triple star in the Trapezium of Orion. This adds double stars and multiple stars (arguably a separate class) to Galileo's list. *Galileo and his contemporaries can thus be credited with the discovery of six new classes of objects in the first few months and years of the telescope as an astronomical instrument – all with telescopes built by Galileo himself.*

However, as telescope dominance passed to Francesco Fontana in Naples, then to Giovanni Domenico Cassini in Paris in the course of the seventeenth century, and thence to others, the incremental increase of telescope size over the next 150 years did not result in any new classes of objects, but rather added new members to classes already known, particularly the class of planetary satellites.[8] The only new classes added during that time were interplanetary dust, inferred by Cassini and Nicolas Fatio de Dullier in 1683–1684 from its visible effects (the zodiacal light), and the terrestrial planets of our solar system, separated from the giant planets in terms of size by micrometer measures. Neither required large telescopes, nor did the largest telescopes in the decades prior to Herschel, James Short's Gregorian reflectors ranging from 15- to 20-inch apertures, discover any new classes. In 1731 John Bevis, using a simple refractor of 24-foot focal length and about 3-inch aperture, made what we would call a pre-discovery observation of the first supernova remnant, the Crab Nebula, an interpretation that came only two centuries later.[9] Messier rediscovered it in 1758 and gave it the nomenclature of M1 in his famous catalogue, which was compiled with small telescopes, the largest of which was a 7.5-inch reflector with a focal length of 32 inches.

It was only with William Herschel's giant reflectors, which he personally constructed, that new possibilities were opened. His "large" 20-foot focal length telescope with its 18.7-inch (0.47-meter) mirror (Figure 9.3), built in 1783, was his favorite instrument, used for most of his nebular searches. It was soon followed by the giant 40-foot telescope with its 48-inch (1.22-meter) mirror, the largest in the world from 1789 to 1815. As described in Chapters 3 and 5, Herschel can claim to have detected five new classes of objects: planetary nebulae, dark nebulae, globular clusters, double and multiple nebulae. Moreover, he materially advanced the interpretation of two others: the diffuse nebulae first detected by Peiresc in 1610 (or Cysat in 1618), and double stars first detected by Castelli and Galileo in 1617–1618. But it is also notable that planetary nebulae were first detected in 1782, and thus with neither of Herschel's two largest telescopes, although the "large 20-foot" instrument of 18.7-inch aperture helped clinch the case that they were a new class, a "most singular phenomenon" with a star in the center of a nebulous mass. All the other new classes were discovered with the 18.7-inch between 1784 and 1789, and in the end the 40-foot proved a failure, too unwieldy and complex to maintain. Although the large 20-foot reverted to the largest in the world from 1815 to 1826, and was used by John Herschel in the brilliant skies of South Africa from 1834 to 1838, it discovered no new classes of objects.[10] Thus, the 20-foot telescopes were most efficacious in cataloguing nebulae, and discovered new classes only in their earlier work, hinting that new instrumentation may have a limited lifetime in terms of discovery.

Similarly, more than a half-century later William Parsons (Lord Rosse), utilizing his 72-inch "Leviathan" mirror constructed at Birr Castle in Ireland (Figure 9.4), found only one new class, spiral nebulae, almost immediately upon turning the telescope toward the heavens in 1845 (Figure 5.5). He did not, of course, realize their extragalactic nature, nor did he discover any other new classes of objects during the telescope's subsequent three decades of operation ending in 1878. During that time, however, Alvan G. Clark did detect the companion to Sirius (Sirius B) while testing what was then the largest refractor in the world, the 18.5-inch destined for Dearborn Observatory. Such a companion had been predicted by Bessel in 1844 due to unexplained perturbations in the motion of Sirius, but again neither knew of its true nature. Because Clark did

Figure 9.3. Herschel's large 20-foot (18.7-inch aperture) telescope, used for most of his discoveries, became operational in October 1783. The instrument is shown here in February 1794, at its height after improvements to the original. The giant 40-foot giant reflector is better known, but most of Herschel's discoveries of new classes of objects were made with this instrument, still the largest of its time. The tube was made of wood, and the mirror was speculum metal. The instrument is on long-term loan to the National Air and Space Museum in Washington, DC, from the National Maritime Museum, London.

Figure 9.4. Lord Rosse's 72-inch telescope, also known as the "Leviathan of Parsonstown," located at Birr Castle in Ireland. The telescope is hung by chains suspended from two 50-foot stone walls. Begun in 1842 and completed in 1845, with this telescope Rosse almost immediately detected the spiral form of nebulae, later determined to be spiral galaxies. His claimed resolution of the Orion Nebula into stars, however, proved spurious, confused by the stars in the Trapezium. From W. H. Smyth, *The Cycle of Celestial Objects* (London, 1860), p. 167.

Figure 9.5. The Mt. Wilson 100-inch Hooker telescope had a hand in detecting elliptical, lenticular, and irregular nebulae, and in determining that they were galaxies of stars in their own right. The image was taken in May, 1928. Courtesy Rockefeller Archive Center, International Education Board, Caltech.

not recognize it as a new class, it also stands as a pre-discovery observation from that point of view. Clark stands as one of the last of the telescope makers able to claim even such a pre-discovery; the division of labor between astronomers and craftsmen was becoming more sharply drawn.

The situation becomes more complex with twentieth-century telescopes, for it was no longer simply a matter of turning the telescope toward the heavens and seeing something new. The 100-inch Hooker telescope on Mt. Wilson (Figure 9.5), unmatched in its light-gathering power from 1917 to 1948, was essential to Edwin Hubble's ability to detect Cepheid variables in the Andromeda Nebula for the first time, enabling him to determine the extragalactic nature of spiral nebulae, in other words, to further interpret the discovery of the class of objects now known as spiral galaxies, first detected by Lord Rosse three quarters of a century earlier. Cepheids were not new, but Hubble's use of them for this purpose was. Moreover, aided by the new era of astronomical photography represented in Figure 9.1, the 100-inch also had a hand in detecting the new classes of elliptical, lenticular, and irregular galaxies. But there is another factor in the efficacy of the 100-inch not depicted in Figure 9.1: its use for spectroscopy. Again and again during its career as the largest telescope in the world, the Mt. Wilson telescope provided data on distances and luminosities that enabled the discovery of new stellar classes of objects, including giant and dwarf stars, supergiants, subgiants, and subdwarfs. And whether determining distances by trigonometric or spectroscopic parallax, or the temperature sequence for spectra, yet another factor was essential: a string of calibrations, calculations, and inferences from this data.[11]

When the Palomar 200-inch assumed the status of the largest telescope in the world in 1948, it took almost fifteen years before a new class of object was found, this time in the form of the famous quasi-stellar objects, or quasars. We recall the sequence of events from Chapter 5: in late 1962 the Australian radio astronomer Cyril Hazard and his colleagues obtained accurate positions of the strong radio source 3C 273; relatively quickly the optical astronomer Maarten Schmidt obtained a spectrum with the 200-inch of the object identified with the radio source; and in February 1963, Schmidt realized the anomalous spectral lines were highly redshifted, revealing their extragalactic nature. The 200-inch did not make this discovery alone, however; it was the extreme outpourings of radio wave radiation from such an apparently small object that soon made astronomers realize they were dealing with a new class of objects, in fact what turned out to be a whole new Family of objects (active galaxies) being revealed with the further opening of the electromagnetic spectrum for astronomy.

More than thirty years later the 200-inch, equipped with a CCD camera, detected what may be another new class of objects, known as Lyman Alpha Blobs, large concentrations of hydrogen gas in the intergalactic medium (IGM), spanning several hundred thousand light-years and emitting light at the "Lyman alpha" wavelength produced when electrons recombine with ionized hydrogen. Astronomers found the first such blobs serendipitously around 1999 during surveys for so-called Lyman break galaxies. Highlighting the uncertainties of first discovery, the authors of one of the teams, led by Charles Steidel, wrote that "We have discovered two objects, which we call 'blobs,' that are very extended (>15"), diffuse, and luminous Ly alpha nebulae, with many properties similar to the giant Lyman alpha nebulae associated with high-redshift radio galaxies . . . we have considered several explanations for the blobs, with no firm conclusion emerging . . . it is unclear whether or not this phenomenon is unusual." They went on to say that "While we refrain from making too much of this result at present, given the many uncertainties involved, it does suggest that the blobs may represent a different class of rare object that, like QSOs [quasars], are preferentially found in rich environments at high redshift."[12] The true nature of these objects remains unknown despite intense study, but they are likely related to the birth to galaxies.

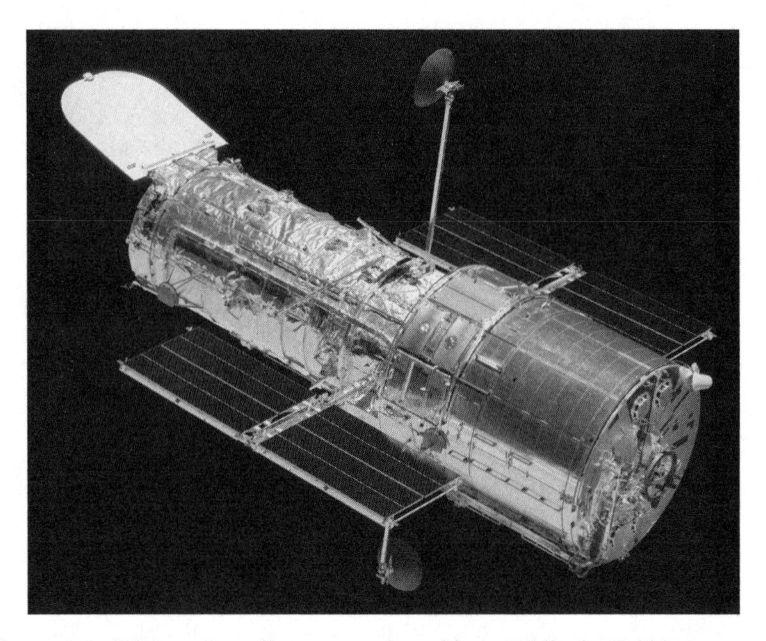

Figure 9.6. "Discovery" applies to more than objects: While the Hubble Space Telescope may claim the detection of only three new classes (protoplanetary systems, galactic black holes, and protogalaxies), it played an important role in discovering the accelerating universe, its age, and aided numerous other discoveries. The telescope is seen here after its release during the last servicing mission in 2009. NASA.

By contrast to Palomar, the Keck 10-meter mirror, the largest in the world from 1992 to 2005, has not as yet discovered any new classes of objects, nor, thus far, has the Large Binocular Telescope with its 12-meter equivalent mirrors, the largest since 2005. And lest we forget the difficulties of these large technologies, the 6-meter reflector in the Russian Caucasus, which technically surpassed Palomar as the largest telescope in the world in 1974, never functioned well enough to do pathbreaking work at all. The fate of Herschel's 40-foot was not only an eighteenth-century phenomenon; technology was always a risky enterprise, as centuries of astronomers can attest through their agonizing experiences of commissioning new telescopes, sometimes successfully and sometimes not. Personal reputations were on the line, as was the ability to make new discoveries.

Finally, at the far end of Figure 9.1, the Hubble Space Telescope (Figure 9.6) stands alone, its 94-inch mirror hardly the largest in the world but having the great advantage of having been lofted above the Earth's atmosphere, with results constituting one of the highlights of the Space Age during its first sixty years. While it has confirmed the accelerating universe, pinpointed its age, provided world-class data and views of a great variety of objects, and constitutes an engine of discovery in many ways, it can claim to have detected only three new classes of objects: protoplanetary systems (proplyds) in 1992, galactic black holes in the early 1990s, and (possibly) protogalaxies in the first decade of the twenty-first century. And the Sloan Digital Sky Survey (SDSS), which through its modest 2.5-meter telescope has provided imagery and redshifts for millions of galaxies, discovered the Sloan Great Wall of galaxies, and mapped vast swaths of the universe in three dimensions, does not claim the discovery of any new classes of objects, even while adding impressively to our knowledge of the nature and extent of the universe.[13]

In summary, while the world's largest telescopes certainly played a role in the detection of new astronomical classes, those classes thus detected constitute fewer than two dozen of the eighty-two classes delineated in Appendix 1, hardly a dominating role. If, as we have argued in Chapter 6, discovery is an extended process consisting of detection, interpretation, and various stages of understanding, the largest telescopes certainly were involved to an even greater extent in the discovery process, if not the initial detection. Moreover, new classes of objects constitute only one of the varieties of discovery made by large telescopes. The 200-inch, for example, in many ways continued the work of the 100-inch, and contributed immensely to cosmology. Already in 1952, Mt. Wilson astronomer Walter Baade used the 200-inch to show that the Andromeda Galaxy was more distant than previously believed, effectively doubling the size of the known universe overnight. In his classic 1961 paper, "The Ability of the 200-inch Telescope to Discriminate Between Selected World Models," Allan Sandage showed how the instrument could be used to determine three key cosmological parameters: the age of the universe, the Hubble constant, and the deceleration parameter. Sandage and his colleagues proceeded to do just that, though with considerable controversy.[14]

Still, it remains true that the initial detection of new classes of astronomical objects based solely on telescope size did not keep pace with the construction of new world-class telescopes. In one sense this is surprising, for not only is sensitivity as depicted in Figure 9.1 dependent on telescope size, so also is angular resolution, the ability to resolve extended objects or systems such as binary stars and globular clusters.[15] Yet this should in no way be construed as a criticism of large telescopes, for other factors are at play. Sensitivity and resolution are only two of the attributes of telescopes, which also include spectral resolution, time resolution, and polarization, quite aside from social factors.[16] And it raises the question of how many more classes of objects remain to be discovered, for the number of such classes is certainly finite, giving rise to the distinct possibility that telescopes already may be bumping up against this finitude.

On the other hand, it is possible that detecting new classes of celestial objects will require some new and currently unimaginable technology.

* * *

How, then, were most of the new classes of astronomical objects discovered, if not by the largest optical telescopes of the world, with the best techniques and detectors? The answer is multifaceted, as we can see by reference to our eighty-two classes delineated in Appendix 1. Some of them, including pulsars and pulsar planets, most active galaxies, and parts of the interstellar medium, were detected as a result of the opening of the electromagnetic spectrum for astronomy, an event of such importance that we reserve it for the next section. Others, including asteroids, Trans-Neptunian objects, and stellar and galactic jets, were detected with telescopes that were not the largest of their time, but that were specialized in other ways. Still others were either *inferred* to be classes only after extended data analysis, including gas giant planets, and dwarf and giant stars. Finally, at least one class was formally declared by vote, the infamous dwarf planets, which (according to the International Astronomical Union) are not planets at all. A few classes, notably energetic particles such as radiation belts, the solar wind, and cosmic rays, were initially discovered by means other than the telescope.

An examination of these modes of discovery is a way of further understanding the strengths and limits of technology as an engine of discovery of new classes of astronomical objects. The detection of the first asteroid is a case in point of a discovery made with a telescope that was one of the smallest, rather than one of the largest, of its time. It is a matter of the historical record that Ceres was discovered, not by Herschel with one of his large telescopes then already available (though he quickly followed up), but with a transit instrument, a very small but stable telescope with an aperture of only a few inches, a standard instrument in virtually all observatories of the time used for accurately determining star positions. It was in the course of observing stars for such positions that Piazzi discovered the uncatalogued object he named Ceres. The detection of other such objects quickly followed, also discovered by relatively small telescopes, though the more versatile equatorial telescopes rather than transit instruments. In all cases, once the hunt was on, modest telescopes were clearly adequate to the task. Nor can we claim that Ceres was discovered after a conscious search; though such searches were under way, Piazzi's discovery was purely serendipitous. Yet, it required the careful measurement of the object's position to detect the motion that set it apart as a class different from the fixed stars, and further observations to separate it from other known classes such as comets.

Similarly, Bevis's lesser-known pre-discovery of the Crab Nebula in 1731 was made with a 3-inch lens; La Caille and Messier's pre-discovery of globular clusters in 1754 was likely made with a 7.5-inch Gregorian speculum reflector, perhaps equivalent to a 3.5-inch refractor; and 1992 QB1, the first unambiguous Trans-Neptunian Object (aside from Pluto), was detected in its eponymous year as a very faint 22nd magnitude object with slow retrograde motion, using the University of Hawaii's 2.2-meter telescope on Mauna Kea.[17] The latter occurred almost a half-century after the Palomar 5-meter came into use, and just before the Keck 10-meter came online. The difference was in part the CCD detector on the 2.2 meter; the bottom line is that if any particular telescope-detector combination can detect an object, any telescope can make a discovery - but only if someone has the motivation, skill, and proper search strategy. With the advance of telescope detector technology, telescopes of relatively small size can (and do) accomplish amazing things.

A case in point involves one of the most spectacular discoveries of recent years: extrasolar planetary systems. For many years, the optimal method for such detection was believed to be the "astrometric" method for detecting the gravitational wobbles they induced in their parent star. But as it turned out, almost all of the more than 1000 planets and thousands of candidate planets found beyond our solar system have been discovered by two methods, Doppler shifts in the radial velocities induced in the star by the planet, and dimming of starlight caused when a planet passes in front of the parent star. These are, respectively, spectroscopic and photometric methods, and most of the discoveries have been made with modest telescopes (though, to be sure the Kepler spacecraft has the advantage of being above the atmosphere). The first unambiguous spectroscopic discovery of an extrasolar planet - 51 Pegasi b - was made in 1995 after eighteen months of precise Doppler measurements with the ELODIE spectrograph on the 1.93-meter telescope of the Observatoire de Haute-Provence; its confirmation, and many more such discoveries, were made with the high-resolution Hamilton spectrometer attached to the 120-inch (3-meter) Shane reflector of the Lick Observatory. The key was not large telescopes (though the Keck 10-meter was later used), but extremely high resolution spectrographs, some built especially for the search, employed on modest-sized telescopes. Most of the photometric discoveries have been made with the Kepler spacecraft, again requiring not large telescope aperture, but high accuracy photometers capable of detecting milli-magnitude changes in light. Astonishingly for one of astronomy's Holy Grails, amateur astronomers can now discover extrasolar planets with their backyard telescopes equipped with CCDs and sensitive photometers.

There is also the historical fact that in retrospect some discoveries could have been made with smaller telescopes, including possibly the Cepheids in the Andromeda Galaxy that lead to the discovery of the extragalactic nature of the spirals and so much else: referring to the 36-inch reflector at Lick Observatory, its former director, C. D. Shane, noted it "would be a close thing for Hubble's Cepheids to have appeared on Crossley plates of the period." But as Robert Smith has pointed out, after H. D. Curtis left Lick in 1921, "no astronomer at Lick in the 1920s particularly wanted to photograph spiral nebulae." As Smith concludes, "while the telescope was at hand at Lick to examine nearby spirals in detail, and perhaps even to have detected Cepheids, the motivation was not present."[18] The same, both in terms of motivation and telescope size, applies to new members of a class; while Asaph Hall used the largest refractor in the world to discover the two moons of Mars, Phobos and Deimos, in 1877, others later saw them with much smaller telescopes, including the 9.6-inch telescope located only a few feet from the 26-inch discovery telescope at the U.S. Naval Observatory. Many who could have looked did not have the motivation – or had not used Hall's method of blocking the light of the planet to reveal any nearby dim objects, another demonstration of the importance of technique at many levels.

In addition to smaller telescopes, another category of discovery stands out. Some new classes of objects were not discovered based on any single observation, but *declared or inferred over an extended period of time, what one might call "discovery by inference" or "discovery by declaration."* Pluto may be considered the most infamous case of discovery by declaration, morphing by decree into a dwarf planet seventy-six years after its discovery, twenty-eight years after its small size was determined with the detection of its moon Charon, and fourteen years after the first discovery of another Trans-Neptunian Object. Even then it required a controversial vote to be officially declared the first in a new class known as "dwarf planet." A formal vote is virtually unique in such matters; a resolution is more often reached by gradual consensus. Terrestrial and giant planets are a case in point; as we saw in Chapter 2, even in the mid-nineteenth century, when Bessel calculated Jupiter's density as 1.35 times the density of water, one-quarter the density of Earth, that average density was no guarantee of the structure of the planet; in the end the term "gas giant" came into use only in the 1950s after incremental evidence showed that a central rocky core must be very small, if indeed it existed at all. And ice giant planets, proposed as a new class only in the late 1990s, are still undergoing this informal process of consensus-building, with some declaring Uranus and Neptune as the first members of this new class, and others still skeptical whether such a separation is necessary or useful.

In a sense, "discovery by declaration," typically by individual declaration or informal consensus, occurs with all classes of objects. Asteroids had to be declared a new class by Herschel and Piazzi, and we have seen how the consensus did not build until many more such objects had been discovered by the mid-nineteenth century. Stellar classes, too, had to be negotiated, formally in the case of the spectral types of the Harvard temperature sequence, and informally in the case of the luminosity classes of the MKK system. Hubble and others had to separate spiral galaxies from other classes of galaxies, at first on the shaky assumption that not every extragalactic object could be resolved into spirals given a large enough telescope, and that not all nebulae could be resolved into stars – the problem that had exercised Herschel and his colleagues in the eighteenth century. Every new class must at some point be declared, often with reputations at stake. Following his detection with the Hubble Space Telescope of circumstellar material in the Orion Nebula, Robert O'Dell stated that it took considerable courage for him to declare a new class of "proplyds" (protoplanetary disks), and that "the competition" has not yet fully accepted their interpretation or the nomenclature.[19]

Finally, a few discoveries were made not with telescopes at all, but with specialized detectors, whether ground-based or space-based. Energetic particles such as solar wind, planetary radiation belts, and cosmic rays are cases in point. We noted in Chapter 7 in connection with "indirect discoveries" that Victor Hess discovered cosmic rays in 1912 after a series of daring balloon flights, which carried not telescopes, but ionization chambers. We have also seen that James van Allen's discovery of the Earth's radiation belts was made with a detector aboard the first U.S. spacecraft, Explorer 1. Van Allen attributed these particles to the solar plasma, but proof of the source remained elusive. Eugene Parker's classic paper in 1958 argued for a "solar wind" of such particles, but its existence was empirically confirmed only by spacecraft in the 1960s, notably Mariner 2, which established the composition, velocity, temperature, and density of the solar wind. Telescopes have since been used to study some of these classes of energetic particles, including arrays of detectors and Cerenkov telescopes such as the High Energy Stereoscopic System (H.E.S.S. in honor of Victor Hess), and (because cosmic rays may produce gamma rays when interacting with gas) the Fermi Gamma Ray Observatory.

All of these factors taken together – large telescopes, specially equipped smaller telescopes, declaration of class by incremental evidence, and even a few non-telescopic discoveries – constitute the largest share of discovery of new objects in the technological realm. But there is one other dominating factor in this realm: the opening of the electromagnetic spectrum for astronomy.

9.2 Discovery and the Opening of the Electromagnetic Spectrum

Thus far we have been focusing on optical astronomy, which for most of history constituted all of the science of the heavens. That would change dramatically in the course of the twentieth century, especially with the advent of high-altitude rocket research and then the Space Age itself, though the potential had been slowly gestating for more than a century. One of the most important developments in the history of science was James Clerk Maxwell's discovery in 1865 that electricity, magnetism, and visible light were manifestations of the same "electromagnetic" phenomena, a discovery he embodied in what soon became known as Maxwell's equations.

Experiments in the early nineteenth century had demonstrated that visible light was only part of a more extended spectrum, though the practitioners of the time had no idea just how extended it would become, and were just beginning to learn of its wavelike character through the experiments of Thomas Young. By placing a thermometer just beyond the red end of the optical spectrum, William Herschel discovered infrared radiation in 1800. Hearing of this observation, the following year the penniless German chemist Johann Wilhelm Ritter found ultraviolet radiation at the violet end of the optical spectrum, made manifest by its discoloration effect on silver chloride. Almost a century passed before Heinrich Hertz produced radio waves via a spark discharge in 1887 through his work at the Technische Hochschule in Karlsruhe, Germany. This was quickly followed by Wilhelm Roentgen's discovery of X-rays in 1895, and Paul Villard's discovery in 1900 of radiation emitted from radium. Ernest Rutherford dubbed the latter gamma rays, a logical extension of the alpha and beta rays he had discovered earlier, but Villard's "radiation" turned out to be particles. These empirical discoveries would barely affect astronomy in the nineteenth century, but – much to the surprise of astronomers – would transform it in the twentieth.[20]

For astronomy, the important implication of these discoveries was that the visible light our eyes and instruments normally detect from the Sun, planets, stars, and other celestial objects might comprise only an extremely tiny part of the radiation they emit. Indeed, we now know that if visible light were an octave on the piano, the rest of the spectrum would comprise at least eighty-one octaves of varying wavelength (Figure 9.7). Put another way, while the wavelength of visible light waves falls between a pinpoint and a bacterium, at the long-wavelength end of the spectrum radio waves can rival the size of a mountain, while at the other end X-rays are the size of an atom, and gamma-rays the size of an atomic nucleus. Stated quantitatively, radio waves may be 10,000 meters or even longer (limited only by the size of the universe), while gamma rays

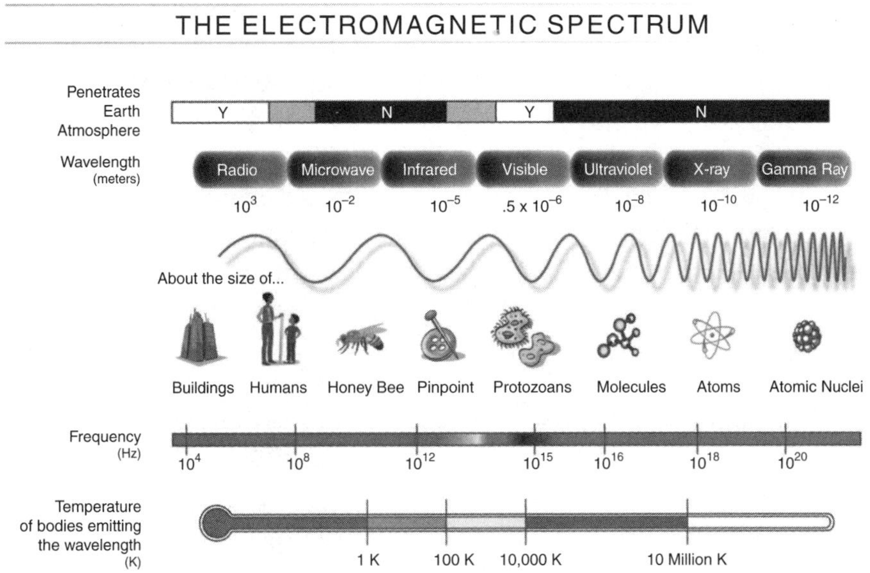

Figure 9.7. The opening of the electromagnetic spectrum proved important for discovering new classes of astronomical objects, as it opened new views of the universe. Radio astronomy can claim to have discovered six new classes of objects over the period 1946–1992, including three classes of active galaxies, giant molecular clouds, pulsars, and pulsar planets. In addition it has assisted in the interpretation and understanding of numerous others. Infrared observations uncovered debris disks around Vega and other Sun-like stars. The opening of the spectrum also aided in the interpretation and understanding of many classes of objects, and led to the discovery of new types of objects below the "class" level in Appendix 1. Courtesy NASA.

may be as small as 10^{-14} meters (limited by the so-called Planck length), a range of 18 orders of magnitude. Light visible to the human eye, presumably having evolved that way because it is an optimal wavelength of sunlight, covers only a range between 380 and 760 nanometers (where one nanometer is 10^{-9} meters), sometimes expressed by astronomers as between 3800 and 7600 Angstroms. It represents the tiniest of slivers of the full electromagnetic spectrum. Ultraviolet radiation, X-rays, and gamma rays represent different wavelengths generated by increasingly energetic processes, X-rays and gamma rays being produced by violent activity involving subatomic particle interactions. Infrared and radio waves, on the other hand, represent more sedate environments – though many wavelengths may be emitted from the same object by many different processes. Little wonder that the universe might look different at varying wavelengths – and that new classes of objects might potentially be discovered.

The discovery of an extended electromagnetic spectrum in the laboratory was one thing, its opening for astronomy was quite another. Despite Herschel's and Ritter's small extensions of the spectrum by measuring the effect of sunlight, it was not at all evident that most astronomical objects would emit at broader wavelengths, or that they could be detected for any but the closest objects even if they did. In order to find out, new techniques had to be developed to detect the generally faint radiations emanating from the depths of outer space. This was made more difficult because, with the exception of radio waves (not even known to exist until the end of the nineteenth century with the work of Hertz), most radiation does not penetrate the Earth's atmosphere, with a few tantalizing exceptions. This situation is good for humans because such penetrating radiation could damage life, as one can see from the sunburn caused by the little UV that does penetrate, not to mention the damage that could be caused by extended exposure to X-rays and gamma rays. But it is bad for astronomy because the opaque windows in the atmosphere necessitate high-altitude balloon flights, high-altitude rocket flights, or, in the ultimate case, instruments lofted entirely above the Earth's atmosphere.

The radiation that did get through the atmosphere was the subject of piecemeal research at first. Following William Herschel's discovery of infrared radiation, some attempts were made at infrared astronomy during the nineteenth century and in the early twentieth century, as when Edison Pettit, Seth Nicholson, and William Coblentz used crude infrared detectors to observe planet or star temperatures. Pioneering astronomers such as Gerard Kuiper extended these researches into the near infrared with detectors developed as a by-product of World War II. Infrared astronomy, however, was largely ignored until the dawn of the Space Age.[21] But as we have seen, a nascent radio astronomy began with the work of Karl Jansky in the 1930s, when he detected the galactic background radiation, and with Grote Reber's maps of celestial radio emission in the 1940s made with a 31-foot telescope dish of his own design (Figure 9.8). This was followed in the late 1940s and 1950s by the first pioneering observations in the ultraviolet and X-ray from rocket flights, including captured German V-2s, as well as Aerobee and Nike sounding rockets.[22] Then, with the beginnings of the Space Age, a flotilla of spacecraft conquered the entire electromagnetic spectrum, including the very energetic gamma ray regime.

To a first approximation the opening of the electromagnetic spectrum, or the "new astronomy" as it has been called in succession to a line of other "new" astronomies, proceeded roughly from the longest wavelengths to the shortest, in other words, from radio to gamma ray. This is no accident in the sense that radio astronomy, and some infrared, could be undertaken from the ground, while most ultraviolet, X-ray, and gamma ray astronomy required high-altitude

Figure 9.8. Grote Reber's radio telescope, constructed in 1937 in his backyard in Wheaton, Illinois. The dish is 31.4 feet in diameter, feeding radio waves to a detector 20 feet above the dish. The telescope has been on display at the National Radio Astronomy Observatory in Green Bank, West Virginia, since the early 1960s. Other types of telescopes have opened other areas of the spectrum for astronomy, revolutionizing the science. NRAO.

flights or spacecraft. We might expect, then, that radio astronomy produced the first discoveries of new classes of objects, and this is indeed the case. As we have seen in Chapter 5, with the opening of the radio spectrum for astronomy, active galaxies - heralded in the optical by the declaration of Seyfert galaxies as a new class in 1943 - were then discovered to consist of a whole family of objects with different apparent characteristics in their radiation at radio wavelengths, including radio galaxies (1946), quasars (1963), and blazars (1972). As the electromagnetic spectrum was opened further for astronomy, these objects were later discovered to radiate also at many other wavelengths, reflecting their internal physical processes. As such, active galaxies stand as the primary example of the effect of the opening of the electromagnetic spectrum on astronomy via new detector technology.

Active galaxies, however, were not the only new classes uncovered by radio astronomy; there were also objects to be found closer to home. The first indication of one such class of objects came when the young engineers Sander

Weinreb, Alan Barrett and their colleagues, using a new "autocorrelator" spectrometer constructed by Weinreb and installed at the 84-foot Millstone Hill radio telescope of the Lincoln Laboratory in Massachusetts, made the discovery of the first interstellar molecules in 1963. The discovery, which came in the form of absorptions lines from the hydroxyl radical (OH) observed in the supernova remnant Cassiopeia A, occurred "almost the first night after we got everything working properly," as Barrett recalled.[23] Such a line had been predicted in the 1950s when Charles Townes and Joseph Shklovskii published their lists of potential spectral lines of possible interest to radio astronomy. The equipment finally caught up with theory, and the discovery was made quickly. The observations opened the way for a whole new field of research on interstellar molecules. But such molecules were not found primarily in already known classes of objects. In 1968, Townes and his colleagues found ammonia (NH_3) emissions toward the center of the Galaxy, followed by water one year later and the first organic molecule, formaldehyde (H_2CO). The discovery of carbon monoxide (CO) emission in 1970 proved especially valuable, because it traces molecular hydrogen, otherwise unobservable in the radio regime at temperatures less than 100 K.[24] Using CO as a tracer, in 1975 N. Z. Scoville and P. M. Solomon, following the first radio survey of CO emission in the galactic plane, reported that a large fraction of interstellar hydrogen is in molecular form. Ever more complex molecules like ethyl alcohol were discovered, and by 1985, sixty-eight interstellar molecules had been reported.[25] These discoveries heralded a new class of object now known as Giant Molecular Clouds, consisting largely of molecular hydrogen but also laced with more complex molecules. Using CO as a tracer, in 1977 astronomers estimated about 3000 Giant Molecular Clouds exist in the Galaxy, with dimensions 30 to 250 light-years.[26] Today a typical spiral galaxy is believed to contain about 1000 to 2000 Giant Molecular Clouds and many more, smaller ones, harboring a considerable amount of the mass of these galaxies, and serving as sites of star formation.

Meanwhile, during the 1960s radio astronomy played a crucial role in the discovery of another new class of object dubbed "pulsars." We need not repeat the story detailed in Chapter 4, but should recall that this discovery too depended on new equipment – an array of 2048 dipole radio antennas covering 4.5 acres, the size of fifty-seven tennis courts. The array was set up to observe the recently discovered class of quasars using the technique of interplanetary scintillation due to plasmas. Because this is most pronounced at long wavelengths, Anthony Hewish and his graduate student Jocelyn Bell and their colleagues optimized their array for a wavelength of 3.7 meters and, crucially, used a short time constant. But instead of finding more quasars, which were by then known to be located far outside our Galaxy, they discovered an entirely new class of object located within our own Galaxy, rotating neutron stars (pulsars), predicted in the 1930s by Walter Baade and Fritz Zwicky, but long forgotten or considered too wild and exotic to actually be produced by nature. But there they were, revealed by radio astronomy in one of the field's most surprising serendipitous discoveries.[27]

Twenty-five years later radio astronomy was to discover one of the most unexpected new classes of objects in the form of pulsar planets. It is true that the existence of pulsar planets was conjectured as far back as 1970, as was the conjecture that such planets would be observable as a change in the frequency of the rotation period of the pulsar. Claims of such observed variation attributed to pulsar planets were also published about the same time.[28] But once again, few believed planets would actually be found in such an exotic and hostile environment as neutron stars. The first confirmed discovery of pulsar planets followed a long period of speculation and spurious claims, including one by Matthew Bailes, Andrew Lyne and colleagues published in *Nature* in 1991 – and a famous retraction at a meeting of the American Astronomical Society at which Lyne was applauded for revealing an honest mistake in the data analysis, the failure to remove the effect of the eccentricity of the Earth's orbit around the Sun. Lyne was part of the Jodrell Bank pulsar group and a director of the famous Jodrell Bank radio telescope, itself a pioneering and controversial landmark of technology associated with the astronomer Bernard Lovell – and with the detection of the first signals from Sputnik in 1957.[29]

Thus, the first confirmed pulsar planets were detected not at Jodrell Bank but during a search for pulsars begun in 1990 with the giant 305-meter radio telescope at Arecibo, Puerto Rico.[30] This telescope, unlike the fully steerable 250-foot Lovell telescope, sits immovable, its structure strung across a natural valley in Puerto Rico, unsteerable except in a very limited way by moving its receiver, a trait only partially compensated by its enormous size. The technique that astronomers Alexandr Wolszczan and Dale Frail used to make the discovery was similar to that of the retracted discovery, and now presented with extreme caution. Since their discovery in 1967, pulsars were known to emit pulsing radio emission with such regularity that they constituted the most precise natural clocks in the universe. Their precise emission is tied to their precise rotation, an astounding 6.22 milliseconds in the case of the planet pulsar, which emits its radio-wave beams like a rotating lighthouse. If a pulsar has one or more orbiting planets, the timing will be off by tiny amounts, ranging from milliseconds for Earth masses to microseconds for asteroid masses, as the pulsar is tugged one way and then another by the planets. Multiple planets can also be separated by this radio-timing technique because they are seen as higher order effects. The pulsar timing method is so powerful that it has been estimated that pulsar planet moons may one day be detected. Although this was believed to be the leading edge of numerous pulsar planets, since the discovery of the first pulsar planetary system, only one other pulsar planet has been confirmed, with 2.5 Jupiter masses in a 100-year orbit around PSR B1620–26 and a companion white dwarf, discovered in 1993 and confirmed by the Hubble Space Telescope in 2003.[31]

Thus, radio astronomy can claim to have discovered six new classes of objects over the period 1946–1992, including three classes of active galaxies, giant molecular clouds, pulsars, and pulsar planets. In addition, it has assisted in the interpretation and understanding of numerous others. For example, as we saw in Section 3.3, although Johannes Hartmann is credited with the discovery of cool atomic clouds in 1904, only gradually were they inferred to consist largely of hydrogen, and only in 1951 did Ewen and Purcell detect the 21cm radiation characteristic of neutral hydrogen, proving the existence of what we now call H I regions. Radio observations have also aided in our understanding of black holes and numerous other objects.

Meanwhile, space-based telescopes were catching up with ground-based astronomy in revealing the extent of electromagnetic radiation from astronomical objects. Some infrared astronomy had taken place from the ground, but it was only with a pioneering infrared satellite that a new class of object was discovered via the infrared.[32] The purpose of the Infrared Astronomical Satellite (IRAS), a joint project of NASA, the Netherlands, and the United Kingdom, was to undertake a survey of the sky in the infrared, a goal that it carried out with complete success – but not without an initial surprise. During its opening calibration tests in 1983, a team of astronomers centered at NASA's Jet Propulsion Laboratory serendipitously found an "infrared excess" around the star Vega, a main sequence star 2.5 times the mass of the Sun but only about 350 million years old. This infrared excess was the thermal signature of the presence of a relatively cool cloud of millimeter-sized solid particles, in this case at a distance of about 85 astronomical units from the star, twice the distance of Pluto from our Sun. At the time the material was interpreted to be the remnant of the cloud out of which Vega had formed. The results, the authors of the discovery paper wrote, "provide the first direct evidence outside of the solar system for the growth of large particles from the residual of the prenatal cloud of gas and dust."[33] Although the scientific paper stopped short of calling the discovery a solar system, a news release from JPL noted that the material could be a solar system at a different stage of development than our own, an interpretation highlighted by the press.

By early 1984, IRAS astronomers had found similar "circumstellar shells" or "protoplanetary disks" around six more stars, including Beta Pictoris,

Fomalhaut, and Epsilon Eridani. By summer of that year, with a total of forty such stars observed, the shells were being reported as a widespread phenomenon. The discoverers in general were careful to emphasize that planets had not been found; instead, "the presumption is that these rings will eventually condense into solar systems like our own; if so, that makes the Vega phenomenon the first semi-direct evidence that planets are indeed common in the universe."[34] We now know that IRAS had found the first direct evidence of debris disks, the material left over after solar systems form, rather than the forming solar systems themselves, discovered a decade later by Robert O'Dell as "proplyds," using the Hubble Space Telescope. Though the initial interpretation proved erroneous, the IRAS finding of debris disks, rather than protoplanetary disks, constituted the discovery of a new class of objects. In general IRAS found that 15 percent of main sequence stars it observed in the solar neighborhood had an infrared excess, and undoubtedly this phenomenon is found throughout the Galaxy.

The environments of some of the IRAS stars have since been observed in much more detail, and illustrate how complex and multifaceted the process of discovery really is. Following up on the IRAS observations, in 1984 astronomers Brad Smith and Richard Terrile used new ground-based CCD technology to image the first debris disk around Beta Pictoris, producing one of the most famous images in the history of twentieth-century astronomy. In a statement that highlights the difficulty of separating protoplanetary disks from debris disks at the time, the authors wrote, "Because the circumstellar material is in the form of a highly flattened disk rather than a spherical shell, it is presumed to be associated with planet formation. It seems likely that the system is relatively young and that planet formation either is occurring now around Beta Pictoris or has recently been completed."[35] Beta Pic, as it is affectionately called, is now categorized as a debris disk rather than a protoplanetary disk, and its planets have even been imaged. Many more disks have been identified, but the identity of many circumstellar disks in terms of their current evolutionary stage is not clear.

Though there have been many more discoveries of infrared radiation from other classes of objects already known, debris disks turn out to be the only new class discovered thus far by infrared observations. Moreover, neither ultraviolet, X-ray, nor gamma-ray satellites have uncovered new classes of objects, despite (among others) the IUE and EUVE spacecraft in the ultraviolet; the Infrared Space Observatory (ISO) and Spitzer Space Telescope in the infrared; Uhuru (1970), HEAO-1 (1977–1979) and HEAO-2 (1978–1981), ROSAT (1990–1999), and Chandra in the X-ray; and OSO-3 (1968), SAS-2 (1972), COS-B (1975–1982), and Fermi in the gamma-ray regime. All of them, however, aided substantially in interpreting and understanding classes of objects already known. Moreover, while they did not discover new objects at the class level, they did discover new types of objects, one level down in the taxonomic hierarchy; for example, not binaries, but X-ray binaries, not supernovae, but X-ray supernovae, not pulsars but X-ray and gamma-ray pulsars such as Geminga, not cosmic rays, but gamma rays caused by cosmic ray interaction with the interstellar medium, and so on. To some extent radio astronomy, with its ground-based head start, had beat other wavelength regimes to the punch. Had X-ray astronomy developed before radio astronomy, it might instead have discovered some of the new classes.

All of this again raises the question whether we have reached the end of discovery in terms of new classes of objects, a question we briefly address at the end of this volume. In any case it is clear that the opening of the electromagnetic spectrum was important for the discovery of new classes, such discoveries often occurring quickly after the new technology was built. It is also clear that the Space Age, in addition to its other beneficial uses such as weather, communications, and navigational satellites, helped open the electromagnetic spectrum for astronomy, with spectacular effect though fewer discoveries of new classes. Even the Hubble Space Telescope, much vaunted for its spectacular imagery in the visible, also provided infrared and ultraviolet data, for its spectral range extended from 1,200 Angstroms in the ultraviolet to 24,000 Angstroms (2.4 microns) in the near infrared. Its successor, the James Webb Space Telescope, will probe even deeper into the infrared, from 0.6 to 28 microns, with some capability from the red to yellow part of the visible spectrum. While Hubble and JWST have revealed, and undoubtedly will reveal, much about cosmology and the early universe, it seems unlikely that they will discover fundamentally new classes of astronomical objects beyond the protoplanetary disks, galactic black holes, and possible proto-galactic clouds already detected by Hubble.

9.3 Theory as Prediction and Explanation in Discovery

The role of theory has long been a perennial theme in science, as well as among its historical, philosophical, and sociological analyzers. Philosophers in the vein of Karl Popper have concentrated on the epistemology of testing theories, arguing they are best judged by their falsifiability rather than their confirmation. Others, such as Norwood Russell Hanson, whose work on the anatomy of discovery helped frame our earlier discussion, have emphasized the role of theory in observation, in particular pointing to the problematic nature of theory-laden observations. Here theory is used in the broader sense of "metaphysically laden" observations. Other philosophers probe the relation of theories to reality, arguing whether they consist of models or represent something closer to the real world. Historians and sociologists each have their own uses for theories and have insisted that the philosophical debates be grounded in real-world science rather than in the formal logic in which they are often couched.[36]

Though fascinating, much of this literature is peripheral to the more specific question we now address: the role of theory in discovery, or more specifically, the role of theory in predicting and driving discoveries in contrast to its role in explaining discoveries. We shall find that, in the extended process of discovery constituted by detection, inference, or declaration, followed by interpretation and understanding, theory historically has been only a weak factor in the prediction of new classes of astronomical objects, and that instead technology has ruled as the driver in the actual detection phase. Theory has been much more successful it its role of explanation in the interpretation and understanding phases of discovery than in the successful prediction for the detection phase.

The problem of whether theory or technology is the most efficient driver of discovery is not just an academic debate, but one that exercises scientists in many fields, often for very personal reasons. Speaking in 1893 of his production of radio waves via the electric spark, the first proof of Maxwell's electromagnetic theory, the experimentalist Heinrich Hertz insisted, "Nor, indeed, do I believe that it would have been possible to arrive at a knowledge of these phenomena by the aid of theory alone. For their appearance upon the scene of our experiments depends not only upon their theoretical possibility, but also upon a special and surprising property of the electric spark which could not be foreseen by any theory."[37] On the other hand, Martin Harwit has recounted the reaction of astronomers at a conference to his conclusion that new instrumentation had proven the best way to make new discoveries. "The audience," he recalled, "got very upset. It was one of the two times I can recall that people literally started screaming in discussions afterwards. Many of them felt discoveries come about through theoretical predictions; those predictions they considered are taken seriously by the observers; the observers go and make measurements and then come back and report to the theorists, who then tell them what to do next." Harwit's event occurred in Holland, where theoreticians had indeed predicted the 21-cm radiation of neutral hydrogen, which observers then proceeded to find after an intense competition described in Chapter 3. Such a course of events, however, appears to be the exception rather than the rule. But Harwit's story does point out that strong opinions exist on both sides when it comes to the drivers of discoveries, which are, after all, the essence of science and the primary goal of its practitioners.[38]

Our discussion of the discovery of new classes of objects in astronomy offers an opportunity to examine the question of the role of theory in a more systematic way, at least for this type of discovery. Our leading question is to what

Figure 9.9. The role of theory: Although Johannes Kepler observed comets, the new star of 1604 and the moons of Jupiter, his life work was as a proto-theorist. Because he never used the telescope in a creative way, he discovered no new classes of objects. Theory, however, plays an important role in the interpretation and understanding of new classes of objects, though not typically in their first detection. This image of Kepler is from an engraving by Jakob von Heyden, 1620/21, based on a painting Kepler gave his friend Matthias Bernegger of Strasbourg in the autumn of 1620. Deutsches Museum, Munich.

extent theory drove discoveries of such new classes; our more focused question is in which component of discovery did theory play a larger role. Is astronomy more like those sciences in which prediction has played a role in discovery – as in chemistry stemming from the periodic table, and physics with respect to elementary particles and the Standard Model – or is it more like biology, in which prediction has played a much smaller role? Put another way, does astronomy more closely resemble the quest for "missing elements" and "missing elementary particles," or the much more general search for "missing links" in an evolutionary scheme? Or none of the above?

If we examine the two great astronomical figures of the early seventeenth century, it should come as no surprise that Galileo the observer trumps Kepler the theorist (to use the term loosely) when it comes to discovering actual objects. Galileo's use of the telescope brought satellites, rings, star clusters, and double stars into the realm of astronomy. Kepler's work, on the other hand, brought the three laws of planetary motion and much else, but no new objects, much less new classes of objects. It is not that Kepler (Figure 9.9) was disinterested in celestial objects; his observations of comets and the "new star" of 1604 attest otherwise. Nor is it the case that Kepler shunned observation in order to theorize in a seventeenth-century ivory tower environment as represented by the Prague castle of his early patron Rudolf II, Emperor of the Holy Roman Empire. To the contrary, it is well known that Kepler coveted Tycho Brahe's observations of Mars, necessary for his work on planetary motion, to the extent that a few daring authors have even suggested (implausibly) that Kepler murdered Tycho in order to obtain them.[39] Rather, one can logically conclude that Kepler discovered no new celestial objects because he never used the telescope in a creative way, despite his *Dissertatio cum Nuncio Sidereo* (Conversation with the Sidereal Messenger) and a brief period from August 30 to September 9, 1610 during which he observed the moons of Jupiter with a Galilean telescope borrowed from the Elector Ernst of Cologne, the Duke of Bavaria.[40] In short, not without reason did Galileo discover new classes of objects rather than Kepler; Galileo was a persistent and skilled observer, Kepler a proto-theorist. And Galileo did not act on any theory; he "simply" pointed his telescope heavenward to see what he could see, even if "seeing" was not always knowing.

Neither the seventeenth nor the eighteenth centuries brought any theoretical predictions of new objects; it is clearly the case that, Kepler notwithstanding, "theory" was not yet a robust activity in astronomy, and even untested theory in the form of hypothesis was shunned, as in Newton's dictum "hypotheses non fingo." No theory guided Charles Messier, Nicolas La Caille, or William Herschel in their reconnaissance of nebulae, nor even in the nineteenth century did it guide Lord Rosse to his spiral nebulae. Those discoveries can be attributed in large part to increasing telescope apertures, along with the social imperatives of discovery such as funding and curiosity. Yet Newton's theory of gravitation had laid the groundwork for prediction, and in fact had been used to predict the return of comets, including what is now known as Halley's comet, which indeed returned in 1758 as predicted, though after Halley's death in 1742.

The first hint of the utility of theory in connection with a new class of objects came with asteroids, but even here it played only a supporting role. At least since the formulation of the Titius-Bode law in 1766 there had been an expectation that a planet might exist in the space between Mars and Jupiter. In 1799, Baron von Zach had organized a coordinated effort to search for such a planet. But in the end the object dubbed Ceres was found serendipitously on January 1, 1801; as we saw in Chapter 2, Piazzi was not part of von Zach's "Celestial Police" effort in this regard. And in any case the Titius-Bode law was only a mathematical progression in planetary distances, not a robust theory. Theory did, however, play a role in the recovery of Ceres when Carl Friedrich Gauss, based on Piazzi's few observations of Ceres over forty-one days and only three degrees of arc, determined an orbit and allowed von Zach to recover the object exactly one year after its initial detection.[41] This feat dramatically demonstrated the utility of Newtonian physics, a demonstration that would be confirmed even more dramatically in the discovery stories of Neptune and (seemingly) Pluto.

Before those discoveries in the solar system, however, gravitational theory proved its usefulness even in the realm of the stars when Friedrich Wilhelm Bessel detected perturbations in the proper motion of Sirius, and announced in 1844 that this could only be due to an unseen object. Less than two decades later, the American telescope-maker Alvan G. Clark proved him right, discovering Sirius B (now known to be a white dwarf) in 1862 while testing the 18.5-inch refractor bound for Dearborn Observatory, then the largest telescope in the world – one of the great testimonies to the power of theory and state-of-the-art technology working together. More precisely, in terms of the structure of discovery, Clark's observation was a pre-discovery observation in the same way there were pre-discovery observations of Uranus and Neptune. We would not say that Clark discovered white dwarfs, any more than seventeenth-century observers discovered Uranus and Neptune. Clark had no idea what he had seen, except a predicted companion. It was only in 1910 that Russell, Pickering, and Fleming *detected* the companion *as a new class of object*, known as a white dwarf, when they determined its extremely low luminosity. The *interpretation* of its physical nature continued with further observations of more white dwarfs. And the mature theoretical *understanding* of white dwarfs as an evolutionary endpoint of low-mass stars like the Sun did not occur until the 1920s and 1930s at the hands of Ralph Fowler and Subrahmanyan Chandrasekhar.[42]

Newtonian physics also played a role in the discovery of various *systems* of objects. The assertion of asteroid groups and meteoroid streams was based on orbit determinations in the mid-nineteenth century. Meteor streams provide the clearer example of the role of theory. We recall that after studying numerous cycles of the Leonids, in 1863 Yale astronomer Hubert A. Newton predicted the November meteors would storm in 1866. When they did, astronomers in four European countries were inspired to solve the riddle of the orbit of the meteor stream and its parent body. The largely independent and almost simultaneous work in 1866–1867 of John Couch Adams in England, Giovanni Schiaparelli in Italy, U. J. J. Le Verrier in France, and Theodor von Oppolzer and C. F. W. Peters in Germany established that orbit. As Adams recounted in 1867, Schiaparelli, the director of the Milan Observatory, showed that the orbits of meteor streams around the Sun are very elongated, as are those of the comets, and that ". . . both these classes of bodies originally come into our system from very distant regions of space." More specifically, in a letter dated December 31, 1866, Schiaparelli remarked on the very close agreement in the orbital elements of the August meteors (now known as the Perseids) and Comet II 1862. To the Italian astronomer thus goes the credit of showing that the comet now known as 109P/Swift-Tuttle is the source of the Perseids. One month later Peters noticed that the orbital elements for the November Leonids agreed closely with Comet I 1866, now known as comet 55P/Tempel-Tuttle, as determined by Oppolzer. Thus, in the space of only one month, the connection we accept today between comets and meteors was established beyond doubt, as was the existence of meteor streams along cometary orbits – and theory via orbit determination had played its role.[43]

The existence of asteroid groups came in a more gradual way, but eventually theory played its role there too. Following Piazzi's discovery of the first object (Ceres), the realization that a "belt" of objects existed between Mars and Jupiter came only gradually as new asteroids were discovered and their distances determined. As early as 1850, an English translation of Alexander von Humboldt's *Kosmos* mentioned in passing the "belt of asteroids," and the idea gradually came into common usage in the 1850s and 1860s, by which time 100 asteroids were known.[44] The popular but erroneous idea that they had originated in the explosion of a planet between Mars and Jupiter hastened their identification as gravitationally connected group.

The realization of more asteroid groups came more than a century after the discovery of the first main belt object. In 1906, the German astronomer Max Wolf discovered 588 Achilles at the L4 Lagrangian point of the Jupiter-Sun system, by definition a gravitational balancing point. This was followed in the next few years by the discovery of 624 Hektor, also at L4, and 617 Patroclus at L5, hinting that this was a real group. Orbital determinations subsequently demonstrated that this was indeed the case. Eleven Trojans were known by 1938, and as of 2010, 2600 were known at L4 and 1500 at L5. The Near-Earth asteroids were the next asteroid group to be discovered, in several subgroups, heralded by 433 Eros, discovered on August 31, 1898. Since then, many more asteroid groups have been discovered, including the Centaurs, their fates in each case bound together by orbital theory. Prediction, however, played little role except in the claim that more asteroids would be found in each group.

Theory therefore first entered the realm of discovery of new objects in the form of Newtonian gravitation, first tentatively with asteroids, followed later in the nineteenth century by white dwarfs, asteroid groups, and meteoroid streams. Its most dramatic demonstration came not with any of these new classes, but with the discovery of Neptune, a new member of the well-known planetary class. And it seemed to be demonstrated once again with the discovery of Pluto. We have seen in Chapter 1, however, that the role of theory in the discovery of Pluto is more complex than the simple story usually told.

It is true that predictions of a new planet beyond Neptune, based on residuals in the motion of Uranus, played a role in motivating the search for such a planet, resulting in the discovery of Pluto. It is now generally agreed, however, that these predictions were a happy accident having little to do with the actual discovery, for Pluto turned out to be much too small to have caused the supposed perturbations. As the astronomer E. W. Brown stated bluntly in May 1930, "The orbit published by the Lowell Observatory for the newly discovered planet shows definitely that it cannot have any connection with that predicted."[45] Writing in his memoir on the fiftieth anniversary of his discovery, Tombaugh himself was clear about his perception of the role of prediction: "The success in finding the new planet was not due to the complex mathematical theory," he wrote, "but to basically simple observational procedure and an enormous amount of painstaking work. Contrary to widespread opinion, the mathematical prediction was of little aid in actually finding the planet, because of earlier negative observational results. Singling out the planet from a sky background, teeming with millions of stars, was like finding a needle somewhere in a large haystack."[46] Tombaugh follows a pattern of discoverers who were loath to assign too much credit to theory.

Smaller Trans-Neptunian objects had also long been hypothesized, though not because of gravitational theory. Theories of planetary accretion, as well as comets and asteroids known to exist in the inner solar system, indicated that similar bodies should be found in the outer solar system. The technology of the time, however, made them very difficult to detect, as the painstaking discovery of Pluto in 1930 proved. Still, both the American astronomer Gerard Kuiper and the Irish astronomer Kenneth Edgeworth noted in the 1940s and 1950s that there was no reason to expect the solar system ended with Neptune or Pluto.[47] But it was advances in technology, in particular charge-coupled device (CCD) detectors, that finally allowed astronomers to image extremely faint objects in the outer solar system. With such a CCD at the focal plane of the University of Hawaii's 2.2-meter telescope on Mauna Kea, in 1992 David Jewitt and Jane Luu spotted the first of these, a 22nd magnitude object dubbed 1992QB1. While this might be considered a case of confirmation of a very general hypothesis, it is hard to credit such a general theory over observations, which would undoubtedly have been made anyway without the theory.

In the realm of the stars, the case of neutron stars provides yet another lesson: theorists may make a prediction that plays no role in the actual detection. This was the case of Walter Baade's and Fritz Zwicky's prediction of neutron stars in 1934, only two years after James Chadwick's discovery of the neutron. Even though the 1968 pulsar discovery paper of Anthony Hewish, Jocelyn Bell and their colleagues cited as a possible explanation a recent theoretical paper by D. W. Meltzer and Kip Thorne about radial pulsations of white dwarfs and neutron stars at the endpoint of stellar evolution, neither that paper nor the discovery paper cited Baade and Zwicky's theoretical prediction made more than three decades earlier. Despite the prediction, astronomers were skeptical that such an object could exist in reality. Prediction therefore did not aid in detection of pulsars, which were discovered serendipitously while searching for the effect of interplanetary scintillation on quasars. But once detected, the earlier prediction did hasten the interpretation of these objects as rotating neutron stars, even though Thomas Gold's 1968 paper identifying neutron stars as the "only theoretically known astronomical object" able to account for such short and accurate periodicities, also did not cite any paper earlier than 1968.[48] The same can be said for black holes, predicted by Oppenheimer and his colleagues in 1939, but too exotic for astronomers to believe, despite Oppenheimer's finding that for a sufficiently massive star, "unless fission due to rotation, the radiation of mass, or the blowing off of mass by radiation, reduce the star's mass to the order of that of the sun, this contraction will continue indefinitely." In other words, a star above a certain mass would collapse completely to an infinitely dense region, what we today term a singularity, or "black hole."[49] Physicists were intrigued, but astronomers aghast, to the extent they listened to physicists at all. Once again, theory did not lead to detection, but hastened interpretation and understanding once the detection had occurred.

This is confirmed by a final example in the form of the galaxies themselves. We recall that in the eighteenth century Kant had said with regard to nebulae, "Here a wide field is open for discovery, for which observation must give the key. The nebulous stars, properly so called, and those about which there is still dispute as to whether they should be so designated, must be examined and tested under the guidance of this theory."[50] Kant's "theory," however, consisted in hypothesizing that different types of nebulae require different explanations, not much of a theory in present-day terms. Though ignored at the time, both Kant and Herschel grasped the basic idea that later became known as the "island universe" theory, which did in fact explain one class of nebulae. The resolution of some nebulae into stars by Lord Rosse in the 1840s tended to support the theory, particularly with the (erroneous) claim of the resolution of the Orion Nebula into stars. But that is a long way from a theoretical prediction. The inside-or-outside our Galaxy "theory" of nebulae was theory only in a very loose, almost colloquial, form – an unfortunate vagueness that still confuses discourse today, as in those who insist that Darwinian evolution by natural selection is "only a theory." By today's standards Kant's conjecture would be called an hypothesis, rather than a well-tested theory. It was not as yet supported by a vast body of evidence. As such it offered no solid prediction.[51]

As we have seen, it was Hubble who finally proved the extragalactic nature of some of the nebulae using Cepheid stars as distance indicators. This too involved no theory, nor did his 1922 and 1926 classification system, the "Hubble sequence" of ellipticals, spirals, and barred spirals, as Hubble himself was at pains to point out. Although his scheme closely followed James Jeans's 1919 sequence based on nebular evolution theory, Hubble held that "a conscious attempt was made to ignore the theory [of Jeans] and arrange the data purely from an observational point of view.[52] In the 1926 paper he continued to insist on his independence from Jeans, while remarking on the suggestive parallels: "Although deliberate effort was made to find a descriptive classification which should be entirely independent of theoretical considerations, the results are almost identical with the path of development derived by Jeans from purely theoretical investigations. The agreement is very suggestive in view of the wide field covered by the data, and Jeans's theory might have been used both to interpret the observations and to guide research. It should be borne in mind, however, that the basis of the classification is descriptive and entirely independent of any theory." Clearly, in Hubble's mind the tuning fork also represented an evolutionary sequence, although he realized the different forms of the ellipticals could be due to projection effects caused by viewing angle.

Perhaps the purest example of prediction leading directly to discovery was the one that exercised Harwit's audience in Holland: the Dutch astronomer Hendrik van de Hulst's prediction in 1944 that neutral hydrogen (H I regions) would emit at a radio wavelength of 21 cm. This led to Harold Ewen and Edward Purcell's discovery of those regions in 1951. Over the next three months two other groups confirmed the 21-cm radiation with their own detections, the Dutch in May and the Australians in June. The American and Dutch results were published in the same issue of *Nature*, as well as a brief note on the Australian results.[53] Similarly, interstellar molecules, predicted in the 1950s by Charles Townes and Joseph Shklovskii, were discovered in 1963, but it was the mid-1970s before astronomers realized they were the trace constituents of Giant Molecular Clouds, which in turn formed a major part of the mass of the Galaxy and were indeed the incubators of the stars.

It is also important to note that while there may have been a general theoretical background behind many discoveries in the twentieth century, rarely did this background actually motivate discovery. Rather, it was the general scientific background against which discoveries were made. The island universe theory, for example, had been around for almost two centuries before Hubble proved its truth. Similarly, the nebular hypothesis had been around for centuries before astronomers found what turned out to be debris disks around stars in 1983, and before Robert O'Dell serendipitously discovered protoplanetary disks in 1992. The idea of a belt of small Trans-Neptunian objects had been around for decades before such objects were actually discovered, and the idea of a more distant "Oort cloud" as a source of long period comets was enunciated more than a half-century ago and still awaits confirmation. Once again, such general "theories" – more appropriately termed untested hypotheses – did not lead directly to the detection of these objects; rather, such detection awaited the technology appropriate to the task.

The role of theory in the prediction of objects that actually led to their discovery is therefore very limited, both from the testimony of the discoverers and by the evidence of history. Although Maxwell's theory did lead Hertz to search for radio waves, he denied it was responsible for his discovery. Although theory led Lowell to instigate a search for a Trans-Neptunian planet, the theory's prediction was flawed, and its discoverer, Clyde Tombaugh, gave no credit to theory as an aid to discovery – a denial widely accepted by both astronomers and historians. And though the Hubble sequence followed the evolutionary sequence of James Jeans, Hubble emphatically distanced both his observations and his sequence from Jeans's theory. Cases could be multiplied, but the conclusion, much to the detriment of theory, is clear for this early stage of discovery. This is true even taking into account that Hubble and Tombaugh perhaps protest too much their distance from theory: Tombaugh would not have found Pluto had Lowell Observatory astronomers not been motivated by a prediction, albeit a bad one. And it seems hard to believe that Hubble could totally segregate his knowledge of Jeans's scheme.

In the next stages of discovery – interpretation and understanding – theory indeed played an extremely important role. This holds in almost whatever class is chosen. No theory was needed for Huygens to interpret Saturn's appendages as rings, but Maxwell's theory of non-solid rotating bodies was required for their understanding as a physically possible rotating system. Similarly, the detection of the classes of stars in the first half of the twentieth century did not involve anything that we would now call theory; rather, the phenomenology of spectral lines and the measurement of distances yielded a Hertzsprung-Russell diagram based almost entirely on observational data – the natural history of the heavens. But the interpretation and understanding of stellar classes certainly did require theory, whether with Saha's theoretical understanding of ionization producing stellar spectra; theories of stellar structure; stellar atomic processes as deduced by Carl von Weisäcker, Hans Bethe, and their colleagues in the late 1930s; or theories of stellar evolution as tracks on the Hertzsprung-Russell diagram became more evident. In the cases of all stellar classes, theory followed discovery as an explainer, not a predictor.[54] The same holds true for nebulae. Theory was the essential ingredient in explaining the physical mechanisms taking place in various forms of nebulae, not in any form of prediction. When in 1926 Donald Menzel and Herman Zanstra independently explained the primary mechanism by which emission lines in a gaseous nebulae are produced as ionization by absorption of stellar ultraviolet radiation followed by recombination – this was by way of interpretation and understanding of objects already detected, not prediction.[55]

Finally, while theory does not often lead to discovery, discovery does sometimes support a very general theory or worldview. Even as he wrote the *Sidereus Nuncius* in February and early March 1610, only weeks after his first observations of the new moons, Galileo was not oblivious to their implications, which at once drew a comparison of the Jovian system of moons and the Earth's Moon, and removed a major argument against the Copernican theory. Two centuries later Herschel's "star gauges," as well as his observations of the distribution of the nebulae, lent support to his idea that the Milky Way was a system of stars, whose form was uncertain but whose structure was nevertheless held together by mutual gravitational attraction. Another century onward, Hubble's proof of the extragalactic nature of spiral galaxies supported the island universe theory of the universe, which was actually more of a cosmological worldview than a robust theory. And in the most general case, all the discoveries we have discussed in this volume have supported the worldview, widely accepted among astronomers only in the last fifty years, that the entire universe and its constituent parts are connected and evolving – in short, the worldview of cosmic evolution. How the synthesis of those discoveries led to the worldview of a dynamic evolving universe is our final subject.

Part V The Synthesis of Discovery

10

Luxuriant Gardens and the Master Narrative

> This method of viewing the heavens seems to throw them into a new kind of light. They now are seen to resemble a luxuriant garden, which contains the greatest variety of productions, in different flourishing beds; and one advantage we may at least reap from it is, that we can, as it were, extend the range of our experience to an immense duration. For, to continue the simile I have borrowed from the vegetable kingdom, is it not almost the same thing, whether we live successively to witness the germination, blooming, foliage, fecundity, fading, withering, and corruption of a plant, or whether a vast number of specimens, selected from every stage through which the plant passes in the course of its existence, be brought at once to our view?
>
> William Herschel, 1789[1]

> Nothing seems to be more important philosophically than the revelation that the evolutionary drive, which has in recent years swept over the whole field of biology, also includes in its sweep the evolution of galaxies and stars, and comets and atoms, and indeed all things material.
>
> Harlow Shapley, 1967[2]

Throughout this volume we have witnessed many narratives of discovery, each groundbreaking in its own way. But one of the greatest achievements of the twentieth century was synthesizing these discoveries into one great master narrative, itself constituting a great discovery. That master narrative is cosmic evolution, the story of the universe from its beginning with the Big Bang 13.7 billion years ago stretching down to the present moment when evolution still continues. The discovery of cosmic evolution was foreshadowed by the luxuriant gardens of William Herschel, who recognized already in the late eighteenth century that individual celestial objects at varying stages of development allow us to "extend the range of our experience to an immense duration." Although he had only the faintest glimpse of the true story, and no idea at all just how immense the duration really was, Herschel was correct that we observe objects in different stages of evolution, stretching back to the beginning of time. In short, not only is the telescope an engine of discovery, it is also a time machine. Because of the finite speed of light, the more distantly we look with our telescopes, the further back we look in time. As Harvard astronomer Harlow Shapley put it in 1967, Darwinian evolution, so controversial among some segments of the population then and now, is seen to be only one small part of the evolutionary narrative in which we all partake. Our exploration in space now takes on the new dimension of an exploration in time.

The evolutionary impulse – part of the penchant of the human mind to fit things into some pattern – has long been active as a survival mechanism among other roles, but has only in the last two centuries achieved a more robust maturity in science. The nebular hypothesis for the origin of the solar system, with a pedigree stretching back to Immanuel Kant and Pierre Simon de Laplace, embraced an evolutionary view of the world that Newtonians had long considered static. Natural history in biology was long seen as a stamp collecting activity even by many of its practitioners until Charles Darwin offered an explanatory framework of evolution by natural selection, the key to explaining the relationship of the parts to the whole. And physicists in the twentieth century, chagrined to find themselves acting as botanists in their attempt to classify a plethora of elementary particles, found solace in insisting that the "continual search for new particles and the structures within them is no mere stamp collecting. Each discovery of a new particle stimulates the human imagination to speculate how it may fit a pattern."[3] The historical interpretation of the universe – embodied today in the concept of "cosmic evolution" – is part of the larger story of the development of the historical interpretation of nature in both the physical and biological sciences.

Just as the patterns in particle physics can only accurately be described by mathematics, and in biology through evolution by natural selection, so classes of astronomical objects may only find their ultimate meaning for the part they play in cosmic evolution, which involves in its deepest sense mathematics and physics, though (so far as we know) no dominant evolutionary principle such as natural selection. In fact we need to use the now-common and canonical term "cosmic evolution" carefully, for "evolution" in astronomy does not imply Darwinian evolution, but development according to the laws of physics, especially gravitational physics, thermodynamics, and hydrostatics. Although a natural selection principle is not at work, those laws are as sweeping in their effect in the physical realm as is natural selection in the biological realm. In short, *the determination of the place of any particular class of object in the scheme of cosmic evolution constitutes the ultimate in mature understanding of the object. That determination is most often a post-discovery activity, occurring decades or even centuries after a basic understanding in terms of physical properties. As such, we suggest it is a step beyond the extended structure of discovery detailed in this volume.*

The full history of the discovery of the relationships among classes of astronomical objects is beyond the scope of this volume. But our examination of the discovery of such classes would be incomplete without a general understanding of what came next, namely, how we came to define the relation of the parts to the whole, an understanding that constitutes what some have called "Genesis for the Third Millennium." That we live in an evolving universe is surely one of the most important and all-encompassing discoveries of the twentieth century, one might say the discovery of all discoveries, making all the more important an understanding of its history. As the discovery not of an object or class of objects, but of the relationship among all objects, it is a different kind of discovery than we have described in most of this volume, but precisely because of that, sheds light on how discoveries can be synthesized and thereby elevated to a new level.

At the same time the discovery of cosmic evolution provides a final opportunity to examine the structure of discovery when the discovery is so all-encompassing as to raise the question of both the nature and the meaning of the universe. We shall find that the discovery of cosmic evolution exhibits some of the same aspects of the extended structure of discovery that we have detailed in the previous chapters. Beyond that, we shall argue that while almost every culture throughout history has had its cosmic worldview, cosmic evolution is the latest, and possibly the greatest, in the sense that it bids fair to actually be "true," or as close as science can come to truth at the present time. As such it may, and should, play a role in culture far beyond science, as have other scientific worldviews in the past.

10.1 Scientific Synthesis: The Discovery of Cosmic Evolution

By what means, with what insights and what motivations did astronomers "discover" the idea of cosmic evolution? Was it, in fact, through the synthesis of many of the discoveries addressed in this volume, or through some other more overarching principle? Does it in fact exhibit the extended structure typical of the classes it embodies, or some different structure? In order to answer these questions, we need to understand that cosmic evolution may be taken in at least three related senses: first, the vast unfolding of cosmic time since the universe began; second, the evolutionary relation among objects in the universe; and third, the evolution of the objects themselves. We can label these, respectively, "universal cosmic evolution" affecting the entire universe at the largest possible scales; "local cosmic evolution," encompassing the relationship among planets, stars, galaxies, and the other classes of objects we have discussed; and "evolution of forms" in terms of individual objects. The three senses are of course related, the vast unfolding of cosmic time providing not only the backdrop but also the necessary and sufficient conditions for the evolution of its constituent parts. In short, local cosmic evolution is the weaving together of the discoveries we have documented in this volume, each with its own evolutionary history (the evolution of forms), and each with a relationship to other classes of objects, but each also a part of universal cosmic evolution. Together, they comprise the master narrative of the universe, in the same way that the evolution of individual forms in biology, and the overall relationship of these forms to each other, provide the master narrative of life on Earth. Whether that master narrative of life on Earth is part of a larger narrative, in which life is common throughout the universe, remains one of the great questions of science. But it is surely the case that biological evolution on Earth, whether rare, unique, or common, is part of the unfolding of cosmic evolution – or so many astronomers have argued over the last half-century.

We argue that, as with the Standard Model in particle physics and evolution by natural selection in biology, there was no one moment when cosmic evolution was "discovered," in either the universal or the local sense. Rather, it was an even more extended process than the discovery of discrete new classes of objects.[4] The Newtonian universe that had ruled since the late seventeenth century was largely static in terms of change over time. Newtonian laws of motion were entirely natural, but in Newton's world God had originally set the particles in motion so as to "produce the harmonious system of mutually adapted structures constituting the present order of nature."[5] Newton's system had no cosmogony and no need for one.

Only in the late eighteenth and early ninteenth centuries, and then haltingly, did this static view of the universe begin to change. For the evolution of forms William Herschel did his part with his observations of the nebulae, which, as we have seen, he believed to be in various states of formation. The French mathematician Pierre Simon Laplace also played his role, with his proposal of the "nebular hypothesis" for the origin of the solar system, whereby planets formed from a rotating gas cloud, providing a link between two previously distinct classes, nebulae, and planets. And the Scottish journalist and intellectual Robert Chambers not only applied evolution to other worlds, but also created a sensation with his anonymous *Vestiges of the Natural History of Creation* (1844), which presented a connected "natural history" of creation from the beginning of the universe to the present - the beginnings of the idea of universal evolution.[6] Later in the century, the philosophy of Herbert Spencer and his followers embraced physical and biological evolution, and even extended it to the evolution of societies. The nebular hypothesis, originally applied to the origin of the solar system, was hijacked to illustrate "general models of the universal progressive development shown in the heavens, on Earth and in human society." By the second half of the century, numerous popular writers had incorporated a sweeping evolutionary epic into their writings.[7] As historian Helge Kragh has concluded, "during the nineteenth century the static clockwork universe of Newtonian mechanics was replaced with an evolutionary worldview. It now became accepted that the world has not always been the same, but is the result of a natural evolution from some previous state probably very different from the present one. Because of the evolution of the world, the future is different from the past - the universe acquired a history."[8]

But the nineteenth century went only so far: "The Victorian conception of the universe was, in a sense, evolutionary, but the evolution was restricted to the constituents of the universe and did not, as in the world models of the twentieth century, cover the universe in its entirety."[9] While popular writers such as Chambers did adopt and spread an evolutionary universe in very general terms, the more scientific idea of universal cosmic development awaited the cosmological worldviews of the early twentieth century. For one thing, the nineteenth century did not as yet have any conception of the vast stretches of time involved, either for the age of the Earth or the age of the universe. In the late nineteenth century, Lord Kelvin was widely criticized for proposing an Earth between 20 and 400 million years old, which seemed to some a huge number, certainly compared to the canonical Christian narrative of Genesis. And although Darwin posited a much longer period necessary for biological evolution, this was unproven, and at the time unprovable, and to his critics represented yet another absurdity of his evolutionary claims. Darwin himself balked at the implied length of the timescale required for natural selection to work.[10]

The problem of age was not resolved until the twentieth century with estimates from radioactivity of the age of the Earth, and various cosmological world models in the case of the universe. Even then, different techniques gave such imprecise results that in 1958, scientific estimates of the age of the universe varied between 13 and 25 billion years. Either way, the universe was conceived as immeasurably older than was thought a century earlier. [11] Only in the late twentieth century did the Hubble Space Telescope narrow the age to within about 1 billion years of its presently accepted value, and it wasn't until the early twenty-first century that the WMAP spacecraft pinned down the number to 13.7 billion years, believed to be accurate within 100 million years. The vast stretches of time evident by the mid-twentieth century gave vast scope for physical evolution, but still did not prove that such evolution had occurred.

Nevertheless, the first three decades of the twentieth century proved crucial for universal and local cosmic evolution, as well as the evolution of forms, which would converge as the master narrative of the universe in the second half of the century. Einstein's theory of general relativity produced the first robust cosmological models, and though Einstein's solution was a static universe (after he fudged a "cosmological constant"), others such as Willem de Sitter and Alexander Friedmann produced solutions indicating an expanding universe.[12] Hubble's publication in 1929 of the velocity-distance relationship gave credence to the idea of fleeing galaxies and an expanding universe, a thought so radical that even Hubble balked at fully embracing it.[13] And in 1931, an obscure Belgian priest, Georges Lemaître, gave the first rendition of what would later become the Big Bang theory.[14]

At the same time, astronomers in the early twentieth century increasingly recognized and advocated the idea of the evolution of celestial forms, mainly in the terms of stellar evolution. Such was the power of the evolutionary impetus inherited from earlier writers that no sooner had stellar spectroscopy sprung up in the late nineteenth century than astronomers began speculating about an evolutionary sequence. The tens of thousands of stellar specimens catalogued in the Harvard spectral sequence were believed almost from the beginning to represent some kind of evolutionary sequence, though (as we have seen in Chapter 4) it took a long time to puzzle out the correct sequence (O, B, A, F, G, K, M) and even longer to determine the underlying physical mechanisms for the stellar classes such as giants and dwarfs. The American astronomer George Ellery Hale's *Study of Stellar Evolution* (1908) spoke only in the most general terms, but Sandage characterized the pathbreaking work of the Mt. Wilson Observatory, founded by Hale, as "breaking the code of cosmic evolution."[15] The scientific basis for stellar evolution was not long in coming: as we have seen, in the decade from 1904 to 1915, the work of Hertzsprung and Russell produced the dichotomy of dwarf and giant stars, and it turned out that the H-R diagram was a veritable roadmap of stellar evolution once enough distances and luminosities were known to populate it not only with giants and dwarfs, but also with subdwarfs, subgiants, supergiants, and white dwarfs. From the 1930s to the 1950s the relationships among these objects were determined, even as Herschel's original evolution of forms of nebulae was parsed into a half-dozen different classes, which in turn were related to the stars.

Meanwhile, the nebular hypothesis, broached in the eighteenth century, had gone through a period of decline in the first half of the twentieth century, due to real and perceived difficulties.[16] The alternative idea that other solar systems were produced by rare stellar collisions implied that solar systems would also be rare. The revival of the nebular hypothesis in the 1940s not only in turn revived the idea of abundant solar systems, but also added one more piece to the evolutionary epic, since the theory not only provided the mechanism for the birth of stars out of rotating gas clouds, but also for the formation of planets as by-products of stellar evolution.[17] Most of the thirty-six classes in the realm of the stars in the Three Kingdom system of Appendix 1 can be tied in some way to stellar evolution, as can many of the classes in the realm of the planets, including not only the planetary classes themselves, but also the leftover debris, including subplanetary, circumplanetary, and interplanetary objects. And, because galaxies consist of stars, an understanding of stellar evolution was also a key to galactic evolution.

Galactic evolution, however, proved more difficult than the sum of its constituent parts. Hubble's "tuning fork" scheme of galaxies (Figure 5.6) turned out not to represent the evolutionary sequence he thought it did. And the formation of any particular galaxy turned out to involve not only gas processes in star formation, but also the central black holes of galaxies, which seemed to be related to galaxy mass. It also involved hierarchical merging of galaxies as evidenced not only by dramatic images of galaxies in collision, but also by dead and dying shredded galaxy streams, a form of cannibalism practiced even today by our own Milky Way Galaxy. And it involved the relationship of the dark matter halo to the stellar halo, bulge, and disk, and ultimately traced back to the irregularities in the cosmic microwave background that provided the seeds for galaxies and clusters of galaxies in the first place. Even with all these clues, galaxy formation is still not well understood and is at the frontier of cosmic evolution research in the sense of the evolution of forms.

Other pieces of the puzzle of cosmic evolution fell into place at intervals. The so-called planetary nebulae first discovered by William Herschel in 1782, for example, remained a mystery well into the twentieth century. In his review of various forms of nebulae in 1933, H. D. Curtis already used the language of cosmic evolution when he wrote, "There seems little doubt that the class of the planetary nebulae must be regarded as an exceptional, and doubtless very rare, branch of cosmical evolutionary development. We must consider them as presumably of catastrophic origin; *lusus naturae* impossible to fit into any orderly fashion within a reasonable gamut of stellar evolution."[18] Because so few planetary nebulae were known (only 150 at the time), Curtis did not believe they could be placed "either at the beginning or the end of the general course of stellar evolution." By 1966, planetary nebulae were seen to be produced by supernovae explosions.[19] The notable feature in both cases is that astronomers were seeking to place a class of objects in the scheme of "stellar evolution" and, more broadly, "cosmical evolution."

This kind of activity is seen in spades during the twentieth century, as astronomers puzzled out the relationships among various classes of objects. In 1933, Curtis was still compelled to admit that "no definite decision can as yet be given as to the precise evolutionary position of the diffuse nebulae." Yet, he could not help trying. Despite the "now discredited" nebular hypothesis by which stars formed out of diffuse nebulae, and admitting that "man's mind demands a logical sequence in such an evolutionary process" even if that sequence may have originated in the mind as a fallacious analogy, Curtis believed that a theory something like the nebular hypothesis (which was soon to be revived) possessed a measure of "sequential harmony" and was to that extent "alluring." He was on the right track. We now believe that stars are born out of molecular clouds, sometimes produce planets out of circumstellar protoplanetary disks, live on the main sequence for millions, billions, or trillions of years, leave it in old age as giants or supergiants, and, depending on their masses, die as supernovae (with accompanying planetary nebulae in some cases), white dwarfs, neutron stars, or black holes. Material from novae and supernovae are recycled into the interstellar medium, from which new generations of stars are born. And on a smaller scale, after being born out of molecular clouds planets also evolve, from amorphous circumstellar material to the terrestrial, gas, or ice giants, and perhaps other forms of planets beyond our solar system. Again, a place in the scheme of local cosmic evolution was seen as the ultimate in the "understanding" component of discovery.

The backdrop to this evolution of forms was universal cosmic evolution. The Big Bang theory had been opposed by the Steady State theory beginning in the 1950s, with Fred Hoyle as its main proponent, but the scientific veracity of the latter gradually declined during the 1960s.[20] With the proof of the Big Bang theory (to the satisfaction of most) through the discovery of the cosmic microwave background in 1965, universal cosmic evolution joined local cosmic evolution to create the dynamically evolving picture of the universe we know today. And in the micro- rather than the macro-direction, by the mid-1950s the chemical evolution of the heavy elements was revealed as nucleosynthesis inside massive stars.[21]

The structure of the discovery of cosmic evolution is thus not only extended, but trifurcated into the separate discoveries of the evolution of forms, the evolution among forms, and universal cosmic evolution, each of which in turn

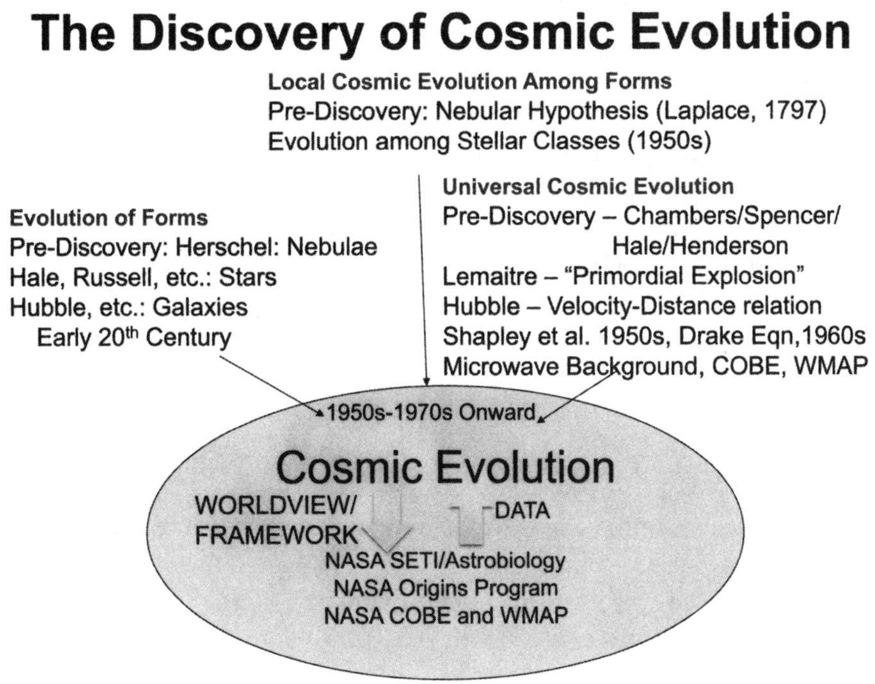

The Discovery of Cosmic Evolution

Local Cosmic Evolution Among Forms
Pre-Discovery: Nebular Hypothesis (Laplace, 1797)
Evolution among Stellar Classes (1950s)

Evolution of Forms
Pre-Discovery: Herschel: Nebulae
Hale, Russell, etc.: Stars
Hubble, etc.: Galaxies
Early 20th Century

Universal Cosmic Evolution
Pre-Discovery – Chambers/Spencer/Hale/Henderson
Lemaitre – "Primordial Explosion"
Hubble – Velocity-Distance relation
Shapley et al. 1950s, Drake Eqn,1960s
Microwave Background, COBE, WMAP

1950s-1970s Onward

Cosmic Evolution
WORLDVIEW/FRAMEWORK
DATA
NASA SETI/Astrobiology
NASA Origins Program
NASA COBE and WMAP

Figure 10.1. The trifurcated structure of the discovery of cosmic evolution. Over the course of several centuries the evolution of forms such as nebulae, stars, and galaxies was uncovered. The relationships among forms were also determined, even as the evolution of the universe as a whole was studied, especially since Lemaître and Hubble in the first half of the twentieth century. Between the 1950s and 1970s all these elements were combined to give the now widely accepted picture known as "cosmic evolution." That worldview provides the framework within which astronomical observations are undertaken in NASA and ground-based research programs, and the results of that research in turn reinforce cosmic evolution, creating an ever more robust picture.

is elaborated in ever more subtle detail (Figure 10.1). Over the course of the twentieth century, the evolution of forms was gradually revealed, the relation of various forms of classes to other classes was determined, and the whole was placed in the context of the unfolding of cosmic time from the Big Bang 13.7 billion years ago. In accordance with the complexity of discovery even (or especially) at this level, no single detection clinched cosmic evolution, no single interpretation revealed it, and our understanding of it is still incomplete. Nevertheless, in stark contrast to a century ago, we have the outlines of what, at present, we take as the "true story" of the universe.

In the fullness of time, we know that, at least on our own planet, life developed, followed by intelligent life and technological civilizations. We do not yet know how often this occurs, but physical, biological, and cultural evolution, incorporating Darwinian evolution and as yet incomplete theories of cultural evolution, rounds out the picture of cosmic evolution as we now understand it. The question remains whether the endpoint of cosmic evolution is commonly planets, stars, and galaxies - the physical universe - or life, mind, and intelligence - the biological universe. In any case this discovery of all discoveries - Genesis for the Third Millennium - is only beginning to seep into cultural worldviews, and bids fair to revolutionize humanity once its implications are fully appreciated.

10.2 Social Integration: The Meaning of Cosmic Evolution

Throughout this volume we have emphasized the nature and structure of discovery, its variety, complexity, and diversity found in the midst of distinct broad patterns. But why are discoveries important? Clearly they are important to scientists, who value them for their own sake, for their potential practical value, and as clues to a larger scientific worldview - the kind of grand synthesis we have seen in the first section of this chapter in the case of astronomy. But discoveries may also acquire a social importance, some more and some less, and nowhere more so than in biology, whose great synthesis is embodied in Darwinian natural selection. Evolution by natural selection, as the Darwin industry has shown us in great detail, is not only a scientific fact, but also part of the social fabric, a driver of considerable controversy over the last 150 years and especially today (unaccountably) in American culture.[22] It imbues life, and individual lives, with meaning that some find uplifting and others degrading. This has been realized for a long time, as exemplified by George Gaylord Simpson's volume *The Meaning of Evolution: A Study of the History of Life and of Its Significance for Man* (1949). Simpson's first "grand lesson" was of the unity of life, the idea that not only are all humans brothers, but all living things are brothers "in the very real, material sense that all have arisen from one source and been developed within the divergent intricacies of one process."[23] Such a vision, which has only been strengthened in the intervening years, has the potential to be transformative in society far beyond the science itself.

Indeed, in the 1960s the leaders of the evolutionary synthesis in biology, including not only Simpson but also Julian Huxley, Theodosius Dobzhansky, and Ernst Mayr, espoused an "evolutionary humanism," a secular progressive vision of the world that, for Huxley at least, was "the central feature of his worldview and of his scientific endeavors." In books and articles, each of

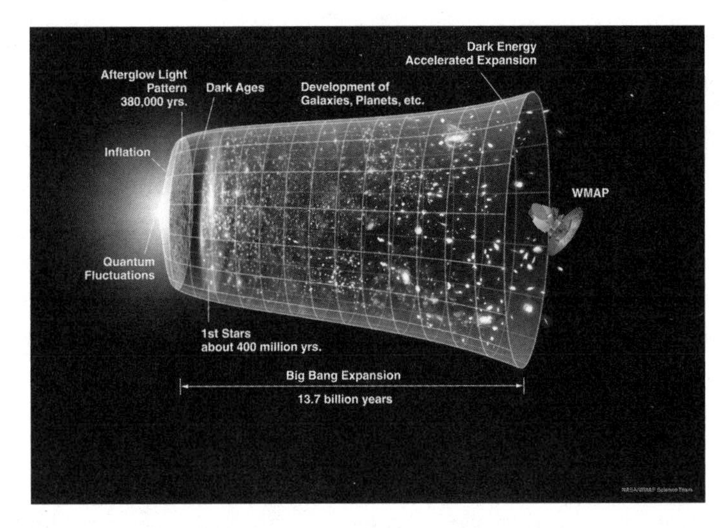

Figure 10.2. Synthesis of discovery: Cosmic evolution as depicted by NASA's Wilkinson Microwave Anisotropy Probe (WMAP) program. Cosmic evolution, the Master Narrative of the Universe, is now believed to have occurred over the last 13.7 billion years plus or minus 100 million years. The current model has the universe beginning with the Big Bang, stars forming within the first few hundred million years, followed by the development of galaxies, planets, and life. The COBE and WMAP spacecraft have observed the radiation remnants of the Big Bang some 380,000 years after its occurrence. In the wake of the Big Bang the universe rapidly expanded, slowed down, and is now accelerating, driven by the mysterious "dark energy." This scientific story of cosmic evolution is increasingly affecting religious and philosophical worldviews. The image was prepared for the public release of three years of WMAP data and was included with a NASA press release on March 16, 2006. Courtesy NASA/WMAP Science Team.

these scientists addressed the future of mankind in evolutionary terms. Huxley (grandson of Darwin's chief defender T. H. Huxley) "offered an inquiry ... into an ethical system, an ethos, grounded in evolution, now a legitimate science, with its fundamental principle of natural selection, verifiable and testable through observation and experiment."[24] Although such a secular worldview does not sit well with many religiously inclined individuals, it must nevertheless be dealt with as part of the larger (and often unnecessarily harsh) dialogue between science and religion.

Cosmic evolution expands this evolutionary worldview into the celestial realm – for the first time placing humans in the context of 13.7 billion years of cosmic history (Figure 10.2). Astronomical discoveries have increasingly had their social impact, as in the public fascination with the imagery from the Hubble Space Telescope on the one hand, and the decision to downgrade Pluto from planet to dwarf planet status on the other. While numerous impacts of astronomy on society could be cited throughout history, they pale in significance to the ever-increasing role of the master narrative of cosmic evolution. Unlike Darwinian evolution, cosmic evolution was not inaugurated or embodied in a single pathbreaking book. Unlike the origin of species, cosmic origins was a stealth revelation whose implications have dawned only gradually and are yet to be fully realized.

But we can confidently predict that a book on *The Meaning of Cosmic Evolution* would cast its shadow far and wide; indeed, the idea has gradually diffused into society to some extent over the past fifty years. Harvard astronomer Harlow Shapley, who has been called a cosmic evolution evangelist, was among the first to realize its scope. He emphasized that in the broadest sense, cosmic evolution integrates humans into the cosmos quite literally. "Mankind is made of star stuff," he wrote in 1963, "ruled by universal laws. The thread of cosmic evolution runs through this history, as through all phases of the universe – the microcosmos of atomic structures, molecular forms, and microscopic organisms, and the macrocosmos of higher organisms, planets, stars, and galaxies. Evolution is still proceeding in galaxies and man – to what end, we can only vaguely surmise."[25] In the 1970s and 1980s Carl Sagan picked up the theme in his books and his wildly popular *Cosmos* television series, viewed by untold millions around the world. The idea that humanity is "at home in the universe," in the felicitous phrase used by several distinguished scientists at the end of the twentieth century, is now common, and made ever more substantial by the increasing knowledge of exactly how humanity and the stars are inextricably linked.[26]

Such a view of our relation to the cosmos naturally has philosophical implications both inside and outside established religions, which have always looked to the heavens for their origins and inspiration. Biologist Ursula Goodenough has argued in *The Sacred Depth of Nature* that cosmic evolution is a shared worldview capable of evoking an abiding religious response. "Any global tradition," she writes, "needs to begin with a shared worldview – a culture-independent, globally accepted consensus as to how things are." She finds this consensus in "our scientific account of Nature, an account that can be called The Epic of Evolution. The Big Bang, the formation of stars and planets, the origin and evolution of life on this planet, the advent of human consciousness and the resultant evolution of cultures – this is the story, the one story, that has the potential to unite us, because it happens to be true."[27] She calls her elaboration of the religious implications "religious naturalism."

Within the Christian tradition, the British biochemist and Anglican priest, Sir Arthur Peacocke, has called cosmic evolution "Genesis for the third millennium." He believes that "any theology - any attempt to relate God to all-that-is - will be moribund and doomed if it does not incorporate this perspective [of cosmic evolution] into its very bloodstream." Christian minister Michael Dowd and science writer Connie Barlow, who also consider themselves "evangelists of cosmic evolution," have proposed "evolutionary Christianity," which embraces what they call "the Great Story" and the "epic of evolution." By now the story has been elaborated by any number of authors, ranging from popular writers to scientists such as E. O. Wilson, Carl Sagan, and Goodenough.[28]

There are cautionary notes to this naturalistic worldview, and not only from the religiously inclined. John Greene, an historian of science who did much to illuminate the history of evolutionary thought in the course of a long career, has reminded us that "the lines between science, ideology, and world view are seldom tightly drawn." Historian John Durant has warned that in the context of the evolutionary worldview, "scientific theology . . . constitutes an attempt to invest the post-Darwinian, scientific world view - and its scientist-custodians - with all the authority traditionally invested in religion and the priesthood." Greene agrees, in no uncertain terms:

> The use of scientific ideas for social and political purposes ranging far beyond the purview of science was not confined to Great Britain or the nineteenth century . . . Ernst Haeckel, Julian Huxley, Ralph Burhoe and Edward O. Wilson all typify the scientist-ideologue bent on saving society by promulgating a new ethics and a new religion claiming the sanction of science. The pattern is depressingly familiar. When will scientists and others learn that naturalism is a philosophical point of view with no more claim to the status of science than any other philosophical viewpoint . . . Scientists have as good a right to expound their philosophical, ethical and religious views as anyone else, but they have no right to palm these off as the findings of science. In doing so they hurt the cause of science by robbing it of such claims to objectivity as it may rightfully assert.[29]

While this is a debate worth having, there is no doubt that cosmic evolution illuminates our place in the cosmos, even in secular terms. It has spawned a movement called "Big History," in which history is taught not as the last 5,000 years of civilization, but as a continuous process extending back to the Big Bang 13.7 billion years ago, emphasizing our origins in stardust. Foreshadowed by Carl Sagan's "cosmic calendar," in which *Homo sapiens* appeared late on December 31 on a scale where the Big Bang occurred one year ago, Big History is gaining broader acceptance in college curricula, along with other cosmic evolution–based courses. Examples could be multiplied of the idea of cosmic evolution being woven into the fabric of society well beyond its scientific content, but enough has been said to indicate the force and reach of the idea and its potential for the future. The ultimate meaning of cosmic evolution is not yet apparent, in part because we do not yet know whether extraterrestrial intelligence exists, or whether there is a deep connection between cosmology and biology.[30]

The discovery of individual classes of objects over the last 400 years of telescopic astronomy, woven together with the discovery of new phenomena such as the expanding and then the accelerating universe, have thus given rise to a cosmic evolutionary worldview in which humans begin to glimpse their true place in the cosmos. This "discovery of discoveries" is likely to become more and more widely influential, perhaps changing the very nature of what it means to be human. We cannot know the end of the story, for it is unfolding with every passing day, as science itself unfolds.

11

The Meaning of Discovery

It would have been fairer, and would send a less distorted message about how this kind of science is actually done, if the [Nobel] award had been made collectively to all members of the two groups [for the discovery of the accelerating universe].

Martin Rees, 2012[1]

Agreeing on what constitutes discovery and assessing the process by which discovery occurs and is accepted are an integral part of accelerating discovery.

Robert Williams[2]

11.1 The Natural History of Discovery

Our natural history of discovery in this volume, analogous to the natural history of the heavens that astronomers have undertaken over the last 400 years, has uncovered some surprising conclusions. Focusing on new classes of astronomical objects – defined in our study by those eighty-two objects listed in Appendices 1 and 2 and as discussed in Chapter 8 – we have found that the discovery of virtually every class of object displays an *extended structure*. Thomas Kuhn and others foreshadowed such a structure already fifty years ago, but unlike Kuhn's anatomy of discovery, which involved anomaly, paradigms, a distinction between normal and revolutionary science, and a subordination of discovery to the concept of revolutions, we place discovery at center stage in science. We find a macrostructure consisting most often of detection, interpretation, and understanding, and a microstructure involving not only the nuanced problems of those three stages and their conceptual content, but also social and technical elements. On occasion the "detection" phase is replaced by an extended "inference" stage (as with gas giant planets, and giant and dwarf stars), itself comprising a chain of reasoning over a long period of time, and necessarily involving interpretation of data *before* the new class is inferred. More rarely, detection or inference are replaced by a formal "declaration" stage, as with Pluto; Pluto was not detected as a dwarf planet, nor even definitively inferred, but was controversially declared to be a dwarf planet by vote of the international astronomical community. Informal declaration of a new class, by the discoverer or someone else, is more common, as in the case of asteroids, spiral nebulae, and Seyfert galaxies. Whether or not it is accepted as a new class depends on community consensus.

These structural elements are present not only in the direct discoveries of the vast majorities of the classes discussed in this volume, but also in the indirect discovery of classes such as cosmic rays, the solar wind, and black holes, seen by their effects. Discoveries of new members of already established classes such as planetary satellites, on the other hand, do not exhibit this structure, or display it in very truncated form, precisely because there is precedent to their discovery and the process of interpretation and understanding has already largely been undertaken earlier. The discovery of new phenomena versus new classes of objects may be considerably more complex, as seen in the case of Hubble's "apparent" velocity-distance relation and its interpretation by others as due to an expanding universe. That discovery was presented and accepted over a few decades; other phenomena, such as cosmic evolution, may have much longer gestation and acceptance periods.

We have found that while no discovery is simple, some are simpler than others when viewed over the extended period of their discovery. Galileo detected, interpreted, and understood – "discovered" – the class of planetary satellites over a period of five days in 1610. They were almost immediately seen as analogous to the Moon. On the other hand, comets, which Tycho demonstrated were celestial objects in 1577, were not understood in the sense of physical composition until the mid-twentieth century. Nor were gas giant planets immediately understood; they were inferred only three centuries after they were first known to be giant planets in the Copernican system. The same can be said for many other classes of objects, ranging from giant and dwarf stars to the many forms of nebulae – including spiral nebulae, which Lord Rosse detected in 1845 but Hubble proved to be extragalactic only in 1924. Techniques such as spectroscopy only made the process of discovery more complicated, interposing more chains of reasoning between the data and the objects themselves. Not surprisingly, the more complex the instrumentation and the more convoluted the chain of reasoning, the more complex the structure of discovery.

Our investigations into the structure of discovery have also led to the concepts of *pre-discovery and post-discovery.* Pre-discovery – casual sightings of astronomical objects with the naked eye, or telescopic observations that go unreported, unrecognized, or undistinguished as new classes of objects – is a concept that applies not only to early observations of meteors, comets, and novae, but also to planets, stars, and galaxies before their classes were distinguished. The realms of the planets and stars were separated in antiquity based on their motion or fixed natures, but their classes were only distinguished after knowledge of their size (in the case of terrestrial versus giant planets) or physical nature (giant and dwarf stars). For those classes inferred rather than directly detected, pre-discovery activities may include a good deal of interpretation of data prior to the discovery of the new class – certainly the case with giant and dwarf stars. The pre-discovery phase also includes theory relevant to the later discovery and classification of phenomenology prior to the discovery of "the thing itself." Altogether, as Appendix 2 indicates, pre-discovery is a surprisingly common phenomenon.

The post-discovery phase, during which the detailed rather than the fundamental properties of the object are discovered, also encompasses its reception among scientists and the public, issues of credit and reward, and the determination of its place in an evolutionary scheme. Pre-discovery and post-discovery bracket the extended process of discovery itself. We have argued that *discoveries end with a basic understanding of the fundamental properties of a class, but before mature understanding, as defined by knowing an object's place in an evolutionary scheme.* Evolutionary understanding with respect to other classes may take decades or centuries (as in planetary nebulae), and the place of all classes in the scheme of cosmic evolution is one of the great achievements of the last half-century of astronomy.

Despite the generally good fit of this scheme of discovery, and its usefulness for analysis, we have argued that we should take care not to shoehorn all discoveries into this structure. There are interesting exceptions, and (as E. C. Pickering is reported to have said with regard to anomalies in astronomy), "it is just such discrepancies which lead to the increase of our knowledge."[3] Occasionally it is difficult to distinguish a pre-discovery stage from the detection of a new class; elliptical galaxies and other nebular forms are good examples. As we have noted, contrary to what one might expect, new classes are not always detected, but sometimes inferred or declared. And the end of discovery can be as indistinct as Peter Galison and others have found to be the case with the end of experiments, in the end a question of community consensus.

Moreover, each of the components of discovery – detection, interpretation, and understanding – has its own gray areas. Again, this is no different from physics, in particular elementary particles, including the Higgs boson, "the most sought-after particle in physics and the key to physicists' explanation of how all particles get their mass." In early 2012, the journal *Science* wrote that scientists at Fermilab reported that "having analyzed all the data they'll ever get, they see hints of the Higgs. The signs are not strong enough to clinch a discovery, but they jibe nicely with hints reported last year by researchers working with Europe's higher-energy Large Hadron Collider."[4] Such "hints" become real discoveries only after sometimes lengthy interpretation and understanding. And even then, consensus is not immediate. The Higgs result, one physicist said, "doesn't make me more convinced, because I'm already convinced . . . But I hope it makes a larger fraction of the audience out there convinced." The "audience" was not the general public but his fellow physicists.

Only a few months later, physicists at CERN's Large Hadron Collider made a more definitive announcement of the Higgs boson, but still laced with caution: "To the layman I now say, I think we have it," said Rolf-Dieter Heuer, director general of CERN . . . we have a discovery. We have discovered a new particle consistent with the Higgs boson. It's a historic milestone today." Two separate experiments at the Large Hadron Collider had indeed detected a signal "just shy of particle physicists' 5-sigma standard for declaring discovery," the journal *Science* reported (where "sigma" represents a measure of uncertainty). While the announcement was met with universal applause, the result was at the lower end of the accepted statistical standard. Others pointed out that it was still unknown whether the new particle was the theorized Higgs of the Standard model, or merely "Higgs-like"; months later *Physics Today* headlined, "The Higgs particle, or something much like it, has been spotted." Properties, such as spin and its interaction with other particles in the manner expected of the theorized Higgs, remained to be determined. Similarly, CERN theorist John Ellis was quoted as saying, "There is no doubt that something very much like the Higgs boson has been discovered."[5] Clearly, the process of detection itself may be extended, mingling with the pre-discovery phase.

Placing the discovery of new classes of objects at the core of our study implies that we have an idea of what it means to be a new class. As we have seen, astronomers have grappled with this problem for centuries, but almost always in the context of their own particular discovery during its extended process, and only very rarely in a more global context, as in Martin Harwit's *Cosmic Discovery*. Unlike Harwit, who defined class as differing in some property by 1000 times or more, *we have advocated the idea of "class" as a taxonomic level in a classification system*, exemplified in Appendix 1. This, we argue, is the optimal way to define "class" most precisely, and thus to delineate new classes of objects as opposed to new types of objects at a taxonomic level below classes, or new families at a level above.

An important characteristic of classes is that their fundamental meaning sometimes evolves, usually over periods exceeding a century, but occasionally over much shorter time scales. The most obvious case involves the discovery that some nebulae are actually located outside of our own galaxy, changing our understanding of their fundamental nature. Thus, with Hubble's announcement in 1925 that spiral nebulae were extragalactic, they became spiral galaxies, eighty years after Lord Rosse first detected them. By analogy, elliptical and irregular nebulae became elliptical and irregular galaxies, in their case only some thirty years after Wolf had detected them as new classes. Simultaneously, the "double nebulae" and multiple nebulae that William Herschel had observed since 1783, and announced in 1811 as physical associations, became double galaxies and multiple galaxies (what we today call galaxy groups). This is to say nothing of the transformation of a non-object (Herschel's "holes in the heavens") into an object composed of dusty clouds (Barnard's dark nebulae). This evolution of our understanding of a particular class of object is not to be confused with the evolution of the object itself, another issue entirely.

Such changes in our understanding, over arcs of history large and small, are caused both by revolutions and (more commonly) by incremental changes in knowledge. Thus, planets were first separated from stars based on their motions against the fixed background. They were then classified as inferior or superior based on their epicyclic motions in the Ptolemaic system, and the terms "inferior" and "superior" were redefined based on their motions in the Copernican system. More important, the latter was a true revolution that made the Earth a planet and the planets potential Earths. Beginning in the seventeenth century, planets were divided into terrestrial and giant planets based solely on their sizes. This classification was in turn gradually refined by 1950 as the giant planets became *gas* giants, and once more by the end of the century as some of the gas giants were seen to be better characterized as *ice* giants. Most astronomers preferred to think of the latter as a new class, rather than as a subclass of giant planets, surely a socially contingent decision.

It should come as no surprise, then, that Pluto should be reclassified from planet to a dwarf planet status as knowledge of its true size and mass became known, and as many more bodies of similar mass were discovered in the outer solar system. The declaration by official vote of members of the International Astronomical Union was indeed unusual, in fact unique, in the annals of astronomical classification, and the incongruous stipulation that a dwarf planet is not a planet was the result of an imperfect community negotiation. The overall reaction revealed both scientific and cultural biases. But the fact of reclassification is not unusual in the history of astronomy, even if our basic understanding of most classes such as satellites, rings, comets, planetary nebulae, and globular clusters, remain stable as their properties are refined.

We have also found that classification plays an important role in all phases of discovery, as well as in pre-discovery and post-discovery. Classification is a component of pre-discovery in terms of phenomenology; it is part and parcel of interpretation and understanding during the discovery process; and (if we accept that discovery indeed ends with a basic understanding of the fundamental properties of a class) classification is also a post-discovery activity in terms of "the thing itself," and as classification systems are continually refined. In other words, *objects are classified according to some observable characteristic, an exercise in "phenomenology"; then discovered during the extended triple process of detection (or inference or declaration), interpretation, and understanding; and finally, classified again in the sense that physically meaningful classes are delineated.*

The stars provide a primary, but far from the only, exemplar of pre-discovery classification of phenomenology and post-discovery classification of the thing itself. They were first distinguished simply by their brightnesses, to which distances were added ever so slowly over time. Spectroscopy entered the scene in the latter half of the nineteenth century, and stars were then distinguished phenomenologically by their spectral lines in many ways, until the Harvard classification system triumphed by its sheer numbers of classified objects and its utility. Only in the first decades of the twentieth century were distances and spectra combined to reveal the existence of dwarfs and giants, and then subdwarfs, subgiants, and supergiants, classes representing the "things in themselves" that could be placed in a scheme of stellar evolution. Hypergiants were added still later. Similarly, the intrepid William Herschel classified the variety of nebulous objects that confronted him only by their appearance; only later, again largely by spectroscopy, were they separated into no less than seven classes, including the gaseous classes of cool atomic clouds (H I regions), hot ionized clouds (H II regions), molecular clouds, and planetary nebulae; the dusty classes of dark nebulae and reflection nebulae; and, last but not least, those objects that turned out to be galaxies.

Finally, while discovery is inevitably socially *influenced*, the concept of class and the development of classification systems are socially *determined*, even while grounded in nature if they are indeed natural systems. It is for this reason, as well as the lack of consistent principles for classification in planetary science, that the Pluto debate was so messy and its conclusion so controversial. There was no "right" answer for Pluto, but some answers might have been better than others if consistent principles and precision of language had been applied. "Dwarf planet" status for Pluto, in which a dwarf planet is not a planet, incorporated neither consistent principles nor precise language. These are among the many lessons astronomers may learn from a long history of classification in biology, which clearly shows the role of community in classification at many levels, from competing comprehensive classification systems between macrobiologists and molecular biologists (Five Kingdoms versus Three Domains), to the classes themselves such as the ambiguities embodied in the controversy over archaeopteryx. Classification in chemistry in the form of the Periodic Table, and in elementary particle physics in the form of the Standard Model, hold similar lessons. Caution is in order, however, in making such comparisons, since there are no transitions between different chemical elements or different subatomic particles - no intermediate states - as there are implied to be between different biological species or different classes of stars.

Because class and classification systems are socially determined - even if they are "natural systems" grounded in nature - it should come as no surprise that controversy has been a constant companion to the discussion of class and classification, whether in astronomy, biology, chemistry, or physics. Thus, from the rings of Saturn in the seventeenth century to Pluto's identity crisis in the twenty-first, controversy is the coin of the realm of discovery. It will likely ever be thus, especially when it comes to class and classification systems that encompass such broad ranges of parameters. If biological classes range from microbes to dinosaurs, astronomical classes range in size from infinitesimal particles like cosmic rays and solar wind to structures that defy the very concept of "object," including superclusters, filaments of galaxies, and their interstitial voids. The difficulties have not kept biologists from devising their classification systems, nor should it deter astronomers, as long as the outcome is pedagogically or scientifically useful.

Finally, we have spoken of the "discovery" of new classes of objects, whereas others might prefer the word "invention." This is much more than semantics. The question of whether an object is discovered or invented goes to the heart of the social construction debate - whether science is an "unveiling" of nature or a human invention of nature. We argue that the discovery of the objects themselves (complex as it is) is an unveiling of nature, while the creation of classification systems is a human invention - not of nature but of the human mind and its way of ordering knowledge. In between is the creation of the classes themselves. Surely the discovery of the majority of new classes of objects such as "rings," "pulsars," and "quasars" is an unveiling of nature's predilection to produce extremely diverse objects. Just as surely, the designation of other objects such as "ice giant planets," "dwarf planets," "bright giant stars," and "lenticular galaxies" - those objects that are declared rather than detected - is less unveiling than invention. Put another way, surely alien minds will have distinguished the former, but not necessarily the latter.

11.2 Beyond Natural History: The Evolution of Discovery

What are the implications of this natural history of discovery? And what is the meaning of discovery thus construed? The implications range from issues of credit, to the drivers of discovery and the evolution of discovery itself, all of intense interest to both discoverers and would-be discoverers. Perhaps most emotional of all from an individual viewpoint, if discovery is an extended process, is the question of who is to be credited with any particular discovery. We have argued that, as Ken Caneva found in physics and Peter Galison in elementary particle physics, discovery in astronomy is best characterized by *collective discovery*. Such a view constitutes a more sophisticated picture of what discovery actually entails as an extended process, and it avoids what are usually counterproductive arguments about the true discoverer. To use only one example we have repeatedly employed in this volume, Galileo detected what we now know to be the rings of Saturn in 1610, Huygens interpreted them as such in 1655, and Maxwell showed how such an object could exist in theory in 1857 – a process encompassing more than two centuries. To say what is often said, that Galileo discovered the rings of Saturn, is to do violence to history, to conflate discovery beyond recognition, and to do a disservice to the beauty and complexity of science and discovery. The same may be said for other classes of astronomical objects.

The concept of collective discovery does not help Nobel Prize committees, but it *would* make them more honest. This is true not only because of the extended nature of discovery over time, but also because of the size of modern research teams. The discovery of the accelerating universe, discussed briefly in Section 7.4, is a prime example. Two rival teams, consisting in total of almost fifty people, were involved in the discovery. By Nobel rules, only three could be selected for the Prize. As British astronomer Sir Martin Rees remarked, "it would have been fairer, and would send a less distorted message about how this kind of science is actually done, if the award had been made collectively to all members of the two groups."[6] Whether in the seventeenth century or the twenty-first, detection, inference, declaration, interpretation, and understanding simply cannot be conflated into a single act of "discovery." The concept of collective discovery, however, is not an easy sell to astronomers, the issue of fairness notwithstanding. At a special session on "Discovery and Classification in Astronomy" at the International Astronomical Union's General Assembly in Beijing, China, in 2012, after an explanation of the idea of collective discovery, astronomers proceeded to argue about who "really" discovered quasars and other new classes of objects. Whether by temperament, training, or human nature, the desire to pinpoint "*the* discoverer" seems to be strongly ingrained among scientists. This, despite heated arguments about credit even in such relatively simple cases such as pulsars, where Anthony Hewish briefly considered suing Fred Hoyle, who claimed Hewish had "filched" the discovery from his graduate student, Jocelyn Bell.[7]

Nevertheless, the idea of collective discovery also sheds light on the public policy question of whether technology or theory drives discovery. The answer of history is clearly that technology drives the detection of new classes of astronomical objects, but theory is essential to their interpretation and understanding. Thus, in the realm of the planets, Galileo's telescope was essential to the detection of what we now know to be rings, but theory was essential to an understanding that they could actually exist, though not in the form of a single solid body. More than a century after Maxwell came to his theoretical conclusion about rings, spacecraft were essential to detecting the properties of the rings, which in turn spurred new theories to explain those properties - both common post-discovery activities. In the realm of the stars, following the phenomenological classification of stellar spectral lines, the delineation of classes of stars in the first decades of the twentieth century depended on distances and spectra - both requiring technological advances - but theory was necessary in the following decades to explain how stars could exist in the form of those classes. And in the realm of the galaxies, though theory had predicted black holes well in advance of the discovery of various classes of active galaxies, that theory played no role in the discovery of those classes, which were distinguished long before their black hole central engines were discovered. These patterns, embracing both observation and theory, repeat again and again in the discovery of new classes of objects over the last four centuries.

Nor is telescope size the only driving factor for discovery; new techniques also play an essential role. Thus, spectroscopy was essential for the discovery of stellar classes such as giants and dwarfs; radio astronomy for radio galaxies, quasars, blazars, and pulsars; various forms of detectors for cosmic rays and the solar wind; and in a few cases (protoplanetary disks and stellar debris disks) space telescopes in the optical or infrared. The first exoplanets were found with relatively small telescopes using new precision radial velocity techniques, and even more were discovered using the photometric techniques of the Kepler spacecraft. Astonishingly, today even small, ground-based telescopes, sometimes run by amateurs, can discover new members of known classes such as exoplanets. Many more techniques are employed in the follow-up observations to the actual detection, in the

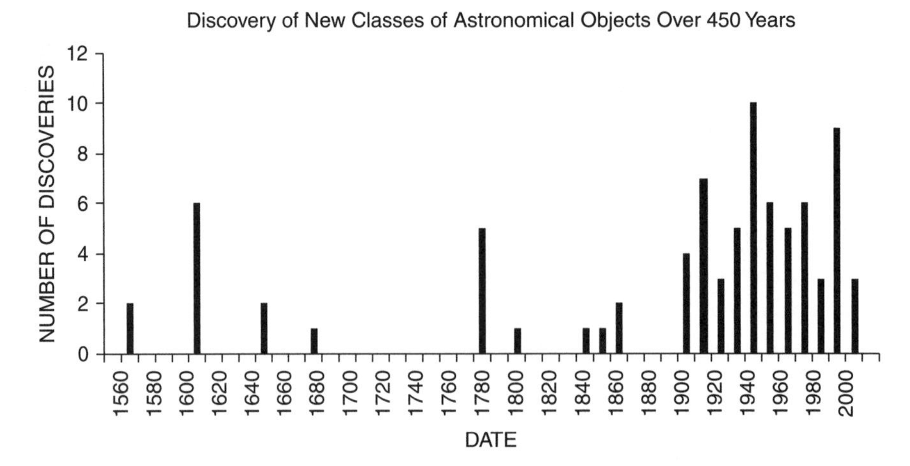

Figure 11.1. Distribution of discoveries of new classes of astronomical objects over the past 450 years. The Galilean and Herschelian peaks are evident around 1610 and the 1780s amid the general desert of discoveries, but the most obvious feature is the mountain of discoveries during the twentieth century. Since discovery is an extended process, what is actually plotted here is the year of first detection, inference, or declaration of each class, as specified in Appendix 2.

processes of interpretation and understanding, and in the post-discovery phase of determining properties of the objects.

Looking back over the last 450 years at the discovery of new classes of astronomical objects as defined in this volume, we can easily see a number of patterns if we plot the date of first detection, inference, or declaration of each classes from the data in Appendix 2. The most obvious feature of the graph (Figure 11.1) is the mountain of discoveries in the twentieth century, three times greater than the sum of the previous 350 years. To put it another way, three-quarters of all discoveries of new classes occurred in the twentieth century. Only a few other peaks stand out as oases, and they are only decadal peaks: the 1610s following Galileo's application of the telescope to the heavens, the 1780s following the construction of William Herschel's large telescopes, and (less noticeably) the 1860s, when both meteors and meteoroid streams were discovered as astronomical phenomena. All else preceding the twentieth century is a virtual desert during which much astronomy was accomplished in the form of theoretical work, data gathering, and the discovery of new types of objects at a level below the class taxonomic level, and therefore by definition of less importance. Notably, to the far left in the histogram are the only pre-telescopic discoveries of new classes: Tycho's determination of the celestial nature of comets and supernovae. What the graph does not show is the grand discovery of cosmic evolution, the combining of classes into the Master Narrative of the Universe that we discussed in Chapter 10. Nor does it show the discovery of cosmic phenomena such as the expanding and accelerating universe. Its strength lies in the systematic definition of classes and the resulting relative homogeneity of the plotted data.

It is true that some of these classes were inferred or declared rather than detected, as was the case of dwarf planets, ice giant planets, and bright giant stars. It is also true that some classes might be questioned, for example, three classes of cosmic rays, one for each Kingdom, even though perhaps justified by their different origin, energy, abundance, and composition – and by aesthetic considerations of symmetry in the Three Kingdom system. And it is true that some classes may turn out to collapse into a single class, as in the case of the active galaxies, which may yet turn out to be the same kind of object viewed from different angles (yet for all that may remain classes if the distinctions prove useful to astronomers). But such declarations or changes in class status will not change the basic form of the histogram. They occur because more information becomes available, never a bad thing. Neither astronomers nor the public should feel cheated when such changes occur; this is a normal part of science.

Finally, our analysis of new classes of astronomical objects informs a question of burning interest to astronomers, if not to historians: Are we at the end of discovery or only the beginning, or somewhere in between? A long history of speculation on this general question has been less than enlightening. One is reminded of the nineteenth-century *fin-de-siecle* declaration that all discoveries in physics had been made, a declaration followed in short order by the revolutionary discoveries of radioactivity, relativity, and quantum physics. Although such speculation is still heard in terms of the aftermath of a unified theory, on general principles it is likely to prove no more accurate than Francis Fukuyama's thesis in *The End of History*.[8]

The question is, however, sharpened when put in terms of new classes of astronomical objects. After all, surely a finite number of such classes must exist. In his book *Cosmic Discovery*, Harwit argued that during the twentieth century astronomical discovery accelerated exponentially, largely due to new technology, and that by 1980 some forty-three cosmic phenomena were known. His definition of discovery included both new classes of objects, defined as those differing in some property by more than 1000 times, and new astronomical phenomena such as the expanding universe. Using a statistical technique, he predicted that the total number of astronomical phenomena is about 130, so that two-thirds remained to be discovered.[9] He envisioned a bell-shaped curve falling off rapidly after the year 2000.

New Zealand astronomer John Hearnshaw, on the other hand, concluded in a much briefer and more cursory study that important astronomical discoveries of the twentieth century, if plotted by decade, showed peaks in the 1920s and 1960s, have declined since 1975, and exhibit only a "dismal number" in the 1990s. His list also included both classes of objects such as quasars and pulsars, and new phenomena, but excluded the solar system. His overall conclusion was that "astronomers are running out of new things to discover," at least things in the same league as the top fifty discoveries of the twentieth century.[10]

We must emphasize several things about these studies, tentative as they are. First, the answer in each case is highly sensitive to the definition of "important discovery," and to the definition of "class." Second, the studies do not take into account the extended nature of discovery. These shortcomings are addressed in some measure by using the classes systematically defined in the Three Kingdom system. We see in Figure 11.1 that there has been no decline of discoveries in the decade of the 1990s, which indeed stands out for its discovery of nine new classes, including such well-known ones as exoplanetary systems, pulsar planets, protoplanetary systems, and the first Kuiper Belt objects. On the other hand, the first decade of the twenty-first century can boast only two new classes (Oort cloud objects and protogalactic clouds) having been discovered, and one (dwarf planets) declared.

We hesitate to repeat the mistake of those in the nineteenth century who proclaimed the end of physics, in particular with the mysteries (we prefer not to say "discoveries") of dark matter, dark energy, and talk of a possible multiverse. It is too soon to come to such an important conclusion based on only the data of the last decade, especially considering the graphical mountain of discoveries in the twentieth century – and the fact that the low number of discoveries in the 1920s, for example, presaged the largest number of discoveries in astronomical history. The argument that there can only be a finite number of classes is compelling from a common-sense point of view, but whether we are entering a temporary valley, or whether the next decades will feature new peaks, only the future will tell. It is difficult to see how a new Kingdom could be discovered, unless it is the universe itself as part of the conjectured multiverse. On the other hand, it is easy to see how many more types of objects could be discovered at taxonomic levels below the class level, the domain of the many types of exoplanets, variable stars, binary stars, and variations on galaxies.

Yet another factor is the changing nature of discovery itself, not only the reality of larger teams mentioned above in the case of the accelerating universe, but the very nature of how science will be done and new discoveries made. An example is the detection of the object known as Hanny's Voorwerp in August 2007 by a twenty-five-year-old Dutch schoolteacher named Hanny van Arkel. Using the "Galaxy Zoo" website, in which the general public was invited to classify some of the million galaxies catalogued by the Sloan Digital Sky Survey, van Arkel discovered a strange blue blob that stumped the astronomers. Several hypotheses have been advanced regarding its nature, including a remnant galaxy or a gas cloud illuminated by a nearby quasar; whether the object is declared a new class of object remains to be seen. Massive projects such as the Sloan Survey and the forthcoming Large Synoptic Survey Telescope (LSST), producing huge amounts of data analyzed by both professional and non-professional astronomers in a new era of "networked science," are only one methodology that may change the face of discovery in the next century, including the discovery of new classes of objects.[11] While the efficiency of a process involving non-professionals seems counter-intuitive, the large number of volunteers (some 200,000 in the case of the Galaxy Zoo) to some extent compensates for the expertise of professional astronomers. This would not have been predicted even a few decades ago, and we cannot predict the future of astronomy's new methodologies.

One thing is sure: the pace of discoveries will require increasingly sophisticated data retrieval systems, for which classification will be crucial. The astronomical community is currently deeply engaged in erecting a Virtual Observatory, a repository of observations spanning all the subdisciplines specializing in the wavelength regime from radio to gamma-ray frequencies, encompassing all the classes listed in Appendix 1. Along with software tools, an intuitive and intelligent system will be necessary to guide researchers rapidly and accurately to the data relevant to a specific project.[12]

In the end, "Accelerating the Rate of Astronomical Discovery," the title of an IAU Symposium in 2009, will remain a goal devoutly to be desired. Whether or not it is reached depends not only on public policy, individual and collective innovation, and networked science, but also on the nature of a universe we do not yet fully understand. The ability of the human mind to comprehend that universe is itself a mystery whose resolution would constitute a landmark discovery, one that will undoubtedly extend over a long period of time. And it is always possible that – as in the case of the electromagnetic spectrum – new views of the universe will be opened up as yet unfathomed.

Appendix 1: Astronomy's Three Kingdoms

As explained in Section 8.2, Appendix 1 constitutes an exercise in constructing a comprehensive classification system for astronomy. The "Three Kingdom" system begins with the three Kingdoms of planets, stars, and galaxies – the three canonical divisions adopted in textbooks for almost a century, since it became clear that galaxies were indeed a separate "realm of the nebulae," as Hubble put it. For each Kingdom six astronomical Families are then delineated, based on the object's origin (Proto-), location (Circum- and Inter-), subsidiary status (Sub-), and tendency to form systems (Systems), in addition to the "Central" Family (planet, star, or galaxy) with respect to which the other Families are defined. These considerations give rise to astronomy's eighteen Families, and the symmetry of the six Families of each Kingdom reflects their physical basis in gravity's action in all three Kingdoms. The system then distinguishes eighty-two classes of objects, a large subset of which are the subject of this volume.

Like biology, the Three Kingdom (3K) system is hierarchical, extending from Kingdom to Family to Class, with the possible extension to further categories lower in the hierarchy such as Type and Subtype. As in biological classification it occasionally adds an intermediate Subfamily level wherever useful. With the benefit of hindsight (and with utility in mind), the system incorporates some classes as they have historically been defined, and adds others as they might be defined in a coherent and consistent system. The criterion adopted for class status is, wherever possible, the physical nature of the object, rather than orbital, dynamical, temperature, morphological, spectral, or any other characteristics. Some of the principles of classification are discussed in Section 8.2.

Clearly, another system might delineate classes differently. Some might question, for example, why there are three classes of cosmic rays, and three classes of "dust," one for each Kingdom, even if based on physical differences and

Astronomy's Kingdoms, Families, and Classes

Kingdom of the Planets

Family: Protoplanetary
- Class P 1: Proplyd (Disk)

Family: Planet
- Class P 2: Terrestrial (rocky)
- Class P 3: Gas Giant
- Class P 4: Ice Giant
- Class P 5: Pulsar Planet

Family: Circumplanetary
- Class P 6: Planetary Satellite
- Class P 7: Planetary Ring
- Class P 8: Planet Radiation Belt

Family: Subplanetary
- Class P 9: Dwarf Planet
- Class P 10: Meteoroid

Subfamily: Small Bodies of Solar System
- Class P 11: Minor Planet/ Asteroid
- Class P 12: Comet
- Class P 13: Trans-Neptunian Objects

Family: Interplanetary Medium
- Class P 14: Gas
- Class P 15: Dust

Subfamily: Energetic Particles
- Class P 16: Solar Wind
- Class P 17: Anomalous Cosmic Ray

Kingdom of the Stars

Family: Protostellar
- Class S 1: Protostar

Family: Star

Subfamily: Pre-Main Sequence
- Class S 2: T-Tauri
- Class S 3: Herbig Ae/Be

Subfamily: Main Sequence (H burning – Luminosity Class V)
- Class S 4: Dwarf
- Class S 5: Subdwarf

Subfamily: Post-Main Sequence (He burning and higher elements)
- Class S 6: Subgiant (Luminosity Class IV)
- Class S 7: Giant (Luminosity Class III)
- Class S 8: Bright Giant (Lumin. Class II)
- Class S 9 Supergiant (Lumin. Class I)
- Class S 10 Hypergiant (Lumin. Class 0)

Subfamily: Evolutionary Endpoints
- Class S 11 Supernova
- Class S 12 White Dwarf
- Class S 13 Neutron Star
- Class S 14 Black Hole

Family: Circumstellar
- Class S 15: Debris disk
- Class S 16: Shell (dying stars)
- Class S 17: Planetary Nebula
- Class S 18: Nova Remnant

Kingdom of the Galaxies

Family: Protogalactic
- Class G 1: Protogalactic Object

Family: Galaxy

Subfamily: Normal
- Class G 2 Elliptical
- Class G 3 Lenticular
- Class G 4 Spiral
- Class G 5 Irregular

Subfamily: Active
- Class G 6 Seyfert
- Class G 7 Radio Galaxy
- Class G 8 Quasar
- Class G 9 Blazar

Family: Circumgalactic
- Class G 10 Galactic Ring
- Class G 11 Galactic Accretion Disk + Jet
- Class G 12 Galactic Halo

Family: Subgalactic
- Class G 13 Subgalactic Object

Family: Intergalactic Medium

Subfamily: Gas
- Class G 14 Warm Hot IGM
- Class G 15 Lyman alpha blobs

Subfamily: Dust
- Class G 16 Dust

Subfamily: Energetic Particles
- Class G 17 Galactic Wind

Kingdom of the Planets (cont.)	**Kingdom of the Stars** (cont.)	**Kingdom of the Galaxies** (cont.)
Family: Systems	Class S 19: Core Collapse Supernova Remnant	Class G 18 Extragalactic Cosmic Rays
Class P 18: Planetary System	Class S 20: Stellar Accretion Disk + Jet	Family: Systems
Class P 19: Asteroid Groups	Class S 21: Herbig-Haro Object	Class G 19 Binary
Class P 20: Meteoroid streams	[See also Planetary System, P 18]	Class G 20 Interacting
Subfamily: Trans-Neptunian Systems	Family: Substellar	Class G 21 Group
Class P 21: Kuiper Belt	Class S 22: Brown Dwarf	Class G 22 Cluster
Class P 22: Oort Cloud	Family: Interstellar Medium	Class G 23 Supercluster
	Subfamily: Gas (99%]	Class G 24 Filaments & Voids
	Class S 23: Cool Atomic Cloud (H I)	
	Class S 24: Hot Ionized Cloud (H II)	
	Class S 25: Molecular Cloud (H_2)	
	Class S 26: White Dwarf Supernova Remnant	
	Subfamily: Dust (1%)	
	Class S 27: Dark Nebulae	
	Class S 28: Reflection Nebulae	
	Subfamily: Energetic Particles	
	Class S 29: Stellar Wind	
	Class S 30: Galactic Cosmic Rays	
	Family: Systems	
	Class S 31: Binary	
	Class S 32: Multiple Star	
	Class S 33: Association (OB)	
	Class S 34: Open Cluster	
	Class S 35: Globular Cluster	
	Class S 36: Population	

considerations of symmetry across the three Kingdoms. Some might question whether a supernova (S 11) is really an object at all, rather than an event (but then when does it cease being an object?). Others might wonder why novae are not included (they are one taxonomic level down from class status, since they are believed to be a type of binary star [S 31]). Still others might question ambiguous Family placement; globular clusters (S 35) could be considered both "systems" of stars and "circumgalactic." Such questions are common in biological taxonomy, and classification principles solve some of them, though ambiguities remain. In the end I would suggest that the majority of these classes would be found in any classification system for astronomy. It is a strength, rather than a shortcoming, of any classification system that it raises interesting questions. Accuracy, simplicity, and utility are the hallmarks of any such system, in whatever science.

Appendix 2: Astronomical Discoveries and Their Extended Structure

(Chronological Order by First Detection, Inference, or Declaration as a New Class of Astronomical Object)

New Object Class	Detection, Inference, or Declaration as A New Class of Astronomical Object	Comments on Pre-Discovery/ Extended Structure of Discovery (Interpretation, Understanding Physical Nature)
Nova/Supernova	Tycho Brahe 1572	**Pre-discovery:** Ancient Chinese believed them to be astronomical, but unproven. **Discovery:** After Tycho's proof of their astronomical nature in 1572 based on a lack of measurable parallax, only in the 1930s are distances and luminosities well enough known for Lundmark and Baade to separate normal novae from much more luminous "upper-class novae," or "Hauptnovae." In 1934, Baade and Zwicky coin the term "supernovae" in print, and theorize they may be a transition stage to neutron stars (pulsars). In 1964, Kraft argues all novae are double stars, thus very different from supernovae. (Section 7.1)

(continued)

Comet	Tycho Brahe 1577	**Pre-discovery:** Ancient Chinese believed them to be astronomical, but unproven. Greeks believed them to be atmospheric phenomena. **Discovery:** After Tycho's proof of their astronomical nature in 1577 based on parallax measurements, by the beginning of the twentieth century some cometary constituents are known, and they are believed to be "dusty sandbanks." Fred Whipple proposes a "dirty snowball" model in the 1950s, subsequently proven by spacecraft. (Section 7.1)
Satellite	Galileo 1610	**Pre-discovery:** Moon is first exemplar (but arguably not yet a class); no method for detecting others prior to telescope. **Discovery:** Galileo detects and, within 5 days, correctly interprets circum-Jovian objects as satellites in 1610; other than analogy with our own Moon and sporadic ground-based observations, their physical nature remains largely unknown until spacecraft. (Sections 2.1, 6.1)
Rings	Galileo 1610	**Pre-discovery:** No theory or precedent **Discovery:** After Galileo's detection in 1610, in 1655 Huygens conjectures they are rings. Only in 1857 does Maxwell determine that dynamically they cannot be a single solid body, but must be composed of separate objects. (Sections 2.1, 6.1)
Diffuse Nebula	Nicolas Claude Fabri de Peiresc 1610? (Orion) and Johann Cysat 1618 (Orion)	**Pre-discovery:** Naked-eye observations of the Orion Nebula **Discovery:** Unseen or unreported by Galileo, after either Peiresc (1610) or Cysat (1618) detects the Orion Nebula, almost two centuries pass before Herschel definitively interprets them as gaseous in 1790. Huggins proves this in 1864. Russell (1921) and Hubble (1922) show the gas is excited by stellar radiation; Menzel and Zanstra (1926) explain the mechanism as ionization by UV radiation; in the 1930s Eddington, Struve, Elvey, and Strömgren show these are what we today call H II regions. (Sections 3.1, 3.2)

Galactic (Star) Cluster	Galileo 1610	**Pre-discovery:** Pleiades and other objects seen with naked eye, mentioned in the Bible and Homer. **Discovery:** After Galileo's detection and resolution of the Praesepe and Orion "nebulosae" [not the Orion Nebula] into stars in 1610, La Caille (1754) and Messier (1781) catalogue a variety of star clusters. In 1767 Michell argues they are gravitationally associated. (Section 3.1)
Double Star (Binary Stars/ Physical Systems after 1802)	Castelli 1616 Galileo 1617 (Mizar A and B)	**Discovery:** After Castelli's and Galileo's observations in 1616–1617 of Mizar as an optical double, in 1767 John Michell argues the probability they are physical systems. In 1802, William Herschel reports observations proving this. (Section 4.5)
Multiple Star (Physical Systems after 1802)	Galileo 1617 Trapezium Huygens 1656	**Discovery:** Galileo shows the Trapezium in Orion consists of three stars, and Huygens in 1656. Herschel records his first "treble star" August 8, 1780, and in 1802 shows observationally they are physical systems.
Terrestrial Planets	Huygens 1659 (apparent planetary diameters) + Cassini 1672 (distances)	**Pre-discovery:** Seen as naked-eye objects, distinguished from fixed stars by their motion. **Discovery** (by inference): Micrometers allowed Huygens to measure the first accurate apparent planetary diameters, published in 1659; distance determinations by Cassini gave absolute diameters. Based on size; this class later evolved to be defined based on composition, after long inference. (Section 2.3)
Giant Planets (Gas Giants after 1940s)	Huygens 1659 (apparent planetary diameters) + Cassini 1672 (distances)	Same as above, but "gas giants" not widely accepted as a class until 1940s/1950s. Illustrates how the fundamental nature of a class can evolve over time. (Section 2.3)

(continued)

Interplanetary Dust (Zodiacal Light)	Cassini 1683 Fatio de Dullier 1684	**Pre-discovery:** Zodiacal light seen since antiquity; called the "false dawn" in reference to the timing of Islamic prayer rituals. **Discovery:** G. D. Cassini investigates the phenomenon of zodiacal light in 1683. In 1684 Fatio de Dullier explains the phenomenon as due to the reflection of sunlight by particles circling the Sun.
Planetary Nebula	Herschel 1782 (Sept. 7)	**Pre-discovery:** Messier's 1781 *Catalogue* contains four objects that turn out to be planetary nebulae; not distinguished as a class from other nebulae. **Discovery:** After Herschel distinguishes them as a class in 1782, in 1790 he conjectures the nebula is a cloud of gas, distinct from its central star. In 1864 Huggins proves this spectroscopically. (Sections 3.1, 6.1)
Double Nebulae (Double Galaxies after 1925)	Herschel 1783–1786	**Pre-Discovery:** None **Discovery:** Detected during Herschel's first sweeps for nebulae, 1783–1786; interpreted as physical systems by Herschel in 1811; determined to be double galaxies by Hubble, 1925. (Section 5.3, 6.1)
Groups of Nebulae (Galaxy Groups after 1925)	Herschel 1784–1786	Ditto (Section 5.3)
Dark Nebulae	Herschel 1785	**Pre-discovery:** Naked-eye dark lanes in the Milky Way. **Discovery:** After Herschel's detection of "holes in the heavens" in 1785, E. E. Barnard interprets them as clouds of unknown material, 1890s–1919. V. Slipher interprets them as dusty clouds in 1916. (Section 3.2)

Globular Cluster	Herschel 1789	**Pre-discovery:** La Caille's 1755 observations; Messier's 1781 catalog contains 28 "round nebulae," observed but arguably not distinguished as a separate class from other nebulae except by morphology. **Discovery:** Herschel interprets them as a new class of gravitationally bound clusters of stars, in 1789. (Sections 3.1, 6.1)
Asteroid/Minor Planet	Piazzi 1801	**Pre-discovery:** In 1779 Messier recorded a 7th magnitude star, now known to be Pallas, on his map of comet C/Bode 1779A1. **Discovery:** On January 1, 1801 Piazzi detected an 8th magnitude object, at first believed to be a comet, then a planet. After Olbers' discovery of a similar object (Pallas) in March 1802, Herschel determined their small size and suggested they were a new class of objects, dubbed "asteroids." Ceres and other such objects were considered planets for a half-century in national ephemerides. (Sections 2.2, 6.1, 8.3)
Spiral Nebulae (Spiral Galaxies after 1925)	William Parsons (Lord Rosse) 1845	**Pre-discovery:** Andromeda "Nebula" is a naked-eye object. Messier and others had seen M51 and other objects, but could not resolve their morphology. **Discovery:** Lord Rosse first detected the spiral form of nebula in 1845 in the form of M51, 14 in total when he published in 1850. Only in 1924 did Hubble determine spiral nebulae were extragalactic star systems of their own. This illustrates how the fundamental meaning of a particular class may change over time. (Section 5.1)
Asteroid Groups (Main Belt, etc.)	1850s–1860s	**Pre-discovery:** None **Discovery:** Humboldt mentions the idea of a "belt of asteroids" in 1850, located between Mars and Jupiter; the idea comes into common usage during 1850s and 1860s as more asteroids are detected. More asteroid groupings are detected in the early twentieth century, including 11 Trojans by 1938 and Near Earth Asteroids beginning in 1898 with Eros.

(continued)

Meteoroids	H. A. Newton 1863	**Pre-discovery**: Records of meteor observations date to at least the 7th century BC. Like comets, Greeks believed them to be atmospheric phenomena, with no celestial connection. Halley proposes a celestial origin in the eighteenth century. **Discovery**: In 1863 H. A. Newton demonstrates that the Leonids have a sidereal period, proving their celestial origin. (Section 7.1)
Meteoroid Streams	Adams Schiaparelli Le Verrier Oppolzer 1866–1867	**Discovery:** In 1866 Schiaparelli shows that the orbital elements of the Perseid meteors closely resemble those of Comet II 1862 (now known as comet 109P/ Swift-Tuttle. In 1867 Peters shows the same for the Leonid meteors and comet I 1866 (comet 55P/Tempel-Tuttle).
Nova Remnant	G. W. Ritchey 1901	**Pre-discovery:** None **Discovery:** In 1901 Ritchey photographs nebulosity around the star GK Persei, which turns out to be the first nova remnant. Nova and supernova were not distinguished before the 1920s, so neither were their remnants.
Cold Atomic Clouds [H I regions]	Johannes Hartmann 1904	**Pre-discovery:** None **Discovery:** Hartmann deduces from stationary absorption lines of ionized calcium in the spectrum of the binary star Delta Orionis that a cloud of gas exists in the line-of-sight in interstellar space. In the 1920s Plaskett, Struve, Eddington prove this, but believe the cloud is composed of calcium. In the 1930s Dunham and Struve identify the gas as neutral hydrogen. In 1944 Van de Hulst predicts neutral hydrogen should produce 21 cm radiation; this is discovered by Ewen and Purcell in 1951. (Section 3.3)

Dwarf Stars	Hertzsprung-Russell 1904–1914	**Pre-discovery:** Most of the stars visible to the naked eye are dwarfs. Classification of spectral line phenomena begins in 1860s. In 1895 Monck suspects two luminosity classes of stars. **Discovery:** In 1905 and 1907 Hertzsprung demonstrates two luminosity classes based on proper motions, and from 1910–1913 Russell infers the giants and dwarfs from parallax data. Bethe, von Weisäcker, and Critchfield work out nuclear reaction in 1938. Saha's ionization theory demonstrates what had been long suspected: that the spectral color sequence was actually a temperature sequence. (Sections 4.1, 6.1)
Giant Stars	Hertzsprung-Russell 1904–1914	**Pre-discovery:** Stars such as Arcturus, Aldebaran, and Pollux are visible to the naked eye without realizing their dimensions. Monck suspects their existence in 1895. **Discovery:** Hertzsprung and Russell demonstrate their existence, 1905–1913. (Section 4.1)
White Dwarf	Fleming Pickering Russell 1910	**Pre-discovery:** William Herschel detects 40 Eridani B in 1793; Alvan G. Clark detects Sirius B in 1862. **Discovery:** After Fleming, Pickering, and Russell realized the anomalous nature of the object 40 Eridani B in 1910, other faint hot stars are found. Luyten coins term "white dwarf" in 1922, popularized by Eddington. Theoretical understanding by Fowler and Chandrasekhar, 1926–1935. (Section 4.3)
Reflection Nebulae	V. M. Slipher 1912	**Pre-discovery:** In 1859 Tempel's discovery of a nebula around one of the stars in the Pleiades. **Discovery:** Slipher's spectroscopically determined in 1912 that the spectrum of the Merope nebula is the same as that of the Pleiades star cluster (a reflection spectrum), and thus composed of dust. (Section 3.2)

(continued)

Galactic Cosmic Rays	Victor Hess 1912	**Pre-discovery:** None **Discovery:** Hess discovers "penetrating radiation" using ionization detectors, one of the few classes not detected with a telescope. The nature of the ionizing agent remains a mystery, resulting in the Millikan-Compton controversy during the 1920s and 1930s. Millikan argues they are rays, Compton that they are charged particles. Compton is proven correct when he finds a latitude effect due to Earth's magnetic field. (Section 7.2)
Subdwarfs	Adams Kuiper 1913–1939	**Pre-discovery:** Seen as faint stars. **Discovery:** In 1913 Adams finds two high-velocity stars, and (with Joy) three more by 1934. In 1934 Kuiper coins the term "subdwarf." Subdwarfs were not one of the official classes of the MKK system. In 1951 Aller finds that subdwarfs have low "metallicity," i.e., a lower content of metals other than hydrogen and helium. (Section 4.2)
Supergiants	W. H. Pickering Cecelia Payne MKK 1917–1943	**Pre-discovery:** Stars such as Rigel and Canopus are seen as naked-eye objects. **Discovery:** In 1917 W. H. Pickering identifies "giants among giants," or "supergiants." (Section 4.1)
Subgiants	Adams KohlschüNewtter (spectroscopic parallaxes) 1917–1935	**Pre-discovery:** Seen as faint stars. **Discovery:** In 1917 Stromberg has early inklings of a separate class, which he dubs "subgiants" in 1930. By 1935 Adams and Kohlschütter publish a catalogue of spectroscopic parallaxes, which shows about 90 out of more than 4000 stars in this separate class. Trig parallaxes for a few stars confirms this in 1936. The class is formalized in Morgan's MKK system in 1943. Their place in stellar evolution is determined in the 1950s. (Section 4.2)

Galactic Jet/ Accretion Disk	H. D. Curtis 1918	**Discovery:** In 1918 Curtis notices "a curious straight ray" near the elliptical galaxy M87. In 1954 Baade and Minkowski identify this as a "straight jet" extending from the nucleus of the galaxy. By mid-1970s Blandford and colleagues show spinning accretion disks will produce jets. (Section 7.2)
Elliptical Nebulae (Elliptical Galaxies after 1925)	Hubble 1922–1926	**Pre-discovery:** Messier (1781) contained what we now know are six ellipticals. Herschel (1811) depicted but did not name them ellipticals. **Discovery:** Wolf (1908) arguably detected ellipticals as a new class, but it was only one of 22 classes depicted, not named. In 1922 Hubble called them "elongated," with "spindle" and "ovate" forms. In 1923 and 1926 he simplified this to "ellipticals." (Sections 5.1, 6.1)
Irregular Nebulae (Irregular Galaxies after 1925)	Hubble 1922–1926	**Pre-discovery:** Magellanic Clouds were seen with the naked eye from the Southern Hemisphere. Prior to Hubble, what we today know as diffuse nebulae within the galaxy were sometimes called "irregular." **Discovery:** In 1922 Hubble applied the term to extragalactic nebulae (those outside the plane of the galaxy), which we now know are outside our galaxy. The term was adopted in his 1926 system. (Sections 5.1, 6.1)
Lenticular Nebulae (Lenticular Galaxies after 1925)	Hubble 1922–1926	**Discovery:** In his 1926 system Hubble declared that extreme ellipticals shaded into "a limiting lenticular figure." Today lenticulars are often defined as a class "S0" between elliptical and spiral, a case of evolution in the meaning of a particular class. (Section 5.1)

(continued)

Trans-Neptunian Objects (TNOs)	Pluto, 1930 (QB1 1992)	**Pre-discovery:** Kuiper and Edgeworth theorize their existence in the 1940s and 1950s. **Discovery:** Tombaugh discovers Pluto as first Trans-Neptunian object. But it is not seen as a new class until more such objects are detected, beginning with Jewitt and Liu's discovery in 1992. (Section 2.4)
Galaxy Clusters	Shapley and Ames 1926 Hubble and Humason, 1931 Zwicky, 1937–1938	**Pre-discovery:** Messier and W. Herschel report concentrations of nebulae in 1780s. **Discovery**: In 1926 Shapley and Ames report report clusters of hundreds of nebulae. In 1931 Hubble and Humason list eight galaxy clusters. Zwicky discusses even more. (Sections 5.3, 6.1)
Circumstellar Shells	Adams and MacCormack 1935	**Discovery:** In 1935 Adams and MacCormack first observe shells around M supergiants. In 1956 Deutsch, also using high-resolution visual spectroscopy, confirms other cases. By 1970s other techniques such as polarimetry and IR photometry yield information on dust in the shells, while microwave spectroscopy gives information on gas, leading to first estimates of mass-loss rates.
Galactic Halo	Shapley and Baade 1938–1958	**Pre-discovery:** In 1920s Shapley shows globular clusters are arranged in a spherical distribution around galactic center. **Discovery:** Rosebrugh, describing Shapley's work, uses the term galactic "halo" in 1938. In 1944 Baade's delineation of two populations of stars in Andromeda helps define the concept. The term comes into wide use between 1955 and 1960, especially after Baade's 1958 lectures describe the halo as one of three structural components of the galaxy.

Core Collapse Supernova Remnant	Mayall 1939	**Pre-discovery:** In 1731 Bevis discovers a new faint nebulosity in the constellation Taurus; Messier rediscovers it in 1758 and labels it M1 in his catalogue. Rosse refers to as the Crab nebula in the mid-nineteenth century, but it is not yet recognized as a distinct class. **Discovery:** In 1939 Mayall argues based on its expansion that a "reasonable working hypothesis" is that it is associated with the supernova of 1054. By 1942 further work by Oort, Mayall, and Duyvendak "leaves hardly any doubt" that it is a supernova remnant.
Giant Molecular Clouds	Andrew McKellar 1940 Weinreb, Barrett et al. 1963 Scoville and Solomon 1975 Bok, 1940–1970s	**Pre-discovery:** In 1930s Merrill, Dunham, and Adams discover spectral lines that cannot be identified with atomic transitions. Swings and Rosenfeld calculate diatomic molecules can occur in interstellar space. **Discovery:** In 1940 McKellar identifies three astrophysical spectral lines with molecular spectra. In 1963 OH is identified, followed by ammonia and many others. In 1975 Scoville and Solomon report a large fraction of interstellar hydrogen is in molecular form. In 1947 Bok draws attention to "Bok globules," which turn out to be small molecular clouds. By 1977 about 3000 Giant Molecular Clouds are estimated to exist in the galaxy. (Section 3.3)
White Dwarf Supernova Remnant	Baade 1943 Hanbury-Brown and Hazard 1952	**Discovery:** In 1943 Baade identifies "nova Ophiuchi of 1604" as a Type I supernova (Kepler's supernova), and announces that the surrounding nebulosity as a remnant of the supernova "seems to be established beyond doubt." In 1952 Hanbury Brown and Hazard observe the remnant of Tycho's supernova as a radio source.

(continued)

Bright Giants	1943	**Pre-discovery:** Seen as naked-eye objects **Discovery:** (Declared) Bright giants were not separated as a distinct class until the MKK system is developed in 1943. (Section 4.2)
Seyfert Galaxies	Carl Seyfert 1943	**Pre-discovery:** In 1908 Fath notices six broad emission lines present in spiral galaxy NGC 1068. **Discovery:** Seyfert observes that a small proportion of spiral galaxies have bright optical nuclei with "high excitation emission lines." He identifies 12 galaxies in "this unusual class of object." In the late 1950s the Burbidges coin the term "Seyfert galaxies." (Section 5.2)
Stellar Populations	Baade 1943–1944	**Pre-discovery:** These stars had been previously seen but not distinguished into populations, though as early as 1926 Oort had found some evidence for this. **Discovery:** In 1943 Baade observes two distinct populations of stars in the Andromeda galaxy and its companions: bright O and B stars in the disks of galaxies and in open star clusters, and older red stars in the spheroidal component of galaxies and globular clusters. (Section 4.5)
T-Tauri Star	Alfred Joy 1945	**Pre-discovery:** Hind discovers T-Tauri itself as a 10th magnitude object in 1852. In 1890 Burnham finds a small nebula around the star, which had dimmed to 14th magnitude. **Discovery:** In 1945 Joy systematically studies the star and names it as the prototype of a new class of 11 other stars known to have similar behavior. In the 1950s astronomers begin to consider them as precursors to low-mass stars. Herbig and others consolidate this view in the 1960s.

Star Associations	Adriaan Blaauw Viktor Ambartsumian 1946–1947	**Pre-Discovery:** These stars had been observed before, but not as associations. **Discovery:** In 1946–1947 Blaauw and Ambartsumian detect OB and T associations. (Section 4.5)
Radio Galaxies	J. S. Hey (Cygnus A) 1946–1952	**Pre-discovery**: Some objects first discovered optically, as Virgo A in 1781. **Discovery:** In 1946 Hey detects Cygnus A; from 1947 to 1949 Bolton and colleagues detect more radio sources and suggest optical identifications for three of them: Taurus A, Virgo A and Centaurus A. In 1952 Baade and Minkowski identify Cygnus A with an 18th magnitude galaxy. (Section 5.2)
Herbig-Haro Objects	Herbig and Haro 1946–1954	**Pre-discovery:** In 1890 Sherburne Burnham first observes nebulosity near the star T-Tauri, but believes it is an emission nebula, not a new class. **Discovery:** In the late 1940s and 1950s Herbig and Haro recognize similar objects in Orion. The similarity to Burnham's T-Tauri observation leads them to suggest these are related to the early stages of star formation. They are formed when gas from a stellar jet collides with the interstellar medium.
Protostar	Bok 1947–1980s	**Pre-discovery:** The basic idea of star formation can be traced back to the nebular hypothesis of Pierre Simon de Laplace in the late eighteenth century, in which a cloud of gas and dust condenses into planets and a central star. The general idea of stars condensing out of gas clouds was in the air in the early twentieth century. **Discovery:** In the 1940s a few astronomers such as Spitzer and Whipple draw attention to the possible formation of stars out of condensations in the interstellar medium, and in 1947 Bok draws attention to small circular dark nebulae as possible sites of such formation. In the late 1970s and 1980s infrared and submillimeter observations begin to penetrate molecular clouds. Infrared and space observatories uncover protostars.

(continued)

Stellar Jets	Ambartsumian 1950s–1980s	**Pre-discovery:** Herbig-Haro objects are the first evidence that jets occur associated with stars. **Discovery:** In the early 1950s Ambartsumian proposes they represent the early stages of T-Tauri stars. By the mid-1970s astronomers realize HH objects have very high proper motions, pointing away from their very young stars. By the early 1980s some of these objects were known to have a highly collimated jet-like nature.
Galaxy Superclusters	Gerard de Vaucouleurs 1953	**Discovery:** In 1953 de Vaucouleurs provides evidence that the Virgo Cluster might be a dominant member of a larger "supergalaxy," referred to as the "Local Supercluster" by 1958. (Sections 5.3, 6.1)
Interacting Galaxies	Boris A. Vonontsov-Velyaminov 1957	**Pre-discovery:** Galaxies in close apparent proximity seen earlier but not considered to be interacting. **Discovery:** In 1957 Vonontsov-Velyaminov describe the two spiral galaxies, dubbed the Mice Galaxies, as an interacting binary galaxy system. Vorontsov-Velyaminov's *Atlas and Catalogue of Interacting Galaxies* is published in 1959. In 1972 the Toomres depict the "Toomre sequence" showing various stages of spiral merging over 500 million years.
Planetary Radiation Belt	Van Allen, spacecraft 1958	**Pre-discovery:** None **Discovery:** In 1958 van Allen places a Geiger counter on board Explorer 1 to observe cosmic rays, but the instrument instead demonstrates the existence of a large population of energetic particles trapped by the Earth's magnetic field. Three months later Explorer 3 shows that the radiation belt is permanent, widespread, and intense. Spacecraft have mapped the nature and extent of the Van Allen belts in more detail ever since.

Solar Wind	Parker 1958 Spacecraft 1960s	**Pre-discovery:** In 1859, after observing a solar flare, Carrington suggests particles emanate from Sun. In 1951 Biermann postulates such particles cause a comet's tail to point away from Sun. **Discovery**: In 1958 Parker presents evidence and coins the term "solar wind." Satellites confirm the solar wind in the 1960s.
Stellar Wind	Deutsch 1956 Cameron 1959 Parker 1960	**Pre-discovery:** Confirmation of the existence of the solar wind in the 1960s was a leading indicator that other stars might have their own stellar winds. **Discovery:** Deutsch reports low-velocity mass loss from cool M supergiants as early as 1956. A full-blown concept of stellar wind was not long in coming after Parker's landmark paper on the solar wind in 1958; Cameron, citing Parker's paper, proposes that stars must lose mass in order to produce white dwarfs and "one method of mass loss is by a 'stellar wind' analogous to the solar wind." In 1960 Parker expanded his concept of solar wind to stellar winds in another landmark paper. In 1967 Morton reported Aerobee rocket-ultraviolet observations of O and B supergiants, showing broad P-Cygni profiles indicating high velocity outflows. Further observations confirm numerous cases.
Herbig Ae/Be Stars	George Herbig 1960	**Pre-discovery:** Ae/Be stars had been observed without distinguishing them as a new class. **Discovery:** Inspired by the work on low-mass T-Tauri main sequence stars, in 1960 Herbig identifies Ae/Be stars as a class of higher mass pre-main sequence stars. Strom and colleagues confirmed the pre-main sequence nature of these objects in 1972.

(continued)

Extragalactic Cosmic Rays	1962	**Pre-discovery:** In 1960 Peters suggests lower-energy cosmic rays are produced predominantly inside our own galaxy, whereas those of higher energy come from more distant sources. **Discovery:** In 1962 Linsley and his team make first discovery of an ECR above 10^{20} eV. Other arrays confirm these results.
Interplanetary Gas	Spacecraft 1960s	**Pre-discovery:** None **Discovery:** In the 1960s detectors aboard the Russian Zond 1 and Venera spacecraft, as well as the American Orbiting Geophysical Observatory and Mariner spacecraft, find hydrogen in interplanetary space based on observations of solar Lyman alpha scattering. In 1970 Paresco, Bowyer and colleagues detect radiation emitted by this neutral helium when a Nike Tomahawk rocket flew three extreme-ultraviolet photometers to an altitude of 264 kilometers to study the 584 Angstrom terrestrial nightglow. Detailed observations begin with spacecraft in 1970s.
Quasars	Maarten Schmidt 1963 (3C 273) Greenstein and Matthews 1963 (3C 48)	**Pre-discovery:** 3C 273 detected optically as 13th magnitude star in 1887; catalogued as radio source in 1959. 3C 48 redshift of 0.37 measured in 1960, but rejected as impossible. **Discovery:** In February 1963 Maarten Schmidt measures redshift of 0.016 in spectrum of 3C 273, identifies it with radio source, indicating extreme distance. In 1964–1965 Salpeter and Zel'dovich suggest accretion by massive objects as the energy source. Only in the late 1970s and 1980s were black holes seen as possible energy sources. (Section 5.2)

Pulsars/Neutron Stars	Hewish and Bell 1967	**Pre-discovery:** Baade and Zwicky theorize neutron stars in 1934. **Discovery:** In November 1967 Hewish and Bell detect an object with a series of radio pulses, dubbed "pulsars." Interpretation takes place over the next few months, concluding that the phenomenon is due to radial pulsations of white dwarfs or neutron stars. Within weeks Gold demonstrates pulsars are the theorized neutron stars. Optical identifications are made by 1969. (Section 4.3)
Hypergiants	1971	**Pre-discovery:** Stars such as Eta Carina seen as naked-eye stars. **Discovery:** In 1956 Feast and Thackeray suggest supergiants brighter than -7 magnitude be called "super-supergiants." In 1971 Keenan suggests a new class, designated 0, to define hypergiants, and this class is incorporated into the next revision of the MK system in 1973.
Galactic Wind	1971	**Pre-discovery:** In 1963 Lynds and Sandage discover an explosion at the center of the galaxy M82, an explosion that drew attention to the possibility of huge outflows of gas. In 1968 J. A. Burke first suggests the presence of a galaxy-scale wind, and by 1971 astronomers speculate that the lack of an interstellar medium in elliptical galaxies was due to galactic winds that had swept them free of gas. **Discovery:** Established as a ubiquitous phenomenon in the 1970s and beyond, based on X-ray emission of hot gas, or optical emission of warm gas.

(continued)

Black Hole	Ground-based and HST observations 1971–1994	**Pre-discovery**: Michell and Laplace glimpse the possibility; in 1916 Schwarzschild provides solution to Einstein's equations of general relativity describing non-rotating black holes. In 1939 Oppenheimer, Volkoff, and Snyder work out theory of black holes. **Discovery:** In 1972 Webster and Murdin theorize a stellar black hole in Cygnus X-1 based on observations. Blandford, Rees, and Lynden-Bell demonstrate accretion disk would create jets, which are observed from centers of some active galaxies. Black hole candidates are proposed in the 1980s. In 1994 HST astronomers claim "seemingly conclusive evidence" for a galactic black hole in M87. (Sections 5.2, 7.2)
Galactic Rings	Olin Eggen 1971 Rodrigo Ibata 1994	**Pre-discovery**: None **Discovery:** In 1971 Eggen identifies "Arcturus Group" based on space motions. In 1994 Ibata launches modern era of galactic stream discoveries with detection of Sagittarius Dwarf Galaxy, in an advanced stage of tidal disruption. Numerous other shredded galaxies, or "rings" found since.
Blazars	Strittmatter 1972	**Pre-discovery**: Some of these objects had been discovered optically earlier, including BL Lac itself in 1929, known as a star varying from 13th to 16th magnitude. In 1968 Schmitt identifies BL Lac as a powerful radio source. **Discovery:** In 1972 Strittmatter and colleagues identify a class called "The BL Lacertae type," characterized by rapid variations in intensity at radio, IR and visual wavelengths, and (unlike quasars) lacking spectral lines. In 1978 Spiegel dubs them "blazars." (Section 5.2)

Anomalous Cosmic Rays	Spacecraft 1972–1973	Whether ACRs should be seen as a class distinct from galactic and extragalactic cosmic rays is open to interpretation, but different origin, composition, and energy.
Circumstellar Debris Disk	IRAS/Aumann 1983 Smith and Terrile 1984	**Discovery:** In 1983 the Infrared Astronomical Satellite (IRAS) discovers an infrared excess around Vega, the thermal signature of a relatively cool cloud of millimeter-sized solid particles. By early 1984 IRAS astronomers find similar shells around six more stars, and a total of 40 by the summer. They are originally interpreted as protoplanetary disks, but later considered to be debris left over after planet formation. Debris disks normally lack large amounts of gas, and may be found around young or old stars, similar to the Kuiper Belt in our solar system.
Galaxy Filaments and Voids	Valerie de Lapparent Margaret Geller and John Huchra 1986	**Pre-discovery:** In 1981 Kirschner and colleagues report the "Boötes Void." Zeldovich and colleagues report other giant voids in 1982. Giovanelli and Haynes report early hints of filaments in 1982. **Discovery:** Based on the first large redshift survey, in 1981 Lapparent, Geller, and Huchra report chains and sheets of galaxies separated by giant voids. More such large-scale structure is reported in subsequent years. (Section 5.3)
Brown Dwarf	Becklin and Zuckerman Latham et al., 1989 T. Nakajima et al. 1988–1994	**Pre-discovery:** Seen as faint stars before their low masses are determined. **Discovery:** Becklin and Zuckerman discover first possible brown dwarf in 1988, but it is now known to be an L dwarf undergoing H fusion; Latham et al. discover a more likely candidate in 1989, HD 114762. In 1994 Nakajima and colleagues image first definitive brown dwarf, Gliese 229B. (Section 4.4)

(continued)

Pulsar Planets	1992	**Pre-discovery:** Conjectured in 1970; claims of observed timing variations attributed to pulsar planets were also published about the same time. Lyne and Bailes make erroneous claim in 1991. **Discovery:** Wolszczan and Frail announce the first two pulsar planets in 1992, orbiting the object known as PSR B1257+12. A third planet is later confirmed in the same system. In 1993 a second system of pulsar planets is announced around PSR B1620–26.
Kuiper Belt	Jewitt and Luu 1992 1992 QB1	**Pre-discovery:** Based on theory, in the 1940s and 1950s Kuiper and Edgeworth suspect a population of objects beyond Neptune and Pluto. **Discovery:** After Jewitt and Luu's detection of 1992 QB1, numerous other Trans-Neptunian objects are discovered. (Sections 1.3, 2.4)
Protoplanetary Disk (Proplyds)	O'Dell 1992	**Pre-discovery:** Laplace's nebular hypothesis. Erroneous interpretations of IRAS infrared excess as protoplanetary disks in the 1980s, now believed to be debris disks after planetary formation. **Discovery:** In 1992 O'Dell and colleagues unexpectedly find circumstellar disks in Orion. Previously thought to be stars, their disks are rendered visible to the Hubble Space Telescope by being near an H II emission nebula. (Section 8.3)

Warm Hot Intergalactic Medium (WHIM)	HST 1992	**Pre-discovery: None** **Discovery:** In 1992 Hubble Space Telescope astronomers reports absorption lines in the ultraviolet spectra of a quasar they interpreted as due to intervening objects such as Lyman alpha blobs and galaxy clusters. Similar absorption lines had been seen for two decades by ground-based telescopes. By 2000 astronomers use Hubble's ultraviolet capabilities to detect, for the first time, highly ionized oxygen between low redshift galaxies by analyzing the light of a quasar. The highly ionized oxygen (O VI) is believed to trace large quantities of hot ionized hydrogen, itself invisible because it is too hot to be seen in visible light, but too cool to be seen in X-rays. The WHIM may account for much of the "missing mass" of cosmology.
Intergalactic Dust	Zaritsky et al. 1994	**Pre-discovery:** In 1957 Zwicky suggests that "the apparent non-uniformities in the apparent distribution of clusters of galaxies are due to the effect of intergalactic (and interstellar) obscuration." **Discovery:** In 1994 Zaritsky and colleagues report dust in galactic halos for two nearby spirals, NGC 2835 and NGC 3521, as seen in the color change in background galaxies whose light passed through these halos. This photometric technique provided the first suggestion that dust surrounds galaxies to distances extending to 200,000 light years. In 1999 and 2005 astronomers provide evidence that galactic winds triggered by supernovae would expel some of the galactic dust into intergalactic space.

Planetary Systems	Mayor and Queloz 1995	**Pre-discovery:** Numerous spurious claims, including van de Kamp's Barnard Star planets. **Discovery:** After Mayor and Queloz's discovery of 51 Pegasi b in 1995, this is confirmed by Marcy and Butler in 1996; numerous other detections follow. Some claim Latham's 1989 discovery of a brown dwarf could in fact be a planet. (Section 4.4)
Subgalactic Objects	HDFs 1995–2004	**Discovery:** Beginning in 1995 the Hubble Deep Field images show high-redshift galaxies smaller and less symmetric in shape than galaxies at lower redshift. These are hypothesized to be objects in a stage prior to galaxy formation, but the evidence is still uncertain.
Ice Giant Planets	Marley 1999	**Pre-discovery:** Detected as planets by Herschel (1781) and Galle (1844), considered as giant planets, then gas giants from 1950s. **Discovery:** Voyager 2 spacecraft observations provide data that models their interiors, different enough from Jupiter and Saturn that some astronomers declare them a new class of planet. (Section 2.3)
Lyman Alpha Blobs	Steidel et al. 1999	**Pre-discovery: None** **Discovery:** Astronomers find the first Lyman alpha blobs serendipitously around 1999 during surveys for so-called Lyman break galaxies. The team led by Steidel writes that "We have discovered two objects, which we call 'blobs,' that are very extended (>15"), diffuse, and luminous Ly alpha nebulae." They go on to say that "While we refrain from making too much of this result at present, given the many uncertainties involved, it does suggest that the blobs may represent a different class of rare object that, like QSOs [quasars], are preferentially found in rich environments at high redshift." Their true nature remains unknown, but they are likely related to the birth of galaxies.

Oort Cloud Objects	Brown (Sedna) 2004	**Pre-discovery:** In the 1950s Oort theorizes a cloud of objects. **Discovery:** Brown argues that Sedna, detected in 2004, may be the first member observed in the Oort Cloud. (Section 2.4)
Dwarf Planet	International Astronomical Union 2006	**Pre-discovery.** Although Tombaugh detected Pluto in 1930, astronomers believed it was another planet, not a new class. **Discovery:** This is a case of discovery by declaration rather than by detection. The IAU declared Pluto a new class (a dwarf planet) in 2006 after its properties were better known, and after other such Trans-Neptunian "Kuiper Belt Objects" were discovered. Ceres was also declared a dwarf planet, giving it dual designation as both an asteroid and a dwarf planet. (Chapter 1, Section 8.2)
Protogalactic Clouds	HST 2006–2010	**Pre-discovery:** Theories of galaxy formation suggest the existence of proto-galactic clouds. **Discovery:** Using results from the Hubble Ultra-Deep Field and the Great Observatories Origins Deep Survey, in 2006 Bouwens and Illingworth find evidence for hierarchical galaxy formation occurring between 700 and 900 million years after the Big Bang, at a redshift of about 7. They find 5 times fewer galaxies at 700 million years than at 900 million years. The following year Stark and colleagues, using the Keck II telescope on Mauna Kea and the technique of gravitational lensing, observed galaxies at a redshift of 9, only 500 million years after the Big Bang. In 2010 Hubble astronomers report finding five galaxies 13 billion light-years distant, dating to 600 million years after the Big Bang. They hypothesize that galaxies began to form between 300,000 years and 600 million years after the Big Bang.

Notes

Introduction: The Natural History of the Heavens and the Natural History of Discovery

1 International Astronomical Union, 2006 resolutions on Pluto at http://www.iau.org/public_press/news/detail/iau0603/; IAU, "The IAU draft definition of 'planet' and 'plutons,'" Press release, August 16, 2006; http://www.iau.org/public_press/themes/pluto/; Owen Gingerich, "The Inside Story of Pluto's Demotion," *Sky and Telescope*, 112 (November 2006), 34ff.; Steven Soter, "What Is a Planet?" *Scientific American* (January 2007).

2 Herschel's biological analogies are discussed in Stephen G. Brush, *Nebulous Earth: The Origin of the Solar System and the Core of the Earth From Laplace to Jeffreys* (Cambridge: Cambridge University Press, 1996), pp. 33–35; see also Richard Panek, "Herschel's Garden," *Natural History*, May 1999, at http://findarticles.com/p/articles/mi_m1134/is_4_108/ai_54574610/. It is also notable that in 1848 the German naturalist and explorer Alexander von Humboldt wrote, "As in our forests we see the same kind of tree in all the various stages of its growth, and are thus enabled to form an idea of progressive, vital development; so do we also in the great garden of the universe recognize the most different phases of sidereal formation. The process of condensation . . . appears to be going on before our eyes" (Humboldt 1848), p. 67.

3 Barbara J. Becker, *Unravelling Starlight: William and Margaret Huggins and the Rise of the New Astronomy* (Cambridge: Cambridge University Press, 2011), p. 5.

1. The Pluto Affair

1 Clyde W. Tombaugh and Patrick Moore, *Out of the Darkness: The Planet Pluto* (New York: New American Library, 1980), p. 126.

2 James W. Christy, Foreword to Tombaugh and Moore, *Out of the Darkness*, p. 1.

3 Michael Brown, *How I Killed Pluto and Why It Had It Coming* (New York: Spiegel & Grau, 2010), pp. 234–35.
4 William G. Hoyt, *Planets X and Pluto* (Tucson: University of Arizona Press, 1980), pp. 76, 83ff. A more succinct, but still insightful, treatment is Gibson Reaves, "The Prediction and Discoveries of Pluto and Charon," in *Pluto and Charon*, S. Alan Stern and David J. Tholen, eds. (Tucson: University of Arizona Press, 1997), pp. 3–25. The Lowell 40-inch was actually a 42-inch mirror, but its full 42-inch aperture could not be used until the tube was rebuilt in 1925. For Lowell in context, see David Strauss, *Percival Lowell: The Culture and Science of a Boston Brahmin* (Cambridge, MA: Harvard University Press, 2001).
5 On Tombaugh see David H. Levy, *Clyde Tombaugh: Discoverer of the Planet Pluto* (Tucson: University of Arizona Press, 1991); for a more concise version, Herbert Beebe, *BEA*, pp. 1145–46. On the Lowell Observatory see William Lowell Putnam, *The Explorers of Mars Hill: A Centennial History of Lowell Observatory, 1894–1994* (West Kennebunk, ME: Phoenix, 1994).
6 Tombaugh and Moore, *Out of the Darkness*, pp. 125–26.
7 V. M. Slipher, "The Discovery of a Solar System Body Apparently Trans-Neptunian," Lowell Observatory *Observation Circular*, March 13, 1930, in Hoyt, *Planets X and Pluto*, pp. 196–97. The April 12th Circular was reprinted, among other places, in "Planet X-Lowell Observatory *Observation Circular*," *JRASC* 24 (1930): 282–84.
8 E. W. Brown, "On the Predictions of Trans-Neptunian Planets from the Perturbations of Uranus," *PNAS* 16 (1930): 364–71, 371; E. C. Bower, "On the Orbit and Mass of Pluto, with an Ephemeris for 1931–32," *Lick Observatory Bulletin* 437, Sept. 26, 1931; E. M. Standish, "Planet X: No Dynamical Evidence in the Optical Observations," *AJ* 105 (1993): 2000–6.
9 David H. DeVorkin, "Pluto: The Problem Planet and its Scientists," in Roger Launius, ed., *Exploring the Solar System: The History and Science of Planetary Exploration* (New York: Palgrave MacMillan, 2013), pp. 323–62: 332.
10 Reaves, "The Prediction and Discoveries of Pluto and Charon," p. 18; Tombaugh and Moore, *Out of the Darkness*, pp. 6–7.
11 Steven J. Dick, *Sky and Ocean Joined: The U. S. Naval Observatory, 1830–2000* (Cambridge: Cambridge University Press, 2003), pp. 425–29; J. W. Christy Oral History Interview, June 16, 1989, copy deposited in U.S. Naval Observatory Library, Washington, DC.
12 J. W. Christy, "The Discovery of Pluto's Moon, Charon, in 1978," in S. A. Stern and D. J. Tholen, eds., *Pluto and Charon* (Tucson: University of Arizona Press, 1997), pp. xvii–xxi; J. W. Christy, *OHI*, 22–23.
13 The discovery paper is R. S. Harrington and J. W. Christy, "The Satellite of Pluto," *AJ*, 83 (1978): 1005–8.
14 A. Stern, "The Pluto-Charon System," *ARAA* 30 (1992): 185–233; Alan Stern and Jacqueline Mitton, *Pluto and Charon: Ice Worlds on the Ragged Edge of the Solar System* (New York: Wiley, 1999); R. M. Marcialis, "The Discovery of Charon: Happy Accident or Timely Find?" *JBAA* 99 (1989): 27–29. On the Hubble images, see http://hubblesite.org/newscenter/archive/releases/1990/14/image/a/; http://hubblesite.org/newscenter/archive/releases/2005/19/.

15 A. J. Dressler and C. T. Russell, "The Pending Disappearance of Pluto," *Eos* 61, no. 44 (1980): 690, reproduced in Tyson, *The Pluto Files*, p. 27.
16 Kenneth Edgeworth, "The Evolution of our Planetary System," *JBAA* 53 (1943): 181–88, and Gerard P. Kuiper, *Astrophysics: A Topical Symposium* (New York: McGraw-Hill, 1951), p. 402; J. McFarland, "Kenneth Essex Edgeworth – Victorian Polymath and Founder of the Kuiper Belt?" *Vistas in Astronomy* 40 (1996): 343–54, 343. On the nomenclature controversy see http://www.cfa.harvard.edu/icq/kb.html, which points out that others also envisioned trans-Neptunian objects beyond Pluto. As with most Americans, in this book we use the term "Kuiper Belt," demonstrating that if classes and classification systems are socially constructed (as we show in due course), nomenclature is even more so.
17 Rex Graham, "Is Pluto a Planet?" *Astronomy* (July 1999): 42–47.
18 Govert Schilling, "Pluto: The Planet that Never Was," *Science* 283 (January 8, 1999): 157; Robert L. Millis, Richard P. Binzel, Joseph A. Burns, Dale P. Cruikshank, William B. McKinnon, Karen J. Meech, and S. Aland Stern, 'Pluto's Planetary Status," *Science* 283 (February 12, 1999): 937.
19 IAU Press Release 01/99, dated February 3, 1999, available at http://nssdc.gsfc.nasa.gov/planetary/text/pluto_iau_pr_19990203.txt, reprinted in Neil deGrasse, *The Pluto Files: The Rise and Fall of America's Favorite Planet* (New York: W. W. Norton, 2009), pp. 67–68.
20 The New York debate is described in Tyson, *Pluto Files*, pp. 68–75.
21 Michael F. A'Hearn, "Pluto: A Planet or a Trans-Neptunian Object?" in H. Rickman, ed., *Highlights of Astronomy*, 12 (San Francisco: International Astronomical Union, 2002), pp. 201–4. Alan Stern and Harold Levison followed A'Hearn's paper with one of their own: "Regarding the Criteria for Planethood and Proposed Planetary Classification Schemes," pp. 205–13. Their first criterion was that the classification be physically based, a criterion that would be violated when dynamical astronomers entered the picture.
22 Alan Boss, *The Crowded Universe: The Search for Living Planets* (New York: Basic Books, 2009), p. 117. The Working Group definition as of 2003 is at http://astro.berkeley.edu/~basri/defineplanet/IAU-WGExSP.htm. The Working Group on Extrasolar Planets was disbanded in August 2006 with the decision of the IAU to create a new commission on extrasolar planets as part of Division III, with the Organizing Committee consisting of most of the same members as the Working Group.
23 Steven Soter, "What is a Planet?" *AJ* 132 (2006): 2513–19, and a popular version in "What is a Planet?" *Sci Am* (January 2007): 34–41; online at http://burro.case.edu/Academics/USNA229/whatisaplanet.pdf.
24 D. W. E. Green, "History and Myth – Trans-Neptunian Objects and their Terminology," *BAAS* 33 (2001): 1363; the situation regarding Trans-Neptunian objects on the eve of the IAU action is well documented in David Weintraub, *Is Pluto a Planet? A Historical Journey through the Solar System* (Princeton: Princeton University Press, 2007).
25 *Reports on Astronomy*, Transactions of the IAU, 36A, Oddbjorn Engvold, ed. (Cambridge: Cambridge University Press, 2006), p. 189.
26 Boss, *Crowded Universe*, pp. 120–21.
27 On the discovery of Trans-Neptunian Objects see Govert Schilling, *The Hunt for Planet X: New Worlds and the Fate of Pluto* (New York: Copernicus Books, 2009). Eris had been discovered in January 2005 on images taken in 2003, thus the designation 2003 UB313. For a first-hand account see Brown, *How I Killed Pluto*.

28 In addition to Gingerich (Chair, USA), Williams (UK), and Sobel (USA), the Committee consisted of André Brahic (France), Junichi Watanabe (Japan), Richard Binzel (USA), and Catherine Cesarsky, president-elect of the IAU.
29 Owen Gingerich, "The Inside Story of Pluto's Demotion," *Sky and Telescope* 112 (November 2006): 34–39.
30 Lars Lindberg Christensen, "The Pluto Affair: When Professionals Talk to Professionals with the Public Watching," in A. Heck and L. Houziaux, eds., *Future Professional Communication in Astronomy* (Memoires Academie Royale Belgique, 2007).
31 Gingerich, "Inside Story," p. 38. See also Gingerich, "Planetary Perils in Prague," *Daedalus* 136, no.1 (2007): 137–40, preceded in the same issue by David Jewitt and Jane X. Luu, "Pluto, Perception and Planetary Politics," pp. 132–36.
32 *Proceedings of the Twenty Sixth General Assembly Prague, 2006, Transactions of the IAU*, volume 26B, Karel A. Van der Hucht, ed. (Cambridge: Cambridge University Press, 2008), pp. 46–47; the resolutions are online at http://www.iau.org/public_press/news/detail/iau0603/.
33 IAU, http://www.iau.org/public_press/themes/pluto/.
34 "Pluto: The Backlash Begins," *Nature* 442 (August 31, 2006): 965–66. The petition is online at http://www.20at.com/assets/pluto/signatures_final.pdf.
35 Editorial page, *Washington Post*, August 27, 2006. For the New Mexico resolution, see http://www.spaceref.com/news/viewsr.nl.html?pid=23558; also S. Zielinski, "New Mexico Declares Pluto a Planet," *Eos Transaction of the American Geophysical Union* 88 (2007): 133. For more on the public reaction, see Tyson, *The Pluto Files*.
36 IAU News Release, June 11, 2008, "Plutoid Chosen as Name for Solar System Objects like Pluto," http://www.iau.org/public_press/news/detail/iau0804/.
37 Lisa Messeri, "The Problem with Pluto: Conflicting Cosmologies and the Classification of Planets," *Social Studies of Science* 40/2 (April 2010): 187–214, 189–90. We discuss Pluto in the context of classification issues in Chapter 8.

2. Moons, Rings, and Asteroids: Discovery in the Realm of the Planets

1 Galileo Galilei, *Sidereus Nuncius, or The Sidereal Messenger*, translated with introduction, conclusion, and notes by Albert van Helden (Chicago: University of Chicago Press, 1989), p. 68.
2 John Heilbron, *Galileo* (Oxford: Oxford University Press, 2010), p. v.
3 Piazzi to Barnaba Oriani, January 24, 1801, in Clifford J. Cunningham, *The First Asteroid: Ceres, 1801–2001* (Surfside, FL: Star Lab Press, 2001), p. 36.
4 On the circumstances of the invention of the telescope and Galileo's role, see the excellent set of articles in Albert van Helden, Sven Dupré, Rob van Gent, and Huib Zuidervaart, eds., *The Origins of the Telescope* (Amsterdam: KNAW Press, 2010). See also Heilbron, *Galileo*, pp. 147–54.

For succinct views of Galileo's career, see Maurice A. Finocchiaro, "Galilei, Galileo," *BEA*, pp. 399–401; Stillman Drake, "Galilei, Galileo," *DSB* 5 (New York: Charles Scribner's Sons), pp. 237–48, and its update, Michele Camerota, "Galilei, Galileo," *NDSB* 3 (New York: Charles Scribner's Sons, 2008), pp. 97–103.

5 Galileo, *Sidereus Nuncius*, pp. 64–86:64.

6 Ibid., pp. 64–66. A letter from Galileo written on January 7, the night of discovery, has been translated in Stillman Drake, "Galileo's First Telescopic Observations," *JHA* 7 (1976): 153–68. Drake also discusses the events surrounding the discovery of the moons of Jupiter in *Galileo at Work: His Scientific Biography* (Chicago and London: University of Chicago Press, 1978), pp. 142–54.

7 Van Helden, *Sidereus Nuncius*, p. 40. Drake, *Galileo at Work*, pp. 151–53, argues that Galileo may not have fully realized until January 15 that the four moons orbited Jupiter.

8 Kepler observed the moons of Jupiter during a brief period from August 30 to September 9, 1610, using a Galilean telescope borrowed from the Elector Ernst of Cologne, the Duke of Bavaria. Kepler published his observations of Jupiter in the *Narratio de observatis a se quatuor Jovis satellitibus erronibus* (Narration on the Four Wandering Satellites of Jupiter Observed by Him). It appeared in late October 1610, with a 1611 imprint, and was only twelve pages in length. In his *Dissertatio cum Nuncio Sidereo*, written in April 1610, Kepler explicitly compares the Jovian satellites with our Moon, with astrology and teleology in mind: "Those four little moons exist for Jupiter, not for us. Each planet in turn, together with its occupants, is served by its own satellites. From this line of reasoning we deduce with the highest degree of probability that Jupiter is inhabited." *Kepler's Conversation with Galileo's Sidereal Messenger*, trans. Edward Rosen (New York and London, 1965), p. 42. For context, see Steven Dick, *Plurality of Worlds: The Origins of the Extraterrestrial Life Debate from Democritus to Kant* (Cambridge: Cambridge University Press, 1982), p. 77.

9 *Galileo on the World Systems*, Maurice A. Finocchiaro, trans. and ed. (Berkeley: University of California Press, 1997), pp. 243–44.

10 Galileo Galilei to Grand Duke Cosimo II, translated in Albert van Helden, "Saturn and his Anses," *JHA* 5 (1974): 105–21, 105.

11 Ibid., 106–7.

12 Van Helden, "Annulo Cingitur: the Solution of the Problem of Saturn," *JHA* 5 (1974): 155–74, 156. Galileo sometimes gets credit for the discovery of the rings, but more precisely his contribution was detection, only the first of three aspects of discovery, not correct interpretation.

13 Van Helden, "Saturn and His Anses," p. 120, and "Annulo Cingitur," 155–74.

14 Van Helden, "Annulo Cingitur," 160–1.

15 Van Helden, "Saturn and His Anses," 105, and A. Van Helden, "Christopher Wren's *De Corpore Saturni*," *Notes and Records of the Royal Society of London* xxiii (1968): 213–29: 220.

16 Stephen G. Brush, Elizabeth Garber, and C. W. F. Everitt, eds., *Maxwell on Saturn's Rings* (Cambridge, MA: MIT Press, 1983). Maxwell's treatise was published in 1857.

17 While a few of these pre-discovery observations were found shortly after the discovery of Uranus, most came to light during the period 1810–1821, when astronomers systematically searched for such earlier observations in order to improve theories of the orbit of Uranus. A. F. O. D. Alexander gives a complete list of twenty-two pre-discovery observations from 1690 to 1771, including a few by Bradley and Tobias Mayer, in *The Planet Uranus: A History of Observation, Theory and Discovery* (New York: American Elsevier, 1965), pp. 80–91. For an update see Eric G. Forbes, "The Pre-Discovery Observations of Uranus," in Garry Hunt, *Uranus and the Outer Planets* (Cambridge: Cambridge University Press, 1982), pp. 67–89. See also Mark Littmann, *Planets Beyond: Discovering the Outer Solar System* (New York: Wiley, 1988), pp. 24–28.

18 William Herschel, "Account of a Comet," *PTRSL* 71 (1781): 492–501, reprinted in part in Bartusiak, *Archives of the Universe* (hereafter Bartusiak), pp. 130–1.

19 Ellis D. Miner, *Uranus: The Planet, Rings and Satellites* (New York: Wiley, 1998), pp. 8–11.

20 William Herschel, *The Scientific Papers of Sir William Herschel*, J. L. E. Dreyer, ed. (London: Royal Society and Royal Astronomical Society, 1912), reprinted in Bartusiak, 132–33.

21 Clifford J. Cunningham, *The First Asteroid: Ceres, 1801–2001* (Surfside, FL: Star Lab Press, 2001), pp. 26–28, and on the Titius Bode Law, pp. 19–26.

22 Giuseppe Piazzi, "Results of the Observations of the New Star Discovered on the 1st of January 1801 at the Royal Observatory of Palermo," English translation excerpts reprinted in Bartusiak, pp. 151–52. The original manuscript (PRO Ref 285) is in the Maskelyne manuscripts of the RGO, now located at Cambridge University. The discovery is described in greatest detail in Clifford J. Cunningham, *First Asteroid*, pp. 35ff.

23 Piazzi's complete letter to Oriani appears in Cunningham, *First Asteroid*, p. 36.

24 "Über einen zwischen Mars und Jupiter längst vermutheten, nun wahrscheinlih entdeckten neuen Haupt-planeten unseres Sonnen-Systems," in von Zach, *Monatliche Correspondenz zur Beförderung der Erd-und Himmels-Kunde*, 3 (1801), 592–623.

25 Gauss, in Cunningham, *First Asteroid*, p. 41. On the debate as to exactly how Gauss made his determination, very close to the current elements, see Cunningham, pp. 40–47.

26 Cunningham, *First Asteroid*, p. 48. Eric G. Forbes, "Gauss and the Discovery of Ceres," *JHA* 2 (1971): 195–99.

27 The Bode paper is reprinted in Bartusiak, 149–51; on Banks's comment see David W. Hughes and Brian G. Marsden, "Planet, Asteroid, Minor Planet: A Case Study in Astronomical Nomenclature," *JAHH* 10 (March 2007): 21–30: 21.

28 A pre-discovery observation of Pallas has recently come to light, made by Messier and recorded in 1779 as a 7th magnitude star on his map of comet C/Bode (1779A1); Rene Bourtembourg, "Messier's Missed Discovery of Pallas in April, 1779," *JHA* 43 (2012): 209–14.

29 William Herschel, "Observations on the Two Lately Discovered Celestial Bodies," *PTRSL* 92 (1802): 213–32, read May 6, 1802, 228; Hughes and Marsden, "Planet, Asteroid," 22. Herschel's paper is

reprinted in full in Cunningham, *First Asteroid*, pp. 228–39: 237. It is also of interest that Herschel turned to the botanist Joseph Banks for help on nomenclature, but he rejected Banks's suggestion (via Stephen Weston) of *aorate*. Banks was greatly influenced by the work of Linnaeus, and had "a lifetime of experience classifying and naming newly-found objects." Clifford J. Cunningham and Wayne Orchiston, "Who Invented the Word Asteroid: William Herschel or Stephen Weston?" *JAHH* 14 (2011): 230–34.

30 Herschel to Piazzi, May 22, 1802, in Hughes and Marsden, "Planet, Asteroid," 22, citing Cunningham, *First Asteroid*; on Bode's letter to Herschel, see Hughes and Marsden, "Planet, Asteroid," 21.

31 Hughes and Marsden, "Planet, Asteroid," 22–30; James Hilton, "When Did the Asteroids Become Minor Planets?" online at http://www.usno.navy.mil/USNO/astronomical-applications/astronomical-information-center/minor-planets.

32 Hilton, "When Did the Asteroids Become Minor Planets?" http://aa.usno.navy.mil/hilton/AsteroidHistory/minorplanets.html; David W. Hughes and Brian G. Marsden, "Planet, Asteroid, Minor Planet: A Case Study in Astronomical Nomenclature," *JAHH* 10 (2007): 21–30. Eric G. Forbes, "Gauss and the Discovery of Ceres," *JHA* 2 (1971): 195–99.

33 Charles T. Kowal and Stillman Drake, "Galileo's Observations of Neptune," *Nature* 287 (25 September 1980): 311–13; David Jamieson, *Australian Physics* 46 (May–June 2009), online at http://www.ph.unimelb.edu.au/~dnj/AP-June-2009-Galileo-Neptune.pdf. Reported online as R. R. Britt, "New Theory: Galileo Discovered Neptune," at http://www.space.com/6941-theory-galileo-discovered-neptune.html.

34 Le Verrier's three memoirs are "Premier memoire sur la theorie d'Uranus," presented to the French Academy of Sciences, November 10, 1846, *Comptes Rendus*, 21 (1845): 1050; "Researches sur les mouvements d'Uranus," presented to the French Academy des Sciences June 1, 1846, *Comptes Rendus* 22 (1846): 907; and "Sur la planete qui produit les anomalies observed dans le mouvement d'Uranus – Determination das a masse, de son orbite et de sa position actuelle," read before the French Academie August 31, 1846, *Comptes Rendus*: 23, 428–38 and 657–62. Le Verrier sent an abstract of the second paper to the AN on September 8, 1846, published as "Letter to the Editor," in *AN* 25 (October 12, 1846), translated in Bartusiak, 162–65.

35 Le Verrier to Galle, September 18, 1846, translated in Morton's Grosser, *The Discovery of Neptune* (Cambridge, MA: Harvard University Press, 1962), pp. 115–16.

36 Encke's first account appeared in a Letter to the *Astronomische Nachrichten* 25 (October 12, 1846), translated in Bartusiak, 165–7, summarized in "Account of the Discovery of the Planet of Le Verrier at Berlin." *MNRAS* 7 (November 13, 1846): 153, as quoted here.

37 William Sheehan, Nicholas Kollerstrom, and Craig B. Waff, "The Case of the Pilfered Planet," *Scientific American* 291 (December, 2004): 92–99; R. W. Smith, "The Cambridge Network in Action: The Discovery of Neptune," *Isis* 80 (1989): 395–422. The literature on the discovery of Neptune is large. Morton Grosser's *The Discovery of Neptune* (Cambridge, MA: Harvard University Press, 1962) needs to be supplemented with more recent research, including Smith (cited above); A. Chapman,

"Private Research and Public Duty: George Biddell Airy and the Search for Neptune," *JHA* 19 (1988): 121–39; and J. G. Hubbell and R. W. Smith, "Neptune in America: Negotiating a Discovery," *JHA* 23 (1992): 261–90.

38 Thomas Kuhn, *The Copernican Revolution: Planetary Astronomy in the Development of Western Thought* (Cambridge, MA: Harvard University Press, 1957), pp. 52–54 for the Ptolemaic ordering; and Robert S. Westman, "Two Cultures or One? A Second Look at Kuhn's The Copernican Revolution," *Isis* 85 (1994): 79–115. In his landmark volume, *The Copernican Question: Prognostication, Skepticism, and Celestial Order* (Berkeley: University of California Press, 2011), Westman argues the importance of the question of planetary re-ordering in the Copernican tradition.

39 Albert van Helden, *Measuring the Universe: Cosmic Dimensions from Aristarchus to Halley* (Chicago: University of Chicago Press, 1985). On measured apparent planetary diameters in the seventeenth century, see Van Helden's table 15, p. 120, and on William Whiston's scheme of distances and planetary diameters as of 1715, see Van Helden's table 19, pp. 155–56. Actual planetary diameters depended on absolute distances, which were still very rough in the seventeenth century. The scale of the entire solar system could be determined by the absolute measure of the distance of one of its bodies, and Van Helden (pp. 142–43) shows that even Cassini's measurement of the parallax of Mars, though yielding distances within 7 percent of the modern value, were swamped by errors and largely a matter of luck. Nevertheless, he concludes that the new quantitative view of the heliocentric solar system came into focus toward the end of the seventeenth century (pp. 160–63). Subsequent refinements did little to change the relative sizes of the planets. For a list of distance determinations in the solar system from Aristarchus to 1976, see Steven J. Dick, *Sky and Ocean Joined: The U. S. Naval Observatory, 1830–2000* (Cambridge: Cambridge University Press, 2003), table 7.1, p. 241.

40 Christiaan Huygens, *The Celestial Worlds Discovered* (London: Frank Cass and Co., 1968), p. 112. This is a facsimile of the first English edition published in 1698. The first Latin edition, titled *Cosmotheoros*, was also published in 1698.

41 Agnes Clerke, *A Popular History of Astronomy During the Nineteenth Century* (London: Adam and Charles Black, 1902; first edition, 1885), p. 288.

42 See Stephen Brush, *Nebulous Earth: The Origin of the Solar System and the Core of the Earth from Laplace to Jeffreys* (Cambridge: Cambridge University Press, 1996), p. 31, table 3.

43 William Whewell, *Of the Plurality of Worlds* (Chicago: University of Chicago Press, 2001), pp. 179–85. This is a facsimile of the first edition (London: J. W. Parker, 1853).

44 Thomas Hockey, *Galileo's Planet* (Bristol and Philadelphia: Institute of Physics Publishing, 1999), pp. 64ff.

45 DeVorkin, "Stellar Evolution and the Origin of the Hertzsprung-Russell Diagram," in Owen Gingerich, ed., *Astrophysics and 20th-Century Astronomy to 1950* (Cambridge: Cambridge University Press, 1984), pp. 90–108: 90–91.

46 George H. Darwin, "On Internal Densities of Planets, and on an Oversight in the *Mécanique Céleste*," *MNRAS* 37 (1876): 77–88: 83; Hockey, *Galileo's Planet*, pp. 168–74.

47 Harold Jeffreys, "The Constitution of the Four Outer Planets," *MNRAS* 83 (1923): 350–54: 353.
48 Mark Marley, *Encyclopedia of the Solar System* (New York: Academic Press, 1999), p. 340; David Leveringon, *History of Astronomy from 1890 to the Present* (London: Springer, 1995), p. 65.
49 C. C. Kiess, C. H. Corliss, and H. K. Kiess, "High-Dispersion Spectra of Jupiter," *ApJ* 132 (1960): 221; Fran Bagenal, Timothy Dowling, and William McKinnon, *Jupiter: The Planet, Satellites and Magnetosphere* (Cambridge: Cambridge University Press, 2004), p. 59; Mark S. Marley, "Interiors of the Giant Planets," in Paul R. Weissman, Lucy-Ann McFadden, and Torrence V. Johnson, *Encyclopedia of the Solar System* (San Diego: Academic Press, 1999), pp. 339–55: 340.
50 "Solar Plexus," in Judith Merril, ed., *Beyond Human Ken*, (1952); R. M. Petrie, "The Atmospheres of the Planets," *JRASC* 34 (1940): 137–45: 143.
51 Mark S. Marley, "Interiors of the Giant Planets," p. 352; also Heidi Hammel, "The Ice Giant Systems of Uranus and Neptune," in Phillippe Blonde and John Mason, eds., *Solar System Update* (New York: Springer, 2006), pp. 251–65: 251.
52 S. A. Stern, "The Third Zone: Exploring the Kuiper Belt," *Sky and Telescope*, 106 (2003): 31–36.
53 John K. Davies et al., "The Early Development of Ideas Concerning the Transneptunian Region," in M. A. Barucci et al., eds., *The Solar System Beyond Neptune* (Tucson: University of Arizona Press, 2008), pp. 11–23.
54 K. E. Edgeworth, "The Evolution of our Planetary System," *JBAA* 53 (1943): 181–88: 186; Edgeworth, "The Origin and Evolution of the Solar System," *MNRAS* 109 (1949): 600–9; and Gerard Kuiper, "On the Origin of the Solar System," in J. A. Hynek, ed., *Astrophysics: A Topical Symposium* (New York: McGraw-Hill, 1951), pp. 357–424: 402.
55 F. Whipple, "Evidence for a Comet Belt Beyond Neptune," *PNAS* 51 (1964): 711–18; J. Fernandez, "On the Existence of a Comet Belt Beyond Neptune," *MNRAS* 192 (1980): 481–91.
56 M. Duncan et al., "The Origin of Short-period Comets," *ApJ* 328 (1988): L69–L73.
57 David Jewitt and Jane Luu, "Discovery of the Candidate Kuiper Belt Object 1992 QB1," *Nature* 362 (1993): 730–32.
58 For the context of these discoveries, see David Weintraub, *Is Pluto a Planet? A Historical Journey through the Solar System* (Princeton: Princeton University Press, 2007), pp. 158ff.; Govert Schilling, *The Hunt for Planet X: New Worlds and the Fate of Pluto* (New York: Copernicus Books, 2009), and Michael Brown, *How I Killed Pluto and Why It Had It Coming* (New York: Spiegel & Grau, 2010). For the Hubble results, see http://hubblesite.org/newscenter/archive/releases/2009/33/.
59 Jane Luu, Brian G. Marsden, David Jewitt et al., "A New Dynamical Class of Object in the Outer Solar System," *Nature* 387 (1997): 573–75.
60 Michael E. Brown, Chadwick Trujillo, and David Rabinowitz, "Discovery of a Candidate Inner Oort Cloud Planetoid," *ApJ* 617 (2004): 645–49; http://www.gps.caltech.edu/%7Embrown/papers/ps/sedna.pdf.
61 Brown, *How I Killed Pluto;* also http://en.wikipedia.org/wiki/List_of_dwarf_planet_candidates.
62 M. Barucci et al., "The Solar System Beyond Neptune: Overview and Perspectives," in M. A. Barucci et al., eds., *Solar System Beyond Neptune*,

pp. 3–10. An official list of Transneptunian objects, updated daily, is found at http://www.cfa.harvard.edu/iau/lists/TNOs.html.
63 M. Fulchignoni et al., "Transneptunian Object Taxonomy," in M. A. Barucci et al., eds., *The Solar System Beyond Neptune*, pp. 181–92.
64 David C. Jewitt, "The Kuiper Belt as an Evolved Circumstellar Disk," in *The Origin of Stars and Planets: The VLT View* (Berlin: Springer-Verlag, 2002), 405–15.

3. In Herschel's Gardens: Nebulous Discoveries in the Realm of the Stars

1 William Herschel, "On Nebulous Stars, Properly so Called," *PTRSL* 81 (1791): 71–88, reprinted in Michael Hoskin, *The Construction of the Heavens: William Herschel's Cosmology* (Cambridge: Cambridge University Press, 2012), pp. 146–56: 153.
2 Herschel, "Catalogue of 500 New Nebulae," *PTRSL* 92 (1802): 477–528; Simon Schaffer, "Herschel in Bedlam: Natural History and Stellar Astronomy," *BJHS* 13 (1980): 211–39: 218.
3 On cosmic distances, see Albert van Helden, *Measuring the Universe: Cosmic Dimensions from Aristarchus to Halley* (Chicago and London: University of Chicago Press, 1985), and Alan W. Hirshfeld, *Parallax: The Race to Measure the Cosmos* (New York: Henry Holt and Company, 2001).
4 For example, in Ptolemy's star catalogue, "the nebulous star in the head of Orion" [gamma, theta 1, and theta 2 Orionis] or "The nebulous mass on the right hand" in Perseus [h and chi Persei], *Ptolemy's Almagest*, translated and annotated by G. J. Toomer (Princeton: Princeton University Press, 1998), pp. 352 and 382, repeated in the catalogue at the end of Book II of Copernicus's *De revolutionibus*. These nebulous stars were later resolved into stars. For details, see Kenneth Glyn Jones, *The Search for the Nebulae* (Chalfont St. Giles: Science History Publications, 1975), pp. 5–7.
5 Galileo Galilei, *Sidereus Nuncius, or The Sidereal Messenger*, translated with introduction, conclusion, and notes by Albert van Helden (Chicago: University of Chicago Press, 1989), p. 62. The same may be said of double and multiple stars, as we shall see in Chapter 4.
6 Owen Gingerich gives this interpretation in "The Mysterious Nebulae, 1610–1924," in *The Great Copernicus Chase* (Cambridge: Cambridge University Press, 1992), pp. 225–37: 229, as does van Helden in his translation of *Sidereus Nuncius*, note 65, pp. 59–60.
7 On Peiresc's ambiguous observations see S. L. Chapin, "The Astronomical Activities of Nicolas-Claude Fabri de Peiresc," *Isis* 1 (1957): 13–29, and Harald Siebert, "Peiresc's Nebel im Sternbild Orion - eine neue Textgrundlage fur die Geschichte on M42," [Peiresc's Nebula in the constellation of Orion, a new textual basis for the history of M42,] *Annals of Science* 66 (2009): 231–46. The latter uses Peiresc's notebooks to question whether he actually observed the nebula, a claim only made in 1916 by the French astronomer Guillaume Bigourdan. See also E. S. Holden, "Monograph of the Central Parts of the Nebula of Orion," appendix I of *Astronomical and Meteorological Observations made during the year 1878 at the United States Naval Observatory* (Washington,

DC: Government Printing Office, 1882). Huygens's results, using his 12- or 23-foot focal length telescope with a lens little more than 2 inches, were reported in his *Systema Saturnium* of 1659. Gingerich, "The Mysterious Nebulae, 1610–1924," p. 230), says the Sicilian astronomer Giovanni Battista Hodierna drew the Orion Nebula for the first time in 1653, but Hodierna's account is exceedingly rare and likely had little influence compared to that of Huygens.

8 Homer, *The Odyssey* (New York: Penguin Books, 1997), Robert Fagles, trans., Book V, 297–300, p. 160; The Old Testament, Book of Job, verse 38:31; Ovid, *Metamorphoses*, Books VI and XIII. The latter reads, referring to Ajax, "He understands nothing of the shield's engraving, Ocean, or earth, or high starry sky; the Pleiades and the Hyades, the Bear, that is always clear of the waters, and opposite, beyond the Milky Way, Orion, with his glittering sword. He demands to bear armour that he does not comprehend!" A. S. Kline translation online at http://www.poetryintranslation.com/PITBR/Latin/Metamorph13.htm.

9 Michell made these claims in his paper, "An Inquiry into the probable Parallax, and Magnitude of the fixed Stars, from the Quantity of Light which they afford us, and the particular Circumstances of their Situation," *PTRSL* 57 (1767): 234–64. On Michell see Laurent Hodges, "Michell," *BEA* 2: 778–9. Herschel was influenced by Michell's argument; see below.

10 La Caille's *Memoire* is translated in Jones, *Search for the Nebulae*, pp. 44–49, where each of the objects is identified according to our knowledge today. Jones also translates (pp. 36–40) another precursor, Philippe Loys de De Cheseaux's "Catalogue of Truly Nebulous Stars," compiled in 1745 or 1746, which had two classes – "star clusters" and "true nebulae." These two works demonstrate the belief in a class of nebulae separate from stars. Jones also documents precursors to La Caille and Messier who observed a few nebulous objects, including Halley in 1715 and Flamsteed in 1929.

11 Charles Messier, *Catalogue des Nébuleuses & des amas d'Étoiles* [Catalogue of Nebulae and Clusters of Stars], *Connaissance des Temps* for 1784 (published 1781), pp. 227–67, with 103 objects, translated into English in Jones, *Search for the Nebulae*, pp. 61–76. A first edition of the *Catalogue* with forty-five objects was published in 1774 in the *Mémoires de l'Académie Royale des Sciences* for 1771, Paris (dated February 16, 1771, published 1774), pp. 435–61, English translation online at http://seds.org/messier/xtra/history/m-cat71.html. A second edition was published in 1780 in the *Connaissance des Temps* for 1783. The *Connaissance* (spelled Connoissance prior to 1779) was an almanac, and therefore published several years ahead of time. On the history of these catalogues, see Owen Gingerich, "Messier and His Catalogue," in John H. Mallas and Evered Kreimer, *The Messier Album* (Cambridge, MA: Sky Publishing, 1978), where the original French of the 1781 third edition [actually a reprint of this edition published in 1784] is reprinted. For an identification of the objects see Kenneth Glyn Jones, *Messier's Nebulae and Star Clusters,* 2nd ed. (Cambridge: Cambridge University Press, 1991), pp. 17ff., also http://seds.org/messier/xtra/history/m-cat.html. Those objects that turned out to be open star clusters are also listed at http://seds.org/MESSIER/open.html, and those that turned out to be globular clusters are at http://cseligman.com/

text/stars/messierglobulars.htm. Wolfgang Steinicke's recent volume, *Observing and Cataloguing Nebulae and Star Clusters: From Herschel to Dreyer's New General Catalogue* (Cambridge: Cambridge University Press, 2010), is also a very valuable work.

12 Michael Hoskin has been the leading scholar on William Herschel over the last two generations. His *William Herschel and the Construction of the Heavens* (London: Oldbourne, 1963), based solely on Herschel's published work, has now been superseded by his volume *The Construction of the Heavens: William Herschel's Cosmology* (Cambridge: Cambridge University Press, 2012), from which passages in the rest of this section are cited. Hoskin's latest research, aside from the extensive introduction to this volume, is found in *Discoverers of the Universe: William and Caroline Herschel* (Princeton: Princeton University Press, 2011). For more succinct biographies see M. A. Hoskin, "Herschel, William," *DSB*, updated in Michael Hoskin, *NDSB* 1, pp. 289–91, and Michael Crowe and Keith Lafortune, "Herschel, (Friedrich) William [Wilhelm]," *BEA*, vol. 1, pp. 494–96. Constance Lubbock's *The Herschel Chronicle: The Life-Story of William Herschel and his Sister Caroline Herschel* (Cambridge: Cambridge University Press, 1933) is still very useful. Woodruff T. Sullivan III is now at work on a full-scale biography of Herschel.

13 The papers "On the Construction of the Heavens" are reprinted in part in Hoskin, *The Construction of the Heavens*, and are available in full in the Astrophysics Data System online. Hoskin originally viewed the first three papers as constituting a first synthesis, in which nebulae were seen as collections of stars held together by gravity, and the latter two papers (plus one "On Nebulous Stars" published in 1791) as the second synthesis, where he discovered true nebulosity. As we shall see below, based on Herschel's papers Hoskin later discovered things were more complicated, requiring three syntheses. The 1789 paper constitutes the introduction to one of Herschel's three catalogues of nebulae and star clusters, where he not only published his observations, but also offered introductory remarks explaining them. These catalogues are "Catalogue of One Thousand New Nebulae and Clusters of Stars," *PTRSL* 76 (1786): 457–99; "Catalogue of a Second Thousand New Nebulae and Clusters of Stars," *PTRSL* 79 (1789): 212–55; and "Catalogue of 500 New Nebulae, Nebulous Stars, Planetary Nebulae, and Clusters of Stars," *PTRSL* 92 (1802): 477–528.

14 Hoskin, *Discoverers*, p. 97. As Steinicke shows, the mean magnitude of Herschel's objects was 12, though many were as faint as 15th magnitude, an amazing achievement considering his 18.7-inch speculum metal mirror was equivalent to a modern 10-inch aluminized glass mirror; see pp. 34–35 in *Observing and Cataloguing Nebulae*, especially figure 2.18.

15 Herschel, 1786, *Catalogue of One Thousand New Nebulae*, p. 466. As Steinicke shows (*Observing and Cataloguing Nebulae*, p. 32), the vast majority of Herschel's eight classes of "nebulae" (2113 out of 2438) were eventually identified as galaxies. Of the remaining, 25 were emission nebulae, 11 reflection nebulae, 34 planetary nebulae, 164 open clusters, and 38 globular clusters.

16 Herschel, "On the Construction of the Heavens," *PTRSL* 75 (1785): 213-66, in Hoskin, *Construction of the Heavens*, p. 115.

17 Herschel, "Catalogue of a Second Thousand of new Nebulae and Clusters of Stars; with a few introductory Remarks on the Construction of the Heavens," *PTRSL* 79 (1789): 212–26, in Hoskin, *Construction of the Heavens*, p. 137.
18 Hoskin, *Construction of the Heavens*, pp. 138–40.
19 Ibid., p. 145. Herschel's biological analogies are discussed in Stephen G. Brush, *Nebulous Earth: The Origin of the Solar System and the Core of the Earth From Laplace to Jeffreys* (Cambridge: Cambridge University Press, 1996), pp. 33–35; see also Richard Panek,"Herschel's Garden," *Natural History*, May 1999, at http://findarticles.com/p/articles/mi_m1134/is_4_108/ai_54574610/.
20 Hoskin, *Construction of the Heavens*, pp. 34–76, and Hoskin, "William Herschel's Early Investigations of Nebulae: A Reassessment," *JHA* x (1979): 165–76; Robert Smith, "Beyond the Galaxy: The Development of Extragalactic Astronomy 1885-1965, Part I," *JHA* 39 (2008): 91–119; Part II, *JHA* 40 (2009): 71–107; and Smith, *The Expanding Universe: Astronomy's Great Debate, 1900–1931* (Cambridge: Cambridge University Press, 1982), pp. 3ff.; Simon Schaffer, "Herschel in Bedlam: Natural History and Stellar Astronomy," *BJHS* 13 (1980): 211–39.
21 Hoskin, *William Herschel*, 61–62.
22 William Herschel, "On Nebulous Stars, Properly so called," *PTRSL* 1791, in Hoskin, *Construction of the Heavens*, pp. 146–156: 146; Hoskin, "William Herschel's Early Investigations of Nebulae: A Reassessment," *JHA* x (1979): 165–76.
23 Hoskin, *Construction of the Heavens*, pp. 146–56: 152–53. The "nebulous star" is a planetary nebula known as NGC 1514. It is notable that of the 77 objects Herschel would eventually place in Class IV as "Planetary Nebulae," only 20 are today classified planetaries; 39 of them are galaxies. See Steinicke, *Observing and Cataloguing Nebulae*, p. 32.
24 Simon Newcomb and E. S. Holden, *Astronomy for High Schools and Colleges* (New York: Henry Holt, 1881), p. 459.
25 Schaffer, "Herschel in Bedlam," 221–22, and 213ff. on the Bath Philosophical Society; Hoskin, *NDSB*, vol. 1, 290. Roy Porter also makes clear the Bath milieu from which Herschel sprang, suggesting that the Bath Philosophical Society, with its emphasis on natural history and classification, "was midwife to Herschel's astronomical revolution." Roy Porter, "Herschel, Bath and the Philosophical Society," in Garry Hunt, ed., *Uranus and the Outer Planets* (Cambridge: Cambridge University Press, 1982), pp. 23–31.
26 Michel Foucault, *The Order of Things: An Archaeology of the Human Sciences* (New York: Random House, 1970).
27 Hoskin, *Construction of the Heavens*, p. 157.
28 Herschel, "Catalogue of 500 new Nebulae," 1802; Schaffer, "Herschel in Bedlam," 218.
29 On the identification of the nebulae in modern terms, see Wolfgang Steinicke, *Nebulae and Star Clusters*, pp. 30–34, especially table 2.14; Bernhard Sticker, "'Artificial' and 'Natural' Classifications of Celestial Bodies in the Work of William Herschel," *Proceedings of the Xth International Congress of the History of Science* (Ithaca, 1962), ii, pp. 729–31.
30 Michael Hoskin, "Rosse, Robinson, and the Resolution of the Nebulae," *JHA* 21 (1990): 331–44: 341. The paper, by the Third Earl of Rosse's son,

is Lord Oxmantown [later fourth Earl of Rosse], "An account of the observations on the Great Nebula in Orion, made at Birr Castle with the 3-feet and 6-feet telescopes, between 1848 and 1867," *PTRSL* 158 (1868): 57–63.

31 William Huggins, "On the Spectra of Some of the Nebulae," *PTRSL* 154 (1864): 437–44.

32 Ibid., 437.

33 Ibid., 442.

34 Bartusiak, *Archives*, p. 219, citing Huggins, "The New Astronomy," *The Nineteenth Century* 41 (June 1897): 907–29: 916–17. The text of the entire essay in reprinted in Barbara Becker, *Unravelling Starlight: William and Margaret Huggins and the Rise of the New Astronomy* (Cambridge: Cambridge University Press, 2011), pp. 328–46. The quote cited here is on p. 335. On the history of astronomical spectroscopy, see John B. Hearnshaw, *The Analysis of Starlight: One Hundred and Fifty Years of Astronomical Spectroscopy* (Cambridge: Cambridge University Press, 1986).

35 Becker, *Unravelling Starlight*, pp. 64–80.

36 Otto Struve, "Recent Progress in the Study of Reflection Nebulae," *PA* 45 (1937): 9–22.

37 V. M. Slipher, "On the Spectrum of the Nebula in the Pleiades," *Lowell Observatory Bulletin* 1 (1912): 26–27; Slipher, "On the Spectrum of the Nebula about Rho Ophiuchi," *PA* 24 (1916): 542–43.

38 Slipher, "On the Spectrum of the Nebula in the Pleiades," p. 26. Slipher cites Clerke's *Problems in Astrophysics*, p. 420.

39 Otto Struve, "Recent Progress in the Study of Reflection Nebulae," *PA* 45 (1937): 9–22:10. This article provides an overview of the status of the subject twenty-five years after Slipher's observations.

40 H. E. Houghton, "Sir William Herschel's 'Hole in the Sky'," *Monthly Notices of the Astronomical Society of South Africa* 1 (1942): 107, transcribed online at http://www.docdb.net/show_object.php?id=barnard_86. Herschel reported the object in "On the Construction of the Heavens," *PTRSL* 75 (1785): 213–66, in Hoskin, *Construction of the Heavens*, p. 128. The object today is most often identified with Barnard 86. Barnard used the term "black hole" to describe it, an early usage that would come to have an entirely different meaning sixty years later. Steinicke, *Observing and Cataloguing Nebulae*, p. 142, notes that the identification with Barnard 86 is doubtful, and thinks it more likely to have been the rho Ophiuchi dark nebula.

41 E. E. Barnard, "On the Vacant Regions of the Sky," *PA* 14 (1906): 579–83: 579; William Sheehan, *The Immortal Fire Within: The Life and Work of Edward Emerson Barnard* (Cambridge: Cambridge University Press, 1995), especially chapter 22.

42 Sheehan, *The Immortal Fire Within*, p. 375; E. E. Barnard, "Some of the Dark Markings on the Sky and What they Suggest," *ApJ* 43 (1916): 1–8: 4.

43 E. E. Barnard, "On the Dark Markings of the Sky, with a Catalogue of 182 Such Objects," *ApJ* 49 (1919): 1–24: 2; E. E. Barnard, *A Photographic Atlas of Selected Regions of the Milky Way*, E. B. Frost and M. R. Calvert, eds. (Washington, DC: Carnegie Institution of Washington,1927).

44 George Ellery Hale, "Barnard's Dark Nebulae," *The Depths of the Universe* (New York: Charles Scribner's Sons, 1924), pp. 37–61: 56; Russell's work is "Dark Nebulae," *PNAS* 8 (1922): 115–18. For subsequent events

see B. T. Lynds, "Dark Nebulae," in Barbara Middlehurst and Lawrence Aller, eds., *Nebulae and Interstellar Matter* (Chicago: University of Chicago Press, 1968), pp. 119–39.

45 E. P. Hubble, "The Source of Luminosity in Galactic Nebulae," *ApJ* 56 (1922): 400–38.

46 Henry Norris Russell, "Dark Nebulae," *PNAS* 8 (1922): 115–18: 117, where he cites his earlier work in *The Observatory* 44 (1921): 72.

47 D. H. Menzel, "The Planetary Nebulae," *PASP* 38 (1926): 295–312; H. Zanstra, *Physical Review* 27 (1926): 644. On Zanstra's work, which Otto Struve described as being based on "a truly revolutionary idea," see Struve and Velta Zebergs, *Astronomy of the 20th Century* (New York: Macmillan, 1962), pp. 386–90; Donald Osterbrock, "Herman Zanstra, Donald H. Menzel, and the Zanstra Method of Nebular Astrophysics," *JHA* 32 (2001): 93–108; R. O. Redman, "The Award of The Gold Medal to Professor Herman Zanstra," *QJRAS* 2 (1961): 109–11, and Zanstra's summary of his own work at "The Gaseous Nebula as a Quantum Counter (George Darwin Lecture)," *QJRAS* 2 (1961): 137–48.

48 E. P. Hubble, "The Source of Luminosity in Galactic Nebulae." *ApJ* 56 (1922): 400–38.

49 Otto Struve and Elvey, "Emission Nebulosities in Cygnus and Cepheus," *ApJ* 88 (1938): 364; Struve and C. T. Elvey, "Observations Made with the Nebular Spectrograph of the McDonald Observatory," series of three articles in *ApJ* 89 (1939): 119ff. and 517ff., and vol. 90, 301ff.; Arthur S. Eddington, "The Density of Interstellar Calcium and Sodium," *MNRAS* 95 (1934): 2–11.

50 Bengt Strömgren, "The Physical State of Interstellar Hydrogen," *ApJ* 89 (1939): 526; Strömgren, "On the Density Distribution and Chemical Composition of the Interstellar Gas," *ApJ* 108 (1948): 242. For the context on how Strömgren came to his ideas about the formation of H II regions, see Lillian Hoddeson and Gordon Baym interview with Strömgren, May 5, 1967, AIP Center for the History of Physics, online at http://www.aip.org/history/ohilist/5070_1.html.

51 Stewart Sharpless, "A Catalogue of H II Regions," *AJ* Supplement, vol. 4 (1959): 257; "A Catalogue of Emission Nebulae Near the Galactic Plane," *ApJ* 118 (1953): 362. The Sharpless 1948 citation is Strömgren, "On the Density Distribution."

52 W. W. Morgan, S. Sharpless, and D. Osterbrock, "Some Features of Galactic Structure in the Neighbourhood of the Sun," *AJ* 57, no. 3 (1952); and *Sky and Telescope* 11, no. 134 (1952).

53 Strömgren, "On the Density Distribution," 242. For background, see AIP interview with Strömgren, cited above.

54 J. Hartmann, "Investigations of the Spectrum and Orbit of delta Orionis," *ApJ* 19 (1904): 268–86: 274.

55 Otto Struve, "On the Calcium Clouds," *PA* 34 (1926): 10.

56 Otto Struve, "The Physical State of Interstellar Gas Clouds," *PNAS* 25 (1939): 36–43; also Otto Struve and Velta Zebergs, *Astronomy of the Twentieth Century*, (New York and London: Macmillan, 1962), 375; and Struve, "Note on Calcium Clouds," *ApJ* (1934). Eddington 's work is "The density of interstellar calcium and sodium," *MNRAS* 95 (1934): 2–11, and Theodore Dunham's is "Interstellar Neutral Potassium and

Neutral Calcium," *PASP* 49 (1937): 26–8, and "Forbidden Transition in the Spectrum of Interstellar Ionized Titanium," *Nature* 139 (1937): 246–47. Saha commented that Dunham's work "forms a landmark in the story of interstellar investigations," *Nature* 139 (1937): 840.

57 H. I. Ewen and E. M. Purcell, "Observation of a Line in the Galactic Radio Spectrum: Radiation from Galactic Hydrogen at 1,420 Mc/sec," *Nature* 168 (1951): 356, followed by C. A. Muller and J. H. Oort, "Observation of a Line in the Galactic Radio Spectrum: The Interstellar Hydrogen Line at 1,420 Mc./sec., and an Estimate of Galactic Rotation," 357. For details, see Woodruff T. Sullivan, III, *Cosmic Noise: A History of Early Radio Astronomy* (Cambridge: Cambridge University Press, 2009), chapter 16, "The 21 cm hydrogen line," 394–417. For recollections by radio pioneers, see Gart Westerhout, "The Pioneers of H I," in A. R. Taylor, T. L. Landecker, and A. G. Willis, eds., *Seeing through the Dust: The Detection of H I and the Exploration of the ISM in Galaxies* (San Francisco: ASP, 2002), vol. 276, p. 3, and the adjacent series of historical articles. This volume also details how the field of 21-cm radio astronomy grew over the next fifty years.

58 J. H. Oort, G. Westerhout, and F. J. Kerr, "The Galactic System as a Spiral Nebula," *MNRAS* 118 (1958): 379–89. The famous map of the galaxy is figure 4. See also the previous 1954 H I map cited in note 5 of this paper. Peter M. W. Kalberla and Jurgen Kerp, "The H I Distribution of the Milky Way," *ARAA* 47 (2009): 27–61: 27.

59 J. M. Dickey and F. J. Lockman, "H I in the Galaxy," *ARAA* 28 (1990): 215–61; Felix J. Lockman, "Discovery of a Population of H I Clouds in the Galactic Halo," *ApJ* 580 (2002): L47–L50.

60 G. Cocconi and P. Morrison, "Search for Interstellar Communications," *Nature* 184 (September 1959): 844. On the context see Steven Dick, *The Biological Universe: The Twentieth Century Extraterrestrial Life Debate and the Limits of Science* (Cambridge: Cambridge University Press, 1996), chapter 8.

61 P. W. Merrill, "Unidentified Interstellar Lines," *PASP* 46 (1934): 206–7, and Merrill, "Stationary Lines in the Spectrum of the Binary Star Boss 6142," *ApJ* 83 (1936): 126–28; T. Dunham Jr., and W. S. Adams, *PASP* 9 (1937): 5; T. Dunham, Jr., "Interstellar Neutral Potassium and Neutral Calcium," *PASP* 49 (1937): 26–28; Dunham, "Forbidden Transition in the Spectrum of Interstellar Ionized Titanium," *Nature* 139 (1937): 246–47. See "The Pioneering Investigations in the Field of the Interstellar Molecules, 1935–1942," *Astrophysics and Space Science* 55 (1978): 263–65.

62 Andrew McKellar, "Evidence for the Molecular Origin of Some Hitherto Unidentified Interstellar Lines," *PASP* 52 (1940): 187–92; P. Swings and L. Rosenfeld, "Considerations Regarding Interstellar Molecules," *ApJ* 86 (1937): 483–86. McKellar cites Otto Struve and C. T. Elvey as having supported the abundance of hydrogen by their observation of H alpha emission in a number of extended Milky Way regions; Struve and Elvey, "Emission Nebulosities in Cygnus and Cepheus," *ApJ* 88 (1938): 364.

63 Kaler, *Cosmic Clouds: Birth, Death and Recycling in the Galaxy* (New York: Scientific American, 1997), 104.

64 S. Weinreb, A. H. Barrett, M. L. Meeks, and J. C. Henry, "Radio Observations of OH in the Interstellar Medium," *Nature* 200 (1963): 829–31. For the competitive environment see Alan H. Barrett, "The Beginnings of Molecular Radio Astronomy," in K. Kellermann and B.

Sheets, *Serendipitous Discoveries in Radio Astronomy* (Green Bank: NRAO, 1983), pp. 280–90.

65 Alan Barrett, in Kellermann and Sheets, *Serendipitous Discoveries*, p. 286.

66 N. Z. Scoville and P. M. Solomon, "Molecular Clouds in the Galaxy," *ApJ* 199 (1975): L105–L109; Lewis Snyder, "The Search for Biomolecules in Space," in K. I. Kellermann and G. A. Seielstad, eds., *The Search for Extraterrestrial Intelligence* (Green Bank: NRAO, 1986), pp. 39–50.

67 Lew Snyder et al., "A Rigorous Attempt to Verify Interstellar Glycine," *ApJ* 619 (2005): 914–30. The original discovery paper was Yi-Jehng Kuan et al., "Interstellar Glycine," *ApJ* 593 (2003): 848–67.

68 Eric Herbst and Ewine F. van Dishoeck, "Complex Organic Interstellar Molecules," *ARAA* 47 (2009): 427–80, including a list of those with more than six atoms (p. 431). An online list with discovery year is available at http://www.daviddarling.info/encyclopedia/I/ismols.html; more official lists are online at http://physics.nist.gov/PhysRefData/Micro/Html/tab1.html and more information at http://www.astrochymist.org/astrochymist_mole.html http://www.astrochymist.org/astrochymist_ism.html.

69 Bart J. Bok and Edith F. Reilly, "Small Dark Nebulae," *ApJ* 105 (1947): 255–57. A. D. Thackeray, "Some Southern Stars Involved in Nebulosity," *MNRAS* 110 (1950): 524; Bo Reipurth, "Star formation in Bok globules and low-mass clouds. I – The cometary globules in the GUM Nebula," *A & A* 117 (1983): 183–98. Russell had proposed dark nebulae as sites of star formation as early as 1924.

70 N. Z. Scoville and P. M. Solomon, "Molecular Clouds in the Galaxy," *ApJ* 199 (1975): L105–L109; P. M. Solomon, D. B. Sanders, and N. Z. Scoville, "Giant Molecular Clouds in the Galaxy: A Survey of the Distribution and Physical Properties of GMC's," *BAAS* 9 (1977): 554; full article in Solomon, Sanders, and Scoville, "Giant Molecular Clouds in the Galaxy – The Distribution of ^{13}CO Emission in the Galactic Plane," *ApJ* 232 (1979): L89–L93. The 1977 article may have been the first use of the term "giant molecular cloud," extending the earlier use of "molecular cloud."

71 Yasuo Fukui and Akiko Kawamura, "Molecular Clouds in Nearby Galaxies," *ARAA* 48 (2010): 547–80.

4. Dwarfs, Giants, and Planets (Again!): The Discovery of the Stars Themselves

1 Hertzsprung to E. C. Pickering, August 17, 1908. Cited in David DeVorkin, *Henry Norris Russell: Dean of American Astronomers* (Princeton: Princeton University Press, 2000), p. 85.

2 Henry Norris Russell, "Giant and Dwarf Stars," *The Observatory* 36 (1913): 324–29: 325–26.

3 A. S. Eddington, *Stars and Atoms* (Oxford: Clarendon Press, 1927), p. 50.

4 David H. DeVorkin, "A Fox Raiding the Hedgehogs: How Henry Norris Russell Got to Mt. Wilson," in Gregory A. Good, ed., *The Earth, the Heavens and the Carnegie Institution of Washington* (New York: American Geophysical Union, 1994), pp. 103–11: 103.

5 Bunsen to the English chemist H. E. Roscoe, November, 1859, cited in A. J. Meadows, "The Origins of Astrophysics," in Owen Gingerich, ed., *Astrophysics and 20th Century Astronomy to 1950*: Part A (Cambridge: Cambridge University Press, 1984), p. 5. This article is a succinct

discussion of the first years of astrophysics; for a more comprehensive treatment, see John B. Hearnshaw, *The Analysis of Starlight: One Hundred and Fifty Years of Astronomical Spectroscopy* (Cambridge: Cambridge University Press, 1986).

6 On the corps of women computers see Bessie Jones and Lyle Boyd, *The Harvard College Observatory (1839–1919)* (Cambridge, MA: Harvard University Press, 1971), especially chapter 11; Peggy Kidwell, "Three Women of American Astronomy" in *American Scientist*, May/June 1990, p. 244. For an annotated illustration dated May 13, 1913, see Barbara L. Welther, "Pickering's Harem," *Isis* 73 (1982): 94. On Fleming, see the Fleming entry by Katalin Keri, *BEA*, p. 374, the *DSB* entry by Owen Gingerich, 5, pp. 33–34, and Dorrit Hoffleit, in *Notable American Women* (Cambridge, MA: Harvard University Press, 1971).

7 On Maury, see the Maury entry by Virginia Trimble, *BEA*, p. 74, and by Owen Gingerich, *DSB*, vol. 9, pp. 194–95. Also Dorrit Hoffleit, "Reminiscences on Antonia Maury and the c-Characteristic," in C. J. Corbally, R. O. Gray, and R. F. Garrison, eds., *The MK Process at 50 Years* (San Francisco: Astronomical Society of the Pacific, 1994), pp. 215–23.

8 On Cannon, see the Cannon entry by Owen Gingerich in *DSB* 3, pp. 49–50; Harry G. Lang, *BEA*, pp. 198–99; Barbara Welther, "Annie J. Cannon: Classifier of the Stars" in *Mercury* (Jan/Feb. 1984): 28; George Greenstein, "The Ladies of Observatory Hill: Annie Jump Cannon and Cecilia Payne-Gaposchkin," in *Portraits of Discovery: Profiles in Scientific Genius* (New York: Wiley, 1998), pp. 7–30.

9 This story has been told many times, in most scientific detail in John B. Hearnshaw, *The Analysis of Starlight*, and in most analytical detail, from the point of view of Henry Norris Russell's work, in David DeVorkin's biography of Russell, *Henry Norris Russell: Dean of American Astronomers* (Princeton: Princeton University Press, 2000). On the Harvard sequence, see p. 132 of the latter; on Saha, pp. 177–78.

10 William Monck, "The Spectra and Colours of Stars," *JBAA* 5 (1895): 416–19: 418; Sandage, 239; David DeVorkin, "Stellar Evolution and the Origin of the Hertzsprung-Russell Diagram," in Owen Gingerich, ed., *Astrophysics and 20th Century Astronomy to 1950* (Cambridge: Cambridge University Press, 1984), pp. 90–108: 96. On the popular literature see, for example, J. E. Gore, *The Worlds of Space* (1894).

11 Ejnar Hertzsprung, "Zur Stralung der Sterne [On the Radiation of Stars]," *Zeitschrift für Wissenschaftliche Photographie* 3 (1905): 429–42, translated into English in Lang and Gingerich, *Source Book in Astronomy and Astrophysics*, pp. 208–211; the second paper was in the same journal, vol. 5 (1907): 86–107. Although Dieter Herrmann (*BEA*, pp. 496–97) says Hertzsprung showed that stars existed in two groups "which he called 'Riesen' (giants) and 'Zwerge' (dwarfs)," and although Russell attributed these terms to Hertzsprung, they do not appear in these two papers. See note 16 below. On Hertzsprung's life and work, see Herrmann's *Ejnar Hertzsprung: Pionier der Sternforschung* (Berlin: Springer-Verlag, 1994), and for a more succinct account by one of his students, K. Aa. Strand, Hertzsprung entry in *DSB*, vol. 6, pp. 350–53.

12 Hertzsprung to Pickering, August 17, 1908, as cited in DeVorkin, *Henry Norris Russell*, p. 85, where the broader context is given, and in DeVorkin, "Stellar Evolution," pp. 90–108: 98.

13 DeVorkin, *Henry Norris Russell*, p. 86, and p. 396, note 39, where he cites Karl Schwarzschild, "Über das System der Fixsterne," *Himmel und Erde* 21 (1909): 433–51, abstract in *Astronomische Jahresbericht*, 179.
14 David DeVorkin, "Stellar Evolution," p. 102. See also Allan Sandage, *The Mount Wilson Observatory*, volume 1 of the Centennial History of the Carnegie Institution of Washington (Cambridge: Cambridge University Press, 2004), p. 240.
15 DeVorkin, "Stellar Evolution," p. 102; Sandage, *Mount Wilson Observatory*, p. 402.
16 Henry Norris Russell, "Giant and Dwarf Stars," *The Observatory* 36 (1913): 324–29: 325–26. This paper is an address given at the Royal Astronomical Society on June 13, 1913. Russell made a similar statement the following year about "the two great classes of stars," the giants and dwarfs, again attributing the terms to Hertzsprung (Lang and Gingerich, *Source Book in Astronomy and Astrophysics*, pp. 212–20: 216). In 1933, after a lengthy discussion of Hertzsprung's two papers, the Swedish astronomer Knut Lundmark concluded regarding the "dwarf" and "giant" terminology that "these names do not originate from Hertzsprung; I have tried to find where they first occur, but so far without success." Lundmark, "Luminosities, Colours, Diameters, Densities, Masses of the Stars," in *Handbuch der Astrophysik*, vol. 5, part 1 (1932), pp. 210–574: 443. Hearnshaw, *Analysis of Starlight*, p. 215, also concludes that Russell's ascription to Hertzsprung of the German equivalent of the terms "giant" and "dwarf" was incorrect, though in December 1908, Schwarzschild referred to some stars as "Giganten" when referring to Hertzsprung's work. It seems likely that Hertzsprung may have used the terms in correspondence with Schwarzschild, rather than in his early published papers.
17 Russell, "Relations between Spectra and Other Characteristics of Stars," *PA* 22 (1914): 275–94, continued at 331–51; the same article appears in *Nature* 93 (May 7, 1914): 227–30, with the diagrams in part 2 of the series, pp. 252–58. See also David DeVorkin, *Henry Norris Russell*, especially chapter 8, "Building a Case for Giants." The *Popular Astronomy* article contains the first appearance of what later became known as the H-R diagram.
18 Sandage, *Mt. Wilson Observatory*, pp. 240ff., has written about the irony of Strömgren's naming of the H-R diagram, since Hertzsprung departed Copenhagen Observatory, directed by Strömgren's father, after accidentally shattering the objective lens of its new 12-inch telescope. Hertzsprung had been a volunteer observer there, just starting what would eventually be a brilliant astronomical career.
19 Arthur S. Eddington, *The Internal Constitution of the Stars* (1st ed.1926; reprinted New York: Dover, 1959), pp. 7–8. At other places in the volume, Eddington seems to question more deeply the distinction between giants and dwarfs: "The evidence is admittedly rough," he wrote, pointing especially to "very little direct evidence" for the mass of giant stars, p. 119. See also DeVorkin, "Stellar Evolution," pp. 105–6.
20 W. S. Adams and A. Kohlschutter, "Some Spectral Criteria for the Determination of Absolute Stellar Magnitudes," *ApJ* 40 (1914): 385–98; *MWC*, 89; Sandage, *Mt. Wilson Observatory*, chapter 15, 239–44.

21 William H. Pickering, "The Distance of the Great Nebula in Orion," *HCO Circular* 205 (1917), 1–8: 5; Pickering refers to supergiants again in connection with the Pleiades in HCO 206, 1–3. The term was first used in a paper title by Cecelia Payne and Carl T. Chase in two articles in 1927, Cecelia Payne and Carl T. Chase, "The Spectra of Supergiant Stars of Class F8," *HCO Circular* 300 (Jan. 1927): 1–10; Cecelia Payne, "Photometric Line Intensities for Normal and Supergiant Stars," *HCO Circular* 306 (1927).
22 Walter S. Adams, Alfred H. Joy, Milton L. Humason, and Ada Margaret Brayton, "The Spectroscopic Absolute Magnitudes and Parallaxes of 4179 Stars," 187–291; Sandage, *Mt. Wilson Observatory*, p. 540.
23 David DeVorkin, "The Changing Place of Red Giant Stars in the Evolutionary Process," *JHA* 37 (2006): 429–69; Karl Hufbauer, "Stellar Structure and Evolution, 1924–1939," *JHA* 37 (2006): 203–32.
24 Gustaf Stromberg, "The Distribution of Absolute Magnitudes among K and M Stars Brighter than the Sixth Apparent Magnitude as Determined from Peculiar Velocities," *ApJ* 71 (1930): 175–90; also *MWC* 142. The earlier inklings came in W. S. Adams and A. H. Joy, "The Luminosities and Parallaxes of Five Hundred Stars," *ApJ* 46 (1917): 313-40. For a more detailed description of the separation of the subgiants from the main sequence and giant stars, see Sandage, *The Mount Wilson Observatory*, pp. 249–50, 363–67, 50; 539–54; also Sandage, Lori Lubin, and Don VandenBerg, "The Discovery of Subgiants (1922–1935)," *PASP* 115 (2003): 1187–1206.
25 Walter S. Adams, Alfred H. Joy, Milton L. Humason, and Ada Margaret Brayton, "The Spectroscopic Absolute Magnitudes and Parallaxes of 4179 Stars," 187–291; Sandage, *Mt. Wilson Observatory*, p. 540.
26 W. W. Morgan, P. C. Keenan, and E. Kellman, *Atlas of Stellar Spectra, with an Outline of Spectral Classification* (Chicago: University of Chicago Press, 1943).
27 On M3, see Sandage, "The Color-magnitude Diagram for the Globular Cluster M 3," *AJ* 58 (1953): 61–74; Sandage, *Mt. Wilson Observatory*, 546.
28 Sandage, *Mt. Wilson Observatory,* 539–54; 541–42, 547, and 548.
29 C. F. von Weizsäcker, "Element Transformation Inside Stars. II," *Physikalische Zeitschrift* 39 (1938): 633–46, and Hans Bethe, "Energy Production in Stars," *Physical Review* 55 (March 1, 1939): 434–56, reprinted in part in Bartusiak, 352–57; H. A. Bethe and C. L. Critchfield, "The Formation of Deuterons by Proton Combination, " *Physical Review* 54 (1938): 248–54; and Bethe's synthetic paper "Energy Production in Stars," *Physical Review* 55 (March 1, 1939): 434–56, reprinted in part in Bartusiak, 352–57.
30 Sandage, *Mt. Wilson Observatory*, pp. 548–49; A. Sandage and M. Schwarzschild, "Inhomogeneous Stellar Models. II. Models with Exhausted Cores in Gravitational Contraction," *ApJ* 116 (1952): 463–76. DeVorkin concludes slightly differently that by 1952, "although no single picture could be described with complete confidence, the basis for consensus had been largely established," and indeed consensus was achieved by 1955; "The Changing Place of Red Giant Stars," p. 461.
31 On M67 see H. L. Johnson and A. R. Sandage, "The Galactic Cluster M67 and its Significance for Stellar Evolution," *ApJ* 121 (1955): 616–27; Sandage, *Mt. Wilson Observatory*, p. 551.

32 Walter S. Adams, "Some Radial Velocity Results," *PASP* 25 (1913): 259–60.
33 Gerard Kuiper, "Two New White Dwarfs; Notes on Proper Motion Stars," *ApJ* 89 (1939): 548–51. Ken Crosswell, "The Rise of the Subdwarfs," chapter 8 in *Alchemy of the Heavens* (New York: Doubleday, 1995), 85–100.
34 Nancy Roman, "A Catalogue of High-Velocity Stars," *ApJ Supplement* 2 (1955): 195. Roman used "sd" for subdwarfs earlier than G0, and VI for stars later than G0. See Hearnshaw, *Analysis of Starlight*, pp. 292–93 on the attempted IAU adoption, citing Keenan and Morgan, *Transactions of the IAU* 11A (New York: Academic Press, 1962), p. 346, and Keenan, in *Stars and Stellar Systems*, 3, *Basic Astronomical Data* (Chicago: University of Chicago Press, 1963), p. 78.
35 Crosswell, *Alchemy*, p. 89; Joseph W. Chamberlain and Lawrence H. Aller, "The Atmospheres of A-Type Subdwarfs and 95 Leonis," *ApJ* 114 (1951): 52–72; reprinted in Abt, *AAS* Centennial Issue, 426–46, with commentary by George Wallerstein, "Chamberlain & Aller's Subdwarf Abundances," pp. 447–49.
36 Henry Norris Russell, "Notes on White Dwarfs and Small Companions," *AJ* 51 (1944): 13–17: 13. *White Dwarfs*, E. Schatzman, ed. (Amsterdam: North-Holland, 1958), p. 1.
37 J. B. Holberg and F. Wesemael, "The Discovery of the Companion of Sirius and its Aftermath," *JHA* 38 (2007): 161–74. U. J. J. Le Verrier made the suggestion of a planet in 1855 even before the observation was made; see J. B. Holberg, "The Discovery of the Existence of White Dwarf Stars: 1860 to 1930," *JHA* 30 (2009): 137–54.
38 Walter S. Adams, "An A-type Star of Very Low Luminosity," *PASP* 26 (1914): 198, and "The Spectrum of the Companion of Sirius," *PASP* 27 (1915): 236–37, both in Bartusiak, 329–30.
39 A. van Maanen, "Two Faint Stars with Large Proper Motions," *PASP* 29 (1917): 258–59.
40 J. B. Holberg, "How Degenerate Stars Came to be Known as White Dwarfs," *BAAS* 37 (2005), p. 150.
41 A. S. Eddington, *Stars and Atoms* (Oxford: Clarendon Press, 1927), p. 50.
42 J. B. Holberg, "The Discovery of the Existence of White Dwarf Stars: 1860 to 1930," *JHA* 30 (2009): 137–54.
43 Ralph Fowler, "On Dense Matter," *MNRAS* 87 (Dec. 1926): 114–22, in Bartusiak, 331–33.
44 S. Chandrasekhar, "The Maximum Mass of Ideal White Dwarfs," *ApJ* 74 (1931): 81–82. Edmund Stoner had come to a similar conclusion several years earlier; see Michael Nauenberg, "Edmund C. Stoner and the Discovery of the Maximum Mass of White Dwarfs," *JHA* 39 (2008): 297–312.
45 Holberg, "The Discovery of the Existence of White Dwarf Stars: 1860 to 1930"; Arthur Miller, *Empire of the Stars* (Boston: Houghton-Mifflin, 2005), especially pp. 102, 132–33. Matthew Stanley, *Practical Mystic: Religion, Science and A. S. Eddington* (Chicago: University of Chicago Press, 2007). It is relevant to note that, exotic as they are, white dwarfs have proven increasingly important to astronomical research over the last few decades. White dwarf supernovae (today known as "Type Ia") were used in the late 1990s to show that the expansion of the universe is

accelerating, giving rise to the concept of a repulsive force known as dark energy, a revival of Einstein's idea of a cosmological constant. Faint white dwarfs are also one of the candidates for baryonic dark matter known to exist from the anomalous rotation rates of galaxies. White dwarfs have been observed in globular clusters and in other galaxies, both from the ground and by spacecraft.

46 Walter Baade and Fritz Zwicky, "Remarks on Super-Novae and Cosmic Rays," *Phys. Rev.* 46 (193): 76–77.

47 Anthony Hewish, "Pulsars and High Density Physics," Nobel Lecture, December 12, 1974, online at http://nobelprize.org/nobel_prizes/physics/laureates/1974/hewish-lecture.pdf.

48 S. Jocelyn Bell Burnell, "Little Green Men, White Dwarfs or Pulsars," *Cosmic Search* 1, online at http://www.bigear.org/vol1no1/burnell.htm, originally presented as an after-dinner speech with the title of "Petit Four" at the Eighth Texas Symposium on Relativistic Astrophysics, published in the *Annals of the New York Academy of Science* 302 (1977): 685–89. A more extensive first-person account is Jocelyn Bell Burnell, "The Discovery of Pulsars," in Ken Kellermann and B. Sheets, eds., *Serendipitous Discoveries in Radio Astronomy* (Green Bank: NRAO, 1983), 160–70. Hewish's detailed account is given in his Nobel lecture, cited above. Another detailed account, presented as part of the American Philosophical Society symposium on "Discovery in Astronomy," is Malcolm Longair, "The Discovery of Pulsars and the Aftermath," *Proceedings of the APS* 155 (2011): 147–57, online at http://www.amphilsoc.org/sites/default/files/4Longair1550204.pdf.

49 Hewish, "Pulsars and High Density Physics," Nobel Lecture, p. 178.

50 Anthony Hewish, S. Jocelyn Bell, John D. H. Pilkington et al., "Observation of a Rapidly Pulsating Radio Source," *Nature* 217 (February 24, 1968): 709–13; reprinted in Bartusiak, 515–18.

51 Thomas Gold, "Rotating Neutron Stars as the Origin of the Pulsating Radio Sources," *Nature* 218 (1968): 731–32; reprinted in Bartusiak, 518–21; Hewish, "Pulsars and High Density Physics," Nobel Lecture, pp. 178–79; David H. Staelin and Edward C. Reifenstein III, "Pulsating Radio Sources near the Crab Nebula," *Science* 162 (1968): 1481–83; W. J. Cocke, M. J. Disney, and D. J. Taylor, "Discovery of Optical Signals from Pulsar NP 0532," *Nature* 221 (Feb. 8, 1969): 525–27.

52 Kaj Strand, "61 Cygni as a Triple System," *PASP* 55 (Feb. 1943): 29–32; for context, and even earlier suggestions of planets, see Dick, *The Biological Universe: The Twentieth Century Extraterrestrial Life Debate* (Cambridge: Cambridge University Press, 1996), p. 186.

53 Dirk Reuyl and E. Holmberg, "On the Existence of a Third Component in the System 70 Ophiuchi," *ApJ* 97 (1943): 41–45; Dick, *Biological Universe*, p. 187.

54 Russell, "Planet Companions," *SciAm* (June 1943): 260–61; Russell, "Anthropocentrism's Demise," *SciAm* (July 1943): 18–19; Russell, "Physical Characteristics of Stellar Companions of Small Mass," *PASP* 55 (April 1943): 79–86.

55 Van de Kamp, "Stars or Planets?" *Sky and Telescope* 4 (December 1944): 5–7, 22; Russell, "Physical Characteristics," 79–86, and "Notes on White Dwarfs and Small Companions," *AJ* 52 (June 1944): 13–17: 17; Dick, *Biological Universe*, p. 189.

56 Shiv S. Kumar, "The Structure of Stars of Very Low Mass," *ApJ* 137 (1963): 1121–25. Kumar describes his work on low-mass stars and "black dwarfs" in Shiv S. Kumar, "The Bottom of the Main Sequence and Beyond: Speculations, Calculations, Observations, and Discoveries (1958–2002), online at http://arxiv.org/pdf/astro-ph/0208096.

57 Jill Tarter, "The Interaction of Gas and Galaxies within Galaxy Clusters," PhD dissertation, University of California, Berkeley, 1975; J. C. Tarter, "An Historical Perspective: Brown is Not a Color," Minas C. Kafatos, Robert S. Harrington, Stephen P. Maran, eds., *Astrophysics of Brown Dwarfs* (Cambridge: Cambridge University Press, 1986), pp. 121–38. A search in the Astrophysics Data System shows that the first cases of the use of "brown dwarf" in paper titles are by Tarter in 1976 and 1978, with increasing use during the 1980s.

58 E. E. Becklin and B. Zuckerman, "A Low-temperature Companion to a White Dwarf Star," *Nature* 336 (1988): 656–58.

59 David W. Latham et al., "The Unseen Companion of HD114762 – A Probable Brown Dwarf," *Nature* 339 (1989): 38–40.

60 T. Nakajima, B. R. Oppenheimer, S. R. Kulkarni et al., "Discovery of a Cool Brown Dwarf," *Nature* 378 (1995): 463–65; the Hubble confirmation is at http://hubblesite.org/newscenter/archive/releases/1995/48/text/.

61 R. Rebolo, M. R. Zapatero Osorio, and E. L. Martin, "Discovery of a Brown Dwarf in the Pleiades Star Cluster," *Nature* 377 (1995): 129–31.

62 Adam J. Burgasser, "The T-type Dwarfs," in Richard O. Gray and Christopher J. Corbally, *Stellar Spectral Classification* (Princeton: Princeton University Press, 2009), pp. 388–440.

63 An archive of L and T dwarfs can be found at http://spider.ipac.caltech.edu/staff/davy/ARCHIVE/index.shtml.

64 Michel Mayor and Didier Queloz, "A Jupiter-mass companion to a Solar-type star," *Nature* 378 (1995): 355–59; Geoff Marcy, Paul Butler et al., "The Planet around 51 Pegasi," *ApJ* 481 (2007): 926–35.

65 The most reliable and continuously updated list of all extrasolar planets is available at http://exoplanets.org, maintained by Jason Wright at Penn State University; see Jason Wright et al., "The Exoplanet Orbit Database," *PASP* 123 (2011): 412–22, preprint online at http://xxx.lanl.gov/abs/1012.5676.

66 For example, Kevin Stevenson et al., "Possible Thermochemical Disequilibrium in the Atmosphere of the Exoplanet GJ 436b," *Nature* 464 (2010): 1161–64.

67 G. Gonzalez, "The Stellar Metallicity-giant Planet Connection," *MNRAS* 285 (1997): 403–12; Debra Fischer and Jeff Valenti, "The Planet-Metallicity Correlation, *ApJ* 622 (2005): 1102–17; A. Sozzetti et al., "A Keck HIRES Doppler Search for Planets Orbiting Metal-Poor Dwarfs. II. On the Frequency of Giant Planets in the Metal-Poor Regime," *ApJ* 697 (2009): 544–56.

68 Raymond Harris, online at: http://www.deepfly.org/TheNeighborhood/PlanetsInPowersOf10.html; J. J. Fortney et al., "Planetary Radii across Five Orders of Magnitude in Mass and Stellar Insolation: Application to Transits," *ApJ* 659 (2007): 1661–72; S. Marchi, "Extrasolar Planet Taxonomy: A New Statistical Approach," *ApJ* 666 (2007): 475–85.

69 Robert Grant Aitken, *The Binary Stars* (New York: D. C. McMurtrie, 1918), pp. 1–9.

70 Leos Ondra, "A New View of Mizar," *Sky and Telescope* 108 (2004): 72ff., online at http://www.leosondra.cz/en/mizar/, contains the original references to Castelli's letter to Galileo regarding Mizar, and Galileo's observations, never published, but located in the Biblioteca Nazionale Centrale Firenze, Ms. Gal. 70 c. 10r. The article also gives the original references in Galileo's *Opere* for Mizar and the other double star observations.

71 As Christopher Graney points out, the sentence occurs in the following context: "The first way to observe the diameters of stars depends on ocular estimate, and on comparison either with the diameter of the moon, or – so its brightness won't be a problem – with the known distance of fixed stars close together to each other . . . by estimating how many such stars would be needed to fill that interval. This conjecture is, of course, liable to error, especially on account of artificial rays enlarging the stars and shrinking the intervals, the way that there appears to be one star in the middle of the Great Bear's tail, when there are actually two, as the telescope reveals." The statement appears in the *Almagestum Novum*, Book VI, chapter 9, "On the Apparent and True Magnitudes of the Fixed Stars," original Latin at http://www.e-rara.ch/zut/content/zoom/140682, translation by Christopher Graney, private communication, September 28, 2012. From this it sounds like the dual nature of Mizar is common knowledge. In his entry on Christian Mayer in the *BEA*, p. 753, Thomas Williams sites two other double stars Riccioli is supposed to have observed. Williams also points out that Ptolemy first applied the name "double star" to a bright naked-eye pair in Sagittarius.

72 Christopher M. Graney, "The Accuracy of Galileo's Observations," *Baltic Astronomy* 16 (2007): 443–49: 445 for Galileo's Trapezium observations.

73 William Herschel, "Catalogue of Double Stars," *PTRSL* 72 (1782): 112–62: 161. On Mayer's catalogues, see Thomas R. Williams, "Mayer, Christian," in *BEA*, p. 753.

74 William Herschel, "Account of the Changes that have happened, during the last Twenty-Five Years, in the relative Situation of Double-stars, with an investigation of the Cause to which they are owing," *PTRSL* 93 (1803): 339–82: 340; Aitken, *Binary Stars*, pp. 2–9. In his entry on Mayer in the *BEA*, p. 753, Tom Williams notes that Mayer detected "some possible orbital motion" when he compared his observations with those of Flamsteed. See also Michael Hoskin, *William Herschel and the Construction of the Heavens* (London: Oldbourne, 1963), pp. 31–40.

75 Brian Mason to Steven Dick, private communication, March 26, 2012. While the appearance of duplicity by happenstance is of little scientific interest (parallax having been determined by other methods), many of the resolved doubles in the Washington Double Star catalogue will undoubtedly turn out to be physical systems. The Catalogue is online at http://ad.usno.navy.mil/wds/

76 Brian Mason to Steven Dick, private communication, March 26, 2012.

77 Adriaan Blaauw, "A Study of the Scorpio–Centaurus Cluster," *Publications of the Kapteyn Astronomical Laboratory at Groningen* 52 (1946 PhD thesis), and V. A. Ambartsumian, in *Stellar Evolution and Astrophysics* (Yerevan: Armenian Academy of Sciences, 1947); Sydney Van den Bergh, "A Study of Reflection Nebulae," *AJ* 71 (1966): 990–98; W. Herbst, "R associations. I – UBV Photometry and MK Spectroscopy of Stars in Southern Reflection Nebulae," *AJ* 80 (1976): 212–26.

78 Walter Baade, "The Resolution of Messier 32, NGC 205, and the Central Region of the Andromeda Nebula," *ApJ* 100 (1944): 137–46; reprinted in part in Lang and Gingerich, *Source Book*, pp. 744–49, and in full in Abt, *AAS* Centennial Issue, 349–58, with commentary by Dmitri Mihalas, pp. 359–61.

5. Galaxies, Quasars, and Clusters: Discovery in the Realm of the Galaxies

1 Earl of Rosse, "Observations on the Nebulae," *PTRSL* 140 (1850): 499–514, reprinted in part in Bartusiak, 191–95.
2 Edwin Hubble, "Photographic Investigations of the Faint Nebulae," *Publications of the Yerkes Observatory* 4, pt. 2 (dissertation completed in 1917, published in 1920), p. 1.
3 Maarten Schmidt, "Discovery of Quasars," in K. Kellermann and B. Sheets, *Serendipitous Discoveries in Radio Astronomy* (Green Bank: NRAO, 1983), pp. 171–74.
4 Thomas Wright of Durham, *An Original Theory or New Hypothesis of the Universe* (1750) facsimile reprint, introduction and transcription by Michael Hoskin (The Netherlands: Elsevier, 1971); Immanuel Kant, *Allgemeine Naturgeschichte und Theory des Himmels* (1755), English edition, *Universal Natural History and Theory of the Heavens*, Stanley L. Jaki, transl. (Norwich: Scottish Academic Press, 1981). Both are discussed in the present context in Richard Berenzden, Richard Hart, and Daniel Seeley, *Man Discovers the Galaxies* (New York: Science History Publications, 1976), pp. 6–10, and in Charles A. Whitney, *The Discovery of Our Galaxy* (New York: Alfred A. Knopf, 1971), pp. 77–86. The passage is here quoted as in Whitney, p. 84. As Whitney points out - and as we saw in the last chapter - the phrase "the nebulous stars, properly so-called" was later used by William Herschel. Hoskin discusses Wright in the introduction to the facsimile edition, and in M. A. Hoskin, "The Cosmology of Thomas Wright of Durham," in *JHA* 1 (1970): 44–52. Relevant parts of Wright and Kant are reprinted in Bartusiak, 168–88.
5 Agnes Clerke, *The System of the Stars*, 1st ed. (London: Longmans, Green and Co., 1890), p. 368, repeated in the second edition of 1905. Much of the history in this section is covered in Robert Smith, *The Expanding Universe: Astronomy's 'Great Debate,' 1900–1931* (Cambridge: Cambridge University Press, 1982), chapter 1, as well as Robert Smith, "Beyond the Galaxy: The Development of Extragalactic Astronomy 1885–1965, Part 1," *JHA* 39 (2008): 91–119; and Part 2, *JHA* 40 (2009): 71–107.
6 On the history of the GC and NGC, see Wolfgang Steinicke, *Observing and Cataloguing Nebulae and Star Clusters: From Herschel to Dreyer's New General Catalogue* (Cambridge: Cambridge University Press, 2010).
7 Allan Sandage, "Classification and Stellar Content of Galaxies," in Allan Sandage, Mary Sandage, and Jerome Kristian, eds., *Galaxies and the Universe* (Chicago and London: University of Chicago Press, 1975), p. 2. Online at http://ned.ipac.caltech.edu/level5/Sandage/paper.pdf.
8 Max Wolf, "Die Klassifizierung der kleinen Nebelflecken," *Publikationen des Astrophysikalischen Instituts Königstuhl-Heidelberg* 3, no. 5 (1909): 109–12.

9 Steven Dick, *The Biological Universe: The Twentieth Century Extraterrestrial Life Debate and the Limits of Science* (Cambridge: Cambridge University Press, 1996), pp. 166–70.
10 The best biography of Hubble is Gale E. Christianson, *Edwin Hubble: Mariner of the Nebulae* (New York: Farrar, Straus and Giroux, 1995).
11 Hubble, "Photographic Investigations of the Faint Nebulae," *Publications of the Yerkes Observatory* 4, pt. 2 (1920): 1. The dissertation has dual pagination, running from pages 1–17 for Hubble's dissertation, corresponding to pages 69–85 of the Yerkes Publications, plus two plates. Christianson, *Edwin Hubble*, pp. 102–3.
12 Hubble, "Photographic Investigations," pp. 3–4.
13 On Hubble's arrival at Mt. Wilson, see Christianson, *Edwin Hubble*, pp. 20ff.
14 Hubble, "A General Study of Diffuse Galactic Nebulae," *ApJ* 56 (1922): 162–99: 166.
15 R. Hart and R. Berenzden, "Hubble's Classification of Non-Galactic Nebulae," *JHA* 2 (1971): 109–19: 113. The Jeans work is *Problems of Cosmogony and Stellar Dynamics* (Cambridge: Cambridge University Press, 1919), updated in 1929 as *Astronomy and Cosmogony*. Hubble's 1926 paper is "Extra-Galactic Nebulae," *ApJ* 64 (1926): 321–72.
16 Hubble himself recalled that planetary and diffuse nebulae were considered galactic, while the spirals had been the subject of the island universe controversy; *The Realm of the Nebulae* (New Haven: Yale University Press, 1936), p. 84. Although we now know we live in a spiral galaxy, this is far from obvious. The Milky Way as seen in the night sky – now known to be the "backbone" of our Galaxy – was one of the first phenomena Galileo observed with his telescope in 1610, resolving it into stars. But the first to speculate that this phenomenon was due to a large, flat disk of stars were Wright and Kant in the eighteenth century. A few decades later, William Herschel was the first to base this conclusion on actual star counts. Even then, the spiral nature of our home galaxy would not be uncovered until the mid-twentieth century, and today it is believed to be a "barred" spiral.
17 Earl of Rosse, "Observations on the Nebulae," *PTRSL* 140 (1850): 499–514, reprinted in part in Bartusiak, 191–95. For context see Michael Hoskin, "Rosse, Robinson and the Resolution of the Nebulae," *JHA* 21 (1990): 331–44, and D. W. Dewhirst and M. Hoskin, "The Rosse Spirals," *JHA* 22 (1991): 257–65.
18 On Scheiner's observations see J. Scheiner, "Über das Spectrum des Andromedanebels," *AN*,148 (1899): 325, English translation in Scheiner, "On the Spectrum of the Great Nebula in Andromeda," *ApJ* 9 (1899): 149–50.
19 Smith, *Expanding Universe*, p. 22.
20 Michael Hoskin, "The 'Great Debate': What Really Happened," *JHA* 7 (1976): 169–82; Smith, *Expanding Universe*, pp. 77–90; Virginia Trimble, "The 1920 Shapley-Curtis Discussion: Background, Issues and Aftermath," *PASP* 107 (1995): 1133–44.
21 Henrietta Leavitt, "1777 Variables in the Magellanic Clouds," *HCO Annals* 60 (1908): 87–108; the 1912 paper is H. Leavitt and E. C. Pickering, "Periods of 25 Variable Stars in the Small Magellanic Cloud,"

HCO Circular 173; see also Smith, *Expanding Universe*, pp. 59–77. In 1908, Leavitt had only determined periods for 16 of the 1777 variables, but it was enough to see the pattern. The 1912 paper reported 25 periods. On Shapley's use of this method, see Owen Gingerich, "Harlow Shapley and the Cepheids," in *The Great Copernicus Chase* (Cambridge: Cambridge University Press, 1992), pp. 238–45.

22 Richard Berenzden, "Hubble's Announcement of Cepheids in Spiral Nebulae," *PASP* 10 (1967), pp. 425–40. Even by the time of the public announcement in the November 23 *New York Times* (p. 6), Hubble's Cepheid work was well known to other astronomers.

23 The M31 and M33 results were reported in Hubble, "Cepheids in Spiral Nebulae," *PA* 33 (1925): 252ff., reprinted in part in Bartusiak, *Archives*, pp. 407–14, and "A Spiral Nebula as a Stellar System, Messier 31," *ApJ* 63 (1926): 236–74. Also Hubble, "NGC 6822, A Remote Stellar System," *ApJ* 62 (1925): 409–33. For further context see Robert Smith, *Expanding Universe*, pp. 111–36, and Smith, "Beyond the Galaxy," Part 2, *JHA* 40 (2009): 71–107.

24 Hubble, "Extra-Galactic Nebulae," *ApJ* 64 (1926): 321–72: note 2, 323. Christianson, *Edwin Hubble*, p. 175; Allan Sandage, *The Mount Wilson Observatory: Breaking the Code of Cosmic Evolution*, vol. 1 of the Centennial History of the Carnegie Institution of Washington (Cambridge: Cambridge University Press, 2004), p. 487. It should also be pointed out that some supposed galaxies actually were nebulosities, and that centers of spirals like M31 had not yet been resolved.

25 Hubble, "Extra-Galactic Nebulae," 321–72: 322. The Magellanic Clouds are today classified as spirals in the de Vaucouleurs system, the LMC as SBm (irregular in appearance with no bulge) and the SMC as Im (highly irregular). Some still classify Im's as irregular (Irr). See Ronald J. Buta, Harold G. Corwin, Jr., and Stephen C. Odewahn, *The de Vaucouleurs Atlas of Galaxies* (Cambridge: Cambridge University Press, 2007), especially p. 16.

26 Hubble, "Extra-Galactic Nebulae," 321–72, reprinted in part in Lang and Gingerich, *Source Book*, pp. 716–24; Hubble, *The Realm of the Nebulae* (New Haven: Yale University Press, 1936), pp. 39–48.

27 Hubble, *The Realm of the Nebulae*, p. 39

28 Ibid., p. 40.

29 Ibid.; Ronald J. Buta, Harold G. Corwin, Jr., and Stephen C. Odewahn, *The de Vaucouleurs Atlas*, pp. 8–9.

30 Hubble, *The Realm of the Nebulae*, 131–37; Buta et al., *The de Vaucouleurs Atlas*, pp. 15–16.

31 C. K. Seyfert, "Nuclear Emission in Spiral Nebulae," *ApJ* 97 (1943): 28–40, reprinted in Helmut Abt, ed., *AAS Centennial Issue*, pp. 324–36, with commentary by Donald E. Osterbrock, pp. 337–38; also reprinted in part in Lang and Gingerich, *Source Book*, pp. 738–43.

32 E. A. Fath, "The Spectra of Some Spiral Nebulae and Globular Star Clusters," *Lick Obs. Bulletin* 5 (1909): 71–77. This is also the title of Fath's PhD dissertation of the same year, and was published in shorter versions under the same title in *PA* 17 (1909): 504–8 and *PASP* 21 (1909): 138–43.

33 Seyfert, "Nuclear Emission," p. 324. In this paper Seyfert reported high dispersion spectrograms for six of twelve of the spirals.

34 Donald E. Osterbrock, "Seyfert Galaxies," in Abt, *AAS Centennial Issue*, pp. 337–38.
35 Osterbrock, "Seyfert Galaxies," has pointed out the gradual way in which Seyfert's name was conferred on a new class of objects, a rarity in astronomical classification, especially at the class level. In 1958, fifteen years after Seyfert declared the new class, Geoffrey Burbidge discussed the galaxies "which Seyfert studied," and the following year Leo Woltjer discussed "objects listed by Seyfert." Also in 1959 Geoffrey Burbidge, Margaret Burbidge and K. H. Pendergast used the term "Seyfert galaxies" for the first time in print in their discussion of NGC 1068, the Messier object that Seyfert had singled out as the prototype in his original paper of 1943. Thereafter, the term "Seyfert galaxies" became commonplace in astronomical literature, but Seyfert barely lived to see it; he died in a car accident in 1960 at the age of forty-nine.
36 J. S. Hey, S. J. Parsons, and J. W. Philips, "Fluctuations in Cosmic Radiation at Radio Frequencies," *Nature* 158 (1946): 234, reprinted in part in Lang and Gingerich, *Source Book*, pp. 774–76. For accounts, see Hey's own book, *The Evolution of Radio Astronomy* (New York: Science History Publications, 1973), pp. 43–46, as well as Woodruff T. Sullivan III, *Cosmic Noise: A History of Early Radio Astronomy* (Cambridge: Cambridge University Press, 2009), pp. 103–5.
37 Sullivan, *Cosmic Noise*, pp. 335–41: 341, details the search for an optical counterpart of Cygnus A. The publication of results is W. Baade and R. Minkowski, "Identification of the Radio Sources in Cassiopeia, Cygnus A and Puppis A," *ApJ* 119 (1954): 206–14, reprinted in part in Lang and Gingerich, *Source Book*, pp. 786–91. See also their adjacent paper, "On the Identification of Radio Sources," *ApJ*, 119 (1954): 215–31, reprinted in Abt, *AAS Centennial Issue*, pp. 538–68.
38 *New York Times*, August 23, 1953, p. E9, as described in Sullivan, *Cosmic Noise*, 341.
39 The discovery of the radio sources is in J. G. Bolton, "Discrete Sources of Galactic Radio Frequency Noise," *Nature* 162 (1948): 141–43; the optical identifications are in John G. Bolton, Gordon J. Stanley, and O. B. Slee, "Positions of Three Discrete Sources of Galactic Radio-Frequency Radiation," *Nature* 164 (1949): 101–2, reprinted in Lang and Gingerich, pp. 777–78. See also Hey, *Evolution*, pp. 46–47, and Sullivan, *Cosmic Noise*, pp. 320–24. It was Bolton who initiated the A nomenclature for the brightest object in the constellation, B for the second brightest, and so on.
40 W. Baade and R. Minkowski, "Identification of the Radio Sources in Cassiopeia, Cygnus A and Puppis A," *ApJ* 119 (1954): 206–14, reprinted in part in Lang and Gingerich, *Source Book*, pp. 786–91; Sullivan, *Cosmic Noise*, pp. 344–48: 346.
41 Hannes Alfven and Nicolai Herlofson, "Cosmic Radiation and Radio Stars," *Physical Review* 78 (1950): 616, reprinted in Lang and Gingerich, *Source Book*, pp. 779–81; G. R. Burbidge, "Estimates of the Total Energy in Particles and Magnetic Field in the Non-Thermal Radio Sources," *ApJ* 129 (1959): 849–51; the quotation is from Suzy Collin, "Quasars and Galactic Nuclei, a Half Century Agitated Story," in Jean-Michel Alimi and Andre Fuzfa, eds., *Albert Einstein Century International Conference* vol. 861 (Melville, NY: American Institute of Physics, 2006), online version only, pp. 587–95: 587; preprint online at http://arxiv.org/abs/astro-ph/0604560.

42 Maarten Schmidt, "Discovery of Quasars," in K. Kellermann and B. Sheets, *Serendipitous Discoveries in Radio Astronomy* (Green Bank: NRAO, 1983), pp. 171–4. See also Maarten Schmidt, "The Discovery of Quasars," *Proceedings of the American Philosophical Society* 155 (2011): pp. 142–6, online at http://www.amphilsoc.org/publications/proceedings/v/155/n/2; and Richard Preston, " Beacons in Time – Maarten Schmidt and the Discovery of Quasars," *Mercury* 17 (1988): 2–11.

43 Schmidt, "Discovery of Quasars,"p. 173. There is some uncertainty in the literature as to whether the date is February 5 or 6. In this 1983 account Schmidt uses February 5, but in his 2011 account he used February 6, as does Preston in his 1988 account. In a private communication Schmidt writes that he vaguely remembers it as a Tuesday, thus February 5. But he goes on to say, "On the other hand, I have a letter to Hazard, dated Feb 8, saying that it happened two days ago. But who dated my letter? Perhaps my secretary on the day I wrote the letter or perhaps the next day? So I am truly uncertain! Fifty years is a long time to reconstruct! . . . It was on the basis of the letter to Hazard that I have been mentioning Feb 6 recently. Yet, my best guess would be Feb 5." Private communication to author, October 26, 2012.

44 Cyril Hazard et al., "Investigation of Radio Source 3C 273 by the Method of Lunar Occultations," *Nature* 197 (1963): 1037–9, followed by Maarten Schmidt, "3C 273: A Star-Like Object with Large Red-Shift," *Nature* 197 (1963): 1040; Jesse Greenstein and Thomas A. Matthews, "Red-Shift of the Unusual Radio Source: 3C 48," *Nature* 197 (1963): 1041–42; and John Oke, "Absolute Energy Distribution in the Optical Spectrum of 3C 273," *Nature* 197 (1963): 1040–41. All are reprinted with commentary and annotations in Lang and Gingerich, *Source Book*, pp. 801–10; see also Bartusiak, pp. 505–7. The first paper with "Quasi-stellar" in the title was published in February 1964, in Maarten Schmidt and Thomas A. Matthews, "Redshift of the Quasi-Stellar Radio Sources 3C 47 and 3C 147," *ApJ* 139 (1964): 781–85.

45 Jesse L. Greenstein and Maarten Schmidt, "The Quasi-Stellar Radio Sources 3C 48 and 3C 273," *ApJ* 140 (1964): 1–34, reprinted in Abt, *AAS Centennial Issue*, pp. 1021–57, with commentary by Greenstein, "Greenstein & Schmidt's Study of the Physics of Quasars," pp. 1058–59. The idea of gravitational redshift in an earlier context has been examined in Norris Hetherington, "Sirius B and Gravitational Redshift – An Historical Review," *QJRAS* 21 (1980): 246–52, reprinted in Hetherington's *Science and Objectivity: Episodes in the History of Astronomy* (Ames: Iowa State University Press, 1988), pp. 65–72.

46 H. J. Smith and D. Hoffleit, "Light Variations in the Superluminous Radio Galaxy 3C273," *Nature* 198 (1963): 650–51; Smith and Hoffleit, "Light Variability and Nature of 3C273," *AJ* 68 (1963): 292–93; Lang and Gingerich, *Source Book*, p. 804.

47 Greenstein and Schmidt, "The Quasi-Stellar Radio Sources," (1964): 1; Jesse L. Greenstein, "Greenstein & Schmidt's Study of the Physics of Quasars," Abt, *AAS* Centennial Issue 525: 1058–59. Greenstein and Schmidt could not have mentioned the words "black hole" in 1964, since John Wheeler did not invent the term until 1967; see below.

48 E. E. Salpeter, "Accretion of Interstellar Matter by Massive Objects," *ApJ* 140 (1964): 796–800; Y. B. Zel'Dovich, "Probability of Quasar Production," *Soviet Astronomy* 9 (1965): 221–23.

49 On this controversy, see especially Halton Arp, *Quasars, Redshifts and Controversies* (Berkeley: Interstellar Media, 1987).
50 Maarten Schmidt, "Discovery of Quasars," in Kellermann and Sheets, *Serendipitous Discoveries*, p. 171.
51 P. A. Strittmatter, K. Serkowski, R. Carswell et al., "Compact Extragalactic Nonthermal Sources," *ApJ* 175 (1972): L7–L13.
52 J. B. Oke and J. E. Gunn, "The Distance of BL Lacertae," *ApJ Letters* 189 (1974): 5.
53 The first "blazar" paper title was J. F. C. Wardle et al., "The Radio Morphology of Blazars and Relationships to Optical Polarization and to Normal Radio Galaxies," *ApJ* 279 (1984): 93–111. The paper reported on "VLA observations of BL Lac objects and highly polarized quasars ('blazars')."
54 Cuno Hoffmeister, "354 Neue Veranderliche," *AN* 236 (1929): 233–44; John L. Schmitt, "BL Lac Identified as Radio Source," *Nature* 218 (1968): 663.
55 Collin, "Quasars and Galactic Nuclei," p. 591.
56 Simon Schaffer, "John Michell and Black Holes," *JHA* 10 (1979): 42–43.
57 S. Chandrasekhar, "The Maximum Mass of Ideal White Dwarfs," *ApJ* 74 (1931): 81–82. Edmund Stoner had come to a similar conclusion several years earlier; see Michael Nauenberg, "Edmund C. Stoner and the Discovery of the Maximum Mass of White Dwarfs," *JHA* 39 (2008): 297–312.
58 J. B. Holberg, "The Discovery of the Existence of White Dwarf Stars: 1860 to 1930," *JHA* 30 (2009): 137–54; Arthur Miller, *Empire of the Stars* (Boston: Houghton-Mifflin, 2005), especially pp. 102, 132–33.
59 J. R. Oppenheimer and G. M. Volkoff, "On Massive Neutron Cores, "*Physical Review* 55 (1939): 374–81; Oppenheimer and Hartland Snyder, "On Continued Gravitational Contraction," *Physical Review* 56 (1939): 455–59. Caltech physicist Kip Thorne tells the story in detail in *Black Holes and Time Warps: Einstein's Outrageous Legacy* (New York: W. W. Norton, 1994), pp. 209–57.
60 Collin, "Quasars and Galactic Nuclei," pp. 590–91.
61 D. Lynden-Bell, "Galactic Nuclei as Collapsed Old Quasars," *Nature* 223 (1969): 690; Martin J. Rees, "Dissipative Processes, Galaxy Formation and 'Early' Star Formation," Symposium on Quasars and Active Nuclei of Galaxies, Copenhagen, Denmark, June 27–July 2, 1977, *Physica Scripta* 17 (1978): 371–6; John Archibald Wheeler, *Geons, Black Holes, and Quantum Foam: A Life in Physics* (New York: W. W. Norton, 1998), p. 296.
62 On the origins and early years of X-ray astronomy, see Richard F. Hirsh, *Glimpsing an Invisible Universe: The Emergence of X-Ray Astronomy* (Cambridge: Cambridge University Press, 1983).
63 B. Louise Webster and Paul Murdin, "Cygnus X-1-a Spectroscopic Binary with a Heavy Companion?" *Nature* 235 (1972): 37–38.
64 Thorne, *Black Holes*, pp. 314–21: 317. Thorne's confidence estimate twenty years later had risen to 95 percent. As he pointed out, there was no unequivocal signal of a black hole, unlike neutron star pulsars with their regular pulses.
65 Ibid., 326.

66 Ibid., 346–51; Lynden Bell, "Galactic Nuclei"; R. D. Blandford and M. Rees, "A Twin-Exhaust Model for Double Radio Sources," *MNRAS* 169 (1974): 395.
67 John Tonry, "Evidence for a Central Mass Concentration in M32," *ApJ* 283 (1984): L27–L30.
68 http://hubblesite.org/newscenter/archive/releases/exotic/black-hole/1994/23/text/ and http://hubblesite.org/newscenter/archive/releases/exotic/black%20hole/1994/23/image/a/.
69 Roeland P. van der Marel, Joris Gerssen et al., "Hubble Space Telescope Evidence for an Intermediate-Mass Black Hole in The Globular Cluster M15," Parts I and II, *AJ* 124 (2002): 3255–69 and 3270–88; NASA Press Release, http://hubblesite.org/newscenter/archive/releases/2002/18/text/.
70 In the stricter systematic terminology of the Three Kingdom system in Appendix 1, these would be in the "Type" taxon, one level below "Class."
71 John Bachall, Hubble press release, November 19, 1996, at http://hubblesite.org/newscenter/archive/releases/1996/35/text/.
72 "Catalogue of One Thousand new Nebulae and Clusters of Stars," *PTRSL* 76 (1786): 457–99; "Catalogue of a Second Thousand new Nebulae and Clusters of Stars," *PTRSL* 79 (1789): 212–55; and "Catalogue of 500 new Nebulae, Nebulous Stars, Planetary Nebulae, and Clusters of Stars," *PTRSL* 92 (1802): 477–528. The telescope used in the first catalogue had an 18.7-inch aperture and 20-foot focal length, giving a field of view of only 15 arcminutes at 157 power.
73 William Herschel, "Astronomical Observations Relating to the Construction of the Heavens, Arranged for the Purpose of a Critical Examination, the Result of Which Appears to Throw Some New Light upon the Organization of the Celestial Bodies," *PTRSL* 101 (1811): 269–336: 285–90. Parts of this paper are reprinted in Hoskin, *The Construction of the Heavens: William Herschel's Cosmology* (Cambridge: Cambridge University Press, 2012), pp. 169–86: 173–76, where some of the multiple nebulae are identified in modern terms. Among those who cite this paper as the pioneering work on double and multiple galaxies is the Swedish astronomer Erik Holmberg, *A Study of Double and Multiple Galaxies,* in *Annals of the Observatory of Lund* (Lund: Lund Observatory, 1937), pp. 1–173. See also Steinicke, *Observing and Cataloguing Nebulae and Star Clusters,* pp. 44–45, where the author notes that nine out of fifteen of Herschel's category of "double nebulae with joined nebulosity" (defining the closest pairs) are today classed as double galaxies included in modern catalogues.
74 Messier's description follows the entry for M91 in the catalogue he published in the *Connaissance des Temps* for 1784. See John H. Mallas and Evered Kreimer, *The Messier Album* (Cambridge, MA: Sky Publishing, 1978), p. 26.
75 Hubble, *Realm of the Nebulae*, pp. 58–82, especially 59–60, 77–82. As Smith has pointed out ("Beyond the Galaxy," p. 92), the idea of clustering was proposed already in the eighteenth century by natural philosophers such as Johann Lambert. But their idea was not based on any real data. On the interpretation of redshifts, see Smith, pp. 80–92.

76 Edwin Hubble, *Realm of the Nebulae*, pp. 124–25.
77 Hubble, "The Local Group," chapter 6 in *Realm of the Nebulae*, pp. 124–51; Baade, "The Globular Cluster NGC 2419," *ApJ* 82 (1935): 396–412: 412. Although Hubble had an entire chapter titled "The Local Group" in his 1936 volume, the first paper with the term in its title was Baade, "NGC 147 and NGC 185, "Two New Members of the Local Group of Galaxies," *ApJ* 100 (1944): 147–50.
78 Harlow Shapley and Adelaide Ames, "A Study of a Cluster of Bright Spiral Nebulae," *HCO Circular*, vol. 294 (1926): pp. 1–8: 1–2. They also noted that the estimation of distances by angular diameters does not work in this case, since the dispersion in diameters was much too great to be the result of a distance effect.
79 E. Hubble and M. Humason, "The Velocity-Distance Relation among Extra-Galactic Nebulae," *ApJ* 74 (1931): 43–80, especially pp. 60ff.; reprinted in Abt, *AAS Centennial Issue*, pp. 214–51, followed by Allan Sandage's commentary, "Hubble & Humason's Evaluation of the Cosmological Expansion," pp. 252–54.
80 Hubble and Humason, "Velocity-Distance Relation," p. 58, in Abt, *AAS Centennial Issue*, p. 229. The velocity-distance relation has now been shown valid to at least 700 megaparsecs, more than 2 billion light-years, revealing the phenomenon of the accelerating universe.
81 The Shapley-Ames Catalogue was published in 1932 as "A survey of the external galaxies brighter than the thirteenth magnitude," *Annals of Harvard College Observatory*, vol. 88, pp. 41–76: Allan Sandage and G. A. Tammann published a revised version in 1981, online at http://nedwww.ipac.caltech.edu/level5/Shapley_Ames/frames.html.
82 Harlow Shapley, "Luminosity Distribution and Average Density of Matter in Twenty-five Groups of Galaxies," *PNAS* (1933): 19, 591–96: 592.
83 Hubble, *Realm of the Nebulae*, p. 77.
84 Fritz Zwicky, "On the Masses of Nebulae and of Clusters of Nebulae," *ApJ* 86 (1937): 217–46, reprinted in part with commentary in Lang and Gingerich, *Source Book*, pp. 729–37, and in Abt, *AAS Centennial Issue*, pp. 267–96, with Jeremiah Ostriker's commentary "Discovery of 'Dark Matter' in Clusters of Galaxies," following on pp. 297–98. The virial theorem derives its name from the Latin "vis," meaning force or energy, and the idea as applied here was borrowed from nineteenth-century statistical mechanics. The brief section on "Nebulae as Gravitational Lenses" appears on pages 237–38 of the original article. Sidney van den Bergh, "The Early History of Dark Matter," *PASP* 111 (1999): 657–60, shows that Zwicky already had published the idea of dark matter in German in 1933.
85 F. Zwicky, "On the Formation of Clusters of Nebulae and the Cosmological Time Scale," *PNAS* 25 (1939): 604–9.
86 Zwicky, "On the Masses on Nebulae," 217–18, 243.
87 Zwicky, "On the Clustering of Nebulae," *PASP* 50 (1938): 218–20.
88 George Abell, "The Distribution of Rich Clusters of Galaxies," *ApJ* Supplement, 3 (1958): 211; G. Abell, H. G. Corwin, and R. P. Olowin, "A Catalog of Rich Clusters of Galaxies," *ApJ Supplement*, 70 (1989): 1–138.

89 Milton L. Humason, Nicholas U. Mayall, and Allan R. Sandage, "Redshifts and Magnitudes of Extra-Galactic Nebulae," *AJ* 61 (1956): 97–162, reprinted in part in Lang and Gingerich, pp. 753–62.
90 G. A. Tammann, in O.-G. Richter and B. Binggeli, eds., *The Virgo Cluster, ESO Workshop Proceedings* No. 20 (Garching: ESO 1985), p. 3. As this volume went to press, data from the Planck spacecraft indicated a Hubble constant of about 67 (km/sec)/mega-parsec.
91 J. H. Lambert, *Cosmologische Briefe über die Einrichtung des Weltbaues* (Augsburg, 1761), translated by Stanley L. Jaki as *Cosmological Letters on the Arrangement of the Universe* (New York: Science History Publications, 1976); Edward R. Harrison, *Darkness at Night: A Riddle of the Universe* (Cambridge, MA: Harvard University Press, 1987), 153–54; on Charlier's arguments, "How an Infinite World May Be Built Up," *Archiv für matematik, astronomi, ffysik*, xvi (1922): 1–34, and "An Infinite Universe," *PASP* 37 (1925): 177–91; de Vaucouleurs, "The Local Supercluster of Galaxies," *Astronomical Society of India Bulletin* 9 (1981): 1–23: 6, online at http://adsabs.harvard.edu/abs/1981BASI....9....1D gives a fascinating history, from his point of view, of the idea of superclustering.
92 Gerard de Vaucouleurs, "Evidence for a Local Supergalaxy," *AJ* 58 (1953): 30–32; de Vaucouleurs, "Further Evidence for a Local Super-cluster of Galaxies: Rotation and Expansion," *AJ* 63 (1958): 253–66.
93 In his brief but intense 1981 article for the Astronomical Society of India, de Vaucouleurs offered an interesting aside on nomenclature. According to him, the terms "metagalaxy" or "metagalactic system" were introduced by Knut Lundmark in 1927 and used by Shapley as early as 1957 to refer to the observable part of the universe, not any particular system. "Supergalaxy" was originally used by Shapley in 1930 to describe the idea of the extended environment of our own Galaxy. De Vaucouleurs adopted it in 1953 for the idea of second-order clustering as in the "Local Supergalaxy," but by 1958 adopted "Local Supercluster" after (as de Vaucouleurs put it), "the discovery of the universality of the superclustering phenomenon by Abell, Kiang, and others in the late 1950s and early 1960s." De Vaucouleurs held the term "Virgo Supercluster" as inappropriate because it was not a gooddescription of a system that encircles the whole sky, and that in any case could cause confusion with the Virgo Cluster.
94 C. D. Shane and C. A. Wirtanen, "The Distribution of Galaxies," *Publications of the Lick Observatory* 22 (1957): 1–60; M. Seldner, B. Siebars, E. J. Groth, and P. J. E. Peebles, "New Reduction of the Lick Catalog of Galaxies," *AJ* 82 (1977): 249–56.
95 Zwicky Rudnicki, "Zur Nichtexistenz von Haufen von Galaxienhaufen," *Zeitschrift für Astrophysik* 64 (1966): 246–55; Maria Karpowicz, "On the Non-Existence of Clusters of Clusters of Galaxies. II," *Zeitschrift für Astrophysik* 66 (1967): 301–7, and subsequent papers in the series also find "no indications ... for the existence of any systematic clustering of clusters of galaxies." For context, see de Vaucouleurs, "The Local Supercluster of Galaxies."
96 M. Joeveer and J. Einasto, "Has the Universe the Cell Structure?" in M. S. Longair and J. Einasto, eds., *The Large Scale Structure of the Universe*

(Dordrecht: Reidel, 1978), pp. 241–51; J. Einasto, "Dark Matter and Large Scale Structure," in *Historical Development of Modern Cosmology*," in PASP Conference series, vol. 252, ed. V. J. Martinez, V. Trimble, and M. J. Pons-Bordeia (San Francisco: ASP, 2001), pp. 85–107.

97 De Vaucouleurs, "Who Discovered the Local Supercluster of Galaxies?" letter in *The Observatory* 109 (1989): 237–38.

98 De Vaucouleurs, "The Local Supercluster of Galaxies," p. 3.

99 De Vaucouleurs, "Who Discovered," p. 237.

100 De Vaucouleurs, "Evidence for a Local Supergalaxy," *AJ* 58 (1953): 30–42, where the plane is recognized from data in the Shapley-Ames catalogue. For a current view of the supergalactic system see O. Lahav et al., "The Supergalactic Plane Revisited with the Optical Redshift Survey," *MNRAS* 312 (2000): 166–76.

101 Harlow Shapley, "Note on a Remote Cloud of Galaxies in Centaurus," *HCO Bulletin* 874 (1930): 9–12; Somak Raychaudhury, "The Distribution of Galaxies in the Direction of the 'Great Attractor'," *Nature* 342 (1989): 251–55.

102 Hubble, *Realm of the Nebulae*, pp. 81–82.

103 M. J. Geller and J. P. Huchra, "Mapping the Universe," *Science* 246 (1989): 897–903: 897, reprinted in Bartusiak, pp. 585–90.

104 R. P. Kirshner et al., "A Million Cubic Megaparsec Void in Boötes," *ApJ* 248 (1981): L57–L60; Ia. B. Zeldovich et al., "Giant Voids in the Universe," *Nature* 300 (1982): 407–13; precursor to the Boötes Void include S. A. Gregory and L. A. Thompson, "The Coma/A1367 Supercluster and Its Environs," *ApJ* 222 (1978): 784–99. On early filament hints, see R. Giovanelli and M. P. Haynes, "The Lynx-Ursa Major Supercluster," *AJ* 87 (1982): 1355.

105 Valerie de Lapparent, Margaret J. Geller, and John P. Huchra, "A Slice of the Universe," *ApJ* 302 (1986): L1–L5.

106 Linda S. Sparke and John S. Gallagher, *Galaxies in the Universe: An Introduction* (Cambridge: Cambridge University Press, 2000), p. 281.

107 R. P. Kirshner et al., "A Survey of the Boötes Void," *ApJ* 314 (1987): 493.

108 M. J. Geller and J. P. Huchra, "Mapping the Universe," *Science* 246 (1989): 897–903.

109 N. Cross et al., "The 2dF Galaxy Redshift Survey: The Number and Luminosity Density of Galaxies," *MNRAS* 324 (2001): 825.

110 J. Richard Gott III et al., "A Map of the Universe," *ApJ* 624 (2005): 463–84.

111 A list and map of some of the well-known voids and filaments can be found at http://en.wikipedia.org/wiki/Void_(astronomy) and http://en.wikipedia.org/wiki/Galaxy_filament.

112 A list of Local Group members may be found at http://www.ast.cam.ac.uk/~mike/local_members.html.

6. The Structure of Discovery

1 Thomas S. Kuhn, *The Structure of Scientific Revolutions* (Chicago: University of Chicago Press, 1962, expanded edition, 1970), p. 55.

2 Norwood Russell Hanson, "An Anatomy of Discovery," *Journal of Philosophy* 64 (June 8, 1967): 321–52: 324.

3 Ludwik Fleck, *Genesis and Development of a Scientific Fact*, Thaddeus J. Trenn and Robert K. Merton, eds. (Chicago: University of Chicago Press, 1979; translation of *Entstehung und Entwicklung einer wissenschaftlichen Tatsache: Einführung in die Lehre vom Denkstil und Denkkollektiv* (Basel: Schwabe und Co., 1935).

4 Ken Caneva, "'Discovery' as a Site for the Collective Construction of Scientific Knowledge," *HSPS* 35 (2005): 175–291.

5 Hanson, "Anatomy of Discovery," pp. 324–34. Hanson, with historian Edward Grant, founded the History and Logic of Science (later History and Philosophy of Science) Department at Indiana University in 1960, the first in the country. When I arrived there eleven years later, stories still abounded about this colorful figure who was not only at the cutting edge of philosophy, but also flew his own plane, rode a motorcycle to work, and sometimes entered his office through the window rather than the door. On Hanson, see Matthew D. Lund, *N. R. Hanson: Observation, Discovery, and Scientific Change* (New York: Humanity Books, 2010), and Kevin T. Grau, "Force and Nature: The Department of History and Philosophy of Science at Indiana University, 1960–1998," *Isis* 90 (1999): S295–S318.

6 Hanson, "Anatomy of Discovery," pp. 334–42.

7 Theodore Arabatzis, "On the Inextricability of the Context of Discovery and the Context of Justification," in J. Schickore and F. Steinle, eds., *Revisiting Discovery and Justification* (Dordrecht: Springer, 2006), pp. 215–30. Kuhn distinguished two classes of discovery in "The Historical Structure of Scientific Discovery," *Science* 136 (1962): 760–64, reprinted in Kuhn, *The Essential Tension: Selected Studies in Scientific Tradition and Change* (Chicago and London: University of Chicago Press, 1977), pp. 165–77: 166–67. For a recent discussion of Kuhn's influence on the philosophy of science, see K. Brad Wray, "Assessing the Influence of Kuhn's *Structure of Scientific Revolutions*," *Metascience* 21 (2012): 1–10.

8 See, for example, K. Kellermann and B. Sheets, *Serendipitous Discoveries in Radio Astronomy* (Green Bank: NRAO, 1983).

9 Kuhn, *Structure*, pp. 52–53.

10 Martin Harwit, *Cosmic Discovery: The Search, Scope & Heritage of Astronomy* (New York: Basic Books, 1981), p. 57. On the general claim of extended structure by other historians and philosophers, see the review in Caneva, "Discovery," pp. 221–61. Caneva also points out (pp. 222–29) how Kuhn conflated his treatments of discovery and scientific revolutions. While Wray (note 8) argues for more emphasis on Kuhn's post-*Structure* work, we would argue that his early ideas about discovery, almost lost in the *Structure*, need more attention.

11 John Heilbron, *Galileo* (Oxford: Oxford University Press, 2010), pp. 184–86; Albert van Helden, "Galileo and Scheiner on Sunspots: A Case Study in the Visual Language of Astronomy," *Proceedings of the American Philosophical Society* 140 (1995): 358–96: 373. The work of Galileo and Scheiner on this subject has been translated in *On Sunspots*, Eileen Reeves and Albert van Helden, trans. (Chicago: University of Chicago Press, 2010). See also Mario Biagoli, *Galileo's Instruments of Credit: Telescopes, Images, Secrecy* (Chicago: University of Chicago Press, 2006).

12 Galileo Galilei to Grand Duke Cosimo II, translated in Albert van Helden, "Saturn and his Anses," *JHA* 5 (1974): 105–21: 105.

13 Piazzi's complete letter to Oriani is in Clifford J. Cunningham, *The First Asteroid: Ceres, 1801–2001* (Surfside, FL: Star Lab Press, 2001), p. 36.
14 Cunningham, *First Asteroid*, p. 48. Eric G. Forbes, "Gauss and the Discovery of Ceres," *JHA* 2 (1971): 195–99. See also Cunningham, *The Collected Correspondence of Baron von Zach* (Surfside, FL: Star Lab Press), vol. 1 (2004) and vol. 2 (2006).
15 William Herschel, "Observations on the two lately discovered celestial Bodies," *PTRSL* 92 (1802): 213–32, read May 6, 1802, 228; David W. Hughes and Brian Marsden, "Planet, Asteroid, Minor Planet: A Case Study in Astronomical Nomenclature," *JAHH* 10 (March 2007): 21–30: 22.
16 On the emergence of the idea and practice of observation ("*observationes*") from the Middle Ages through the Renaissance and nineteenth century, see *Histories of Scientific Observation*, Lorraine Daston and Elizabeth Lunbeck, eds. (Chicago: University of Chicago Press, 2011).
17 Bartusiak, *Archives*, p. 219, citing Huggins, "The New Astronomy," *The Nineteenth Century* 41 (June 1897): 907–29: 916–17.
18 Such immediate interpretation is bound up with the problems of "perception." See William Sheehan, *Planets and Perception: Telescopic Views and Interpretations, 1609–1909* (Tucson: University of Arizona Press, 1988).
19 It is worth noting, however, that in the more mature stages of science, novelty may emerge against a theoretical background that determines scientists' expectations.
20 On the problematic nature of observation, see Norwood Russell Hanson, *Patterns of Discovery* (Cambridge: Cambridge University Press, 1958) and Daston and Lunbeck, *Histories of Scientific Observation.*
21 Ludwik Fleck, *Genesis*, p. 76; Robert K. Merton, "Science, Technology and Society in Seventeenth Century England," *Osiris*, 4, pt. 2 (1938): pp. 360–632; Michael Polanyi, *Personal Knowledge: Towards a Post-Critical Philosophy* (Chicago: University of Chicago Press, 1958), and *The Tacit Dimension* (Garden City, NY: Doubleday, 1966). Fleck's work was not well known until its English translation in 1979.
22 See, for example, Steven Shapin, *Never Pure: Historical Studies of Science as If It Was Produced by People with Bodies, Situated in Time, Space, Culture, and Society, and Struggling for Credibility and Authority* (Baltimore: Johns Hopkins University Press, 2010), and references therein, as well as Robert E. Kohler's review, "A Naturalizer's Vision," in *Science* 330 (2010): 450–51.
23 Augustine Brannigan, *The Social Basis of Scientific Discoveries* (Cambridge: Cambridge University Press, 1981).
24 In addition to Shapin cited above, see Bruno Latour and Steve Woolgar, *Laboratory Life: The Social Construction of Scientific Facts* (Princeton: Princeton University Press, 1986), and Peter Galison, *How Experiments End* (Chicago: University of Chicago Press, 1987). Interestingly, Latour and Woolgar dropped "social" from their title for the second edition.
25 David Aubin et al., eds., *The Heavens on Earth: Observatories and Astronomy in Nineteenth-Century Science and Culture* (Durham and London: Duke University Press, 2010).

26 Toby E. Huff, *Intellectual Curiosity and the Scientific Revolution: A Global Perspective* (Cambridge: Cambridge University Press, 2011), p. 142. For strong caveats on this book and its definition of "intellectual curiosity," see the review by Sonja Brentjes, *Isis* 103 (2012): 179–80.
27 Xiao-Chun Sun, "The Impact of the Telescope on Astronomy and Society in China," in Donald G. York, Owen Gingerich, and Shuang-Nan Zhang, eds., *The Astronomy Revolution: 400 Years of Exploring the Cosmos* (Boca Raton, FL: CRC Press, 2012), pp. 281–90: 290; Tsung-Dao Lee, "From the Language of Heaven to the Rationale of Matter," ibid., pp. 3–12: 7.
28 On social aspects of quasar discovery, see David O. Edge and Michael J. Mulkay, *Astronomy Transformed: The Emergence of Radio Astronomy in Britain* (New York: Wiley, 1976), pp. 204–8, and Harwit, *Cosmic Discovery*, pp. 137–38.
29 Suzy Collin, "Quasars and Galactic Nuclei, a Half Century Agitated Story," in Jean-Michel Alimi and Andre Fuzfa, eds., *Albert Einstein Century International Conference*, vol. 861 (Melville, NY: American Institute of Physics, 2006), online version only, pp. 587–95: 587; preprint online at http://arxiv.org/abs/astro-ph/0604560. On gravitational waves, see Harry Collins, *Gravity's Ghost: Scientific Discovery in the Twenty-first Century* (Chicago: University of Chicago Press, 2011). At low resolution the macrostructure of discovery does not seem to have changed much over time, but the same may not be true of the microstructure. In other words, the tripartite structure of detection, interpretation, and understanding seems to be a constant, but the internal structure has become more complex over time, with more inference, more complex (digital) social interactions, etc.
30 Hanson, "Anatomy," p. 328.
31 Ken Caneva, "Discovery," 175–291: 255–56.
32 Galison, *How Experiments End*, pp. 76, 126–27.
33 Ibid., p. 127.
34 Caneva, "Discovery," p. 175.

7. The Varieties of Discovery

1 Abraham Pais, *Inward Bound: of Matter and Forces in the Physical World* (Oxford: Oxford University Press), p. 130.
2 Sara Schechner, *Comets, Popular Culture, and the Birth of Modern Cosmology* (Princeton: Princeton University Press, 1997).
3 A catalog of naked-eye comets reported through AD 1700 is found in Donald Yeomans, *Comets: A Chronological History of Observation, Science, Myth, and Folklore* (New York: Wiley, 1991), pp. 361–424. For more detail on those in East Asia, see the list in David Pankenier, Zhentao Xu and Yaotiao Jiang, *East Asian Archaeoastronomy: Historical Records of Astronomical Observations of China, Japan, and Korea* (New York and London: Cambria Press, 2008), chapter 5; the first edition was CRC Press, 2000. The comet of 1059 BC is listed first in both Yeomans and Pankenier.
4 John T. Ramsay and A. Lewis Licht, *The Comet of 44 B.C. and Caesar's Funeral Games* (Atlanta: Scholars Press, 1997).

5 *Anglo-Saxon Chronicle*, part 5; opening lines of 1066 entry, "The Online Medieval and Classical Library," http://omacl.org/Anglo/, utilizing *The Anglo-Saxon Chronicle* (London: Everyman Press, 1912).
6 J. R. Christianson, "Tycho Brahe's German Treatise on the Comet of 1577: A Study in Science and Politics," *Isis* 70 (1979): 110–40. The full text of Brahe's German Treatise on the comet is translated here and appended to the article; the quotation here is found on page 134. Though Tycho speaks of the "star Saturn," he would have separated the planets from the "fixed stars" by their motion. Similarly, Galileo indiscriminately used the terms *planetae*, *sidera*, and *stelluale* for the moons of Jupiter.
7 "Detection" in this case involves some form of interpretation, since in order to detect an object as an instance of a class, one needs to form the corresponding concept.
8 Victor E. Thoren, *The Lord of Uraniborg* (Cambridge: Cambridge University Press, 1990), p. 125; C. Doris Hellman, *The Comet of 1577: Its Place in the History of Astronomy* (New York: AMS Press, 1971), reprint of the 1944 edition with addenda, errata, and a supplement to the appendix. Yeomans (pp. 33ff.), points out that both Michael Maestlin and Cornelius Gemma also came to the same conclusions based on fewer and less accurate observations. Hellman and Robert Westman point out that more than 100 authors opined on the comet in 1578 alone, many of them in an astrological context; Westman, *The Copernican Question: Prognostication, Skepticism and Celestial Order* (Berkeley and Los Angeles: University of California Press, 2011), pp. 250–58. On Galileo's skepticism, see J. L. Heilbron, *Galileo* (Oxford: Oxford University Press, 2010), p. 237.
9 Fred L. Whipple, "A Comet Model I. The Acceleration of Comet Encke," *ApJ* 111 (1950): 375–94, reprinted in part in Bartusiak, *Archives*, 445–48.
10 Jan H. Oort, "The Structure of the Cloud of Comets Surrounding the Solar System and a Hypothesis Concerning Its Origin," *BAIN* 11 (1950): 91–110, reprinted in part in Bartusiak, *Archives*, 441–45.
11 Thoren, *Lord of Uraniborg*, pp. 52–53; Christianson, "Tycho Brahe's German Treatise," pp. 121–22.
12 David H. Clark and F. Richard Stephenson, *The Historical Supernovae* (Oxford: Pergamon Press, 1977); F. R. Stephenson and D. A. Green, *Historical Supernovae and Their Remnants* (Oxford: Clarendon Press, 2002).
13 Thoren, *Lord of Uraniborg*, pp. 55–73. Tycho was not the first to see the star; in Europe Thomas Digges and John Dee observed it, but failed to make accurate measurements. Others also observed it, some measuring parallax, others not. Westman, *Copernican Question*, pp. 230–35.
14 *De Stella Nova in Pede Serpentarii* (1606) has been translated into German and French, but not English. The best description of Kepler's observations in English is A. Lombardi, "Kepler's Observations of the Supernova of 1604," in M. Turatto et al., eds., *1604–2004: Supernovae as Cosmological Lighthouses, PASP* 342 (2005): 21–29. On priority disputes over discovering this object, see Westman, *Copernican Question*, 382–401.
15 Baade and Zwicky, "On Supernovae," *PNAS* 20 (May 15, 1934): 254–59, reprinted in Bartusiak, *Archives*, 341–43; Baade and Zwicky, "Supernovae and Cosmic Rays," *Physical Review* 45 (1934): 138. Donald E. Osterbrock, "Who Really Coined the Word Supernova/Who First

Predicted Neutron Stars," *BAAS* 33 (2001): 1330–31. Osterbrock's answer to both questions is Baade and Zwicky, who used the term "supernovae" in seminars and courses at Caltech in 1931. See also Hilmar Duerbeck, "Novae: An Historical Perspective," in Michael F. Bode and Adeurin Evans, eds., *Classical Novae* (Cambridge: Cambridge University Press, 2nd ed.,
2008), p. 2.

16 Fritz Zwicky, "On the Theory and Observation of Highly Collapsed Stars," *Physical Review* 55 (1939): 726–43.

17 R. Minkowski, "Spectra of Supernovae," *PASP* 53 (1941): 224–25.

18 Virgil, *Georgics*, Book I, lines 365–7. For East-Asian records see Pankenier et al., *Archaeoastronomy*.

19 Mark Littmann, *The Heavens on Fire: The Great Leonid Meteor Storms* (Cambridge: Cambridge University Press, 1998), p. 35.

20 D. Olmsted, "Observations on the Meteors of 13 Nov. 1833," *AJSA* 25 (1834): 363–411 and *AJSA* 26 (1834): 132–74; Steven J. Dick, "Observation and Interpretation of the Leonid Meteors Over the Last Millennium," *JAHH* 1 (1998): 1-20.

21 H. A. Newton, "Evidence of the Cosmical Origin of Shooting Stars Derived from the Dates of Early Star-showers," *AJSA* 36 (1863): 145–49.

22 For background see Michael W. Friedlander, *A Thin Cosmic Rain: Particles from Outer Space* (Cambridge, MA: Harvard University Press, 2000), pp. 8–9; Robert P. Crease and Charles C. Mann, *The Second Creation: Makers of the Revolution in 20th-Century Physics* (New York: MacMillan, 1986); and Per Carlson, "A Century of Cosmic Rays," *Physics Today* 65 (2012): 30–36.

23 Crease and Mann, *The Second Creation*, pp. 147–48; Friedlander, *Thin Cosmic Rain*, pp. 7–8.

24 V. F. Hess, "Über Beobachtungen der durchdringenden Strahlung bei sieben Freiballonfahrt," *Physikalische Zeitschrift* 13 (1912): 1084–91; translated by Brian Doyle, "Concerning Observations of Penetrating Radiation on Seven Free Balloon Flights," in Lang and Gingerich, pp. 13–20: 19. See also Crease and Mann, *The Second Creation*, pp.147–57.

25 W. Kolhörster, *Physicakische Zeitschrift* 14 (1913): 1153; Carlson (2012): 30–36.

26 R. M. Otis and R. A. Millikan, *Phys. Rev.* 23 (1924): 778; cited in Carlson, "A Century of Cosmic Rays."

27 Robert A. Millikan and G. H. Cameron, "The Origin of the Cosmic Rays," *Physical Review* 32 (Oct. 1928): 533–57; for Millikan's experiments on cosmic rays, see Peter Galison, *How Experiments End* (Chicago: University of Chicago Press, 1987), pp. 80ff. Carlson points out (p. 32) that in Europe the names Höhenstrahlung (high-altitude rays) and Ultra-Gammastrahlung became current.

28 For context see Galison, *How Experiments End*, pp. 93ff., and Daniel Kevles, *The Physicists* (New York: Vintage Books, 1971), p. 179.

29 Stuart Clark, *The Sun Kings: The Unexpected Tragedy of Richard Carrington and the Tale of How Modern Astronomy Began* (Princeton: Princeton University Press, 2007).

30 Eugene N. Parker, "Dynamics of the Interplanetary Gas and Magnetic Field," *ApJ* 128 (1958): 664–76; Parker, "Cosmic-ray Modulation by Solar Wind," *Physical Review* 110 (1958): 1445–49; Karl Hufbauer, *Exploring*

the Sun: Solar Science Since Galileo (Baltimore and London: Johns Hopkins University Press, 1991), especially "The Solar Wind, 1957–1970," pp. 213ff.

31 Anthony Hewish, "Pulsars and High Density Physics," Nobel Lecture, December 12, 1974, pp. 174–75, online at http://nobelprize.org/nobel_prizes/physics/laureates/1974/hewish-lecture.pdf; James van Allen, "Interplanetary Particles and Fields," *SciAm* (September 1975), reprinted in *The Solar System* (San Francisco: W. H. Freeman, 1975), pp. 125–34.

32 R. A. Howard, "A Historical Perspective on Coronal Mass Ejections," in N. Gopalswamy, R. Mewaldt, and J. Torsti, eds., *Solar Eruptions and Energetic Particles* (New York: American Geophysical Union, 2006).

33 http://www.lpi.usra.edu/lunar/missions/apollo/apollo_11/experiments/swc/.

34 Steven J. Dick, *Sky and Ocean Joined, The U. S. Naval Observatory, 1830–2000* (Cambridge: Cambridge University Press, 2003); Dick, "Discovering the Moons of Mars," *Sky and Telescope* 76 (Sept. 1988): 242–43.

35 Helge Kragh, *The Moon that Wasn't: The Saga of Venus' Spurious Satellite* (Berlin: Birkhauser Basel, 2008); Kragh, "The Second Moon of the Earth," *JHA* 40 (2009): 1–10. The same is true of searches for an intra-Mercurial planet Vulcan; see Richard Baum and William Sheehan, *In Search of Planet Vulcan: The Ghost in Newton's Clockwork Universe* (New York and London: Plenum, 1997). On Lassell, see Robert Smith and Richard Baum, "William Lassell and the Ring of Neptune: A Case Study in Instrumental Failure," *JHA* 15 (1984): 1–17.

36 J. L. Elliot, E. Dunham, and D. Mink, "The Rings of Uranus," *Nature* 267 (1977): 328–30; Elliot, Dunham, and R. L. Millis, "Discovering the Rings of Uranus," *ST* 53 (1977): 412–16.

37 B. A. Smith, L. A. Soderblom, T. V. Johnson et al., "The Jupiter System Through the Eyes of Voyager 1," *Science* 204 (1979): 951–7, 960–72.

38 B. A. Smith, L. A. Soderblom, D. Banfield et al., "Voyager 2 at Neptune: Imaging Science Results," *Science* 246 (1989): 1422.

39 Harwit, *Cosmic Discovery*, chapter 2.

40 Ibid., p. 16; Bartusiak, *Archives*. On the concept of "phenomena," see Ian Hacking, *Representing and Intervening: Introductory Topics in the Philosophy of Natural Science* (Cambridge: Cambridge University Press, 1983).

41 John North, *The Measure of the Universe: A History of Modern Cosmology* (Oxford: Oxford University Press, 1965); Norris Hetherington, *Hubble's Cosmology: A Guided Study of Selected Texts* (Tucson: Pachart Publishing, 1996), and Hetherington, *The Edwin Hubble Papers: Previously Unpublished Manuscripts of the Extragalactic Nature of Spiral Nebulae* (Tucson: Pachart Publishing, 1990); Kragh, *Cosmology and Controversy* (Princeton: Princeton University Press, 1996); Robert Smith, *The Expanding Universe* (Cambridge: Cambridge University Press, 1982); Smith, "Beyond the Galaxy: The Development of Extragalactic Astronomy, 1885–1965, Part 2," *JHA* 40 (2009): 71–107, especially 80–107. See also Harry Nussbaumer and Lydia Bieri, *Discovering the Expanding Universe* (Cambridge: Cambridge University Press, 2009).

42 Smith, "Beyond the Galaxy," Part 1, *JHA* (2008): 104. On Slipher's redshift measurements see Robert Smith, "Redshifts and Gold

Medals," in William Lowell Putnam, ed., *The Explorers of Mars Hill* (West Kennebunk, ME: Phoenix Publishing), pp. 43–65.

43 Willem de Sitter, "On Einstein's Theory of Gravitation, and Its Astronomical Consequences. Third Paper," *MNRAS* 78 (1917): 3–28: 28, brief excerpt in Bartusiak, *Archives*, p. 316; John North, *The Measure of the Universe*, pp. 92–104; Robert Smith, *The Expanding Universe*, pp. 170–76; Harry Nussbaumer and Lydia Bieri, *Discovering the Expanding Universe*, chapters 6, 12, and 14.

44 Edwin Hubble, "A Relation Between Distance and Radial Velocity Among Extra-Galactic Nebulae," *PNAS* 15 (March 15, 1929): 168–73.

45 Smith, "Beyond the Galaxy," *JHA* 40 (2009): 88.

46 Hetherington, "Philosophical Values and Observation in Edwin Hubble's Choice of a Model of the Universe," *HSPS* 13 (1982): 42–67; Hubble later explicitly addressed the question of the expanding universe in his Sigma Xi lecture, delivered in 1941, published as "The Problem of the Expanding Universe," *American Scientist* 30 (April 1942), reprinted with commentary in Norris Hetherington, *Hubble's Cosmology: A Guided Study of Selected Texts* (Tucson: Pachart Publishing, 1996), pp. 148–209.

47 Smith, "Beyond the Galaxy," *JHA* 40 (2009): 91; Norris Hetherington, "Sirius B and Gravitational Redshift - An Historical Review," *QJRAS* 21 (1980): 246–52, reprinted in Hetherington's *Science and Objectivity: Episodes in the History of Astronomy* (Ames: Iowa State University Press, 1988), pp. 65–72.

48 Helge Kragh and Robert Smith, "Who Discovered the Expanding Universe?" *History of Science* 41 (2003): 141–62.

49 G. Lemaître, "Un univers homogène de masse constante et de rayon croissant rendant compte de la vitesse radiale des nébuleuses extra-galactic," *Annales de sociétés scientifique de Bruxelles* 47 (1927): *Annales sociétés*, 49–56; translated into English as "A homogeneous universe of constant mass and increasing radius," *MNRAS* 91 (1931): 483–90, reprinted in Lang and Gingerich, *Source Book*, pp. 845–48.

50 Adam G. Riess, Alexei Filipchenko, et al., "Observational Evidence from Supernovae for an Accelerating Universe and a Cosmological Constant," *AJ* 116 (1998): 1009–38; Saul Perlmutter et al., "Measurements of Omega and Lambda from 42 High-Redshift Supernovae," *ApJ* 517 (June 1, 1999): 565–86. The Riess article also provides some of the history of Type Ia supernova studies. Proof of the accelerating universe was the "breakthrough of the year" for *Science* magazine in 1998, and Perlmutter, Riess, and Brian Schmidt received the Nobel Prize for their work on the subject. For the inside story of the race between the Perlmutter and Riess teams, see Robert Kirshner, *The Extravagant Universe: Exploding Stars, Dark Energy and the Accelerating Cosmos* (Princeton: Princeton University Press, 2002); Alex Filippenko, "Einstein's Biggest Blunder? High-Redshift Supernovae and the Accelerating Universe," *PASP* 113 (2001): 1441; and Yudhijit Bhattacharjee, "A Week in Stockholm," *Science* 336 (2012): 26–31.

51 The fullest account of the discovery of cosmic radio waves is Woodruff T. Sullivan, III, *Cosmic Noise: A History of Early Radio Astronomy* (Cambridge: Cambridge University Press, 2009), chapter 3 "Jansky and His Star Static," pp. 29–53. Earlier accounts are given in Sullivan, "Karl

Jansky and the Beginnings of Radio Astronomy," in K. Kellermann and B. Sheets, eds., *Serendipitous Discoveries in Radio Astronomy* (Green Bank: NRAO, 1983), pp. 39–56, and Sullivan, *The Early Years of Radio Astronomy: Reflections Fifty Years after Jansky* (Cambridge: Cambridge University Press, 1984).

52 Karl G. Jansky, "Directional Studies of Atmopsherics at High Frequencies," Proceedings of the IRE 20 (Dec. 1932); "Electrical Disturbances Apparently of Extraterrestrial Origin," *Proc. IRE* 31 (Oct.1933); "A Note on the Source of Interstellar Interference," *Proc. IRE* 23 (Oct.1935).

53 Models of the antennae of Jansky and Reber are on display at the National Radio Astronomy Observatory (NRAO) in Green Bank, West Virginia.

54 Arno A. Penzias and Robert W. Wilson, "A Measurement of Excess Antenna Temperature at 4080 Mc/s," *ApJ* 142 (July 1, 1965): 419–21; R. H. Dicke, P. J. E. Peebles, P. G. Roll, and D. T. Wilkinson, "Cosmic Black-Body Radiation," *ApJ* 142 (1965): 416–19. The Penzias and Wilson article is reprinted in Bartusiak, *Archives*, pp. 510–12.

55 George Smoot and Keay Davidson, *Wrinkles in Time* (New York: Avon Books, 1993); John C. Mather and John Boslough, *The Very First Light* (New York: Basic Books, 1996; revised edition, 2008); see p. 221 for the AAS meeting, p. 235 for the APS meeting and news conference. The Hawking quote is admittedly a promotional quote that appears on the cover of Smoot's book, but demonstrates how scientific discovery is sold to the public.

56 Kragh, *Cosmology and Controversy*, p. 343.

8. Discovery and Classification

1 Carolus Linnaeus, *Systema naturae: Facsimile of the First Edition* (1735; Niewwkoop: B. DeGraaf, 1964), p. 19, quoted in Paul L. Farber, *Finding Order in Nature: The Naturalist Tradition from Linnaeus to E. O. Wilson* (Baltimore and London: Johns Hopkins University Press, 2000), pp. 8–9.

2 Allan Sandage, *The Mount Wilson Observatory: Breaking the Code of Cosmic Evolution*, vol. 1 of the Centennial History of the Carnegie Institution of Washington (Cambridge: Cambridge University Press, 2004), pp. 230–31.

3 W. W. Morgan, "Remarks on the MK System," in M. F. McCarthy, A. G. D. Philip, and G. V. Coyne, eds., *Spectral Classification of the Future, Proceedings of IAU Colloq. 47, held in Vatican City, July 11–15, 1978* (Rome: Vatican Observatory, 1979), pp. 59–61.

4 On the most general aspects of classification even beyond the sciences, as well as their social roots and implications, see Geoffrey C. Bowker and Susan Leigh Star, *Sorting Things Out: Classification and Its Consequences* (Cambridge, MA: MIT Press, 1999).

5 Stephen Jay Gould, in Foreword to Lynn Margulis and Karlene V. Schwartz, *Five Kingdoms: An Illustrated Guide to the Phyla of Life on Earth* (New York: W. H. Freeman, 2nd ed., 1988), p. x.

6 Paul L. Farber, *Finding Order in Nature;* Ernst Mayr, *The Growth of Biological Thought: Diversity, Evolution and Inheritance* (Cambridge, MA: Harvard University Press, 1982), especially pp. 185ff.; Jan Sapp, *The New Foundations of Evolution on the Tree of Life* (Oxford: Oxford University Press, 2009).

7 David L. Hull, *Science as a Process: An Evolutionary Account of the Social and Conceptual Development of Science* (Chicago: University of Chicago Press, 1988), pp. 75–110: 109.
8 Hull, *Science as a Process*, pp. 109–10. On Darwin's species concept and its change over his career, see Mayr, *Growth*, pp. 265ff.
9 Ernst Mayr, "Species Concepts and Their Application," chapter 2 of *Populations, Species and Evolution* (Cambridge, MA: Harvard University Press, 1963), reprinted in Marc Ereshefsky, ed., *The Units of Evolution: Essays on the Nature of Species* (Cambridge, MA: MIT Press, 1992), pp. 15–25, elaborated in Ernst Mayr, *Toward a New Philosophy of Biology* (Cambridge, MA: Harvard University Press, 1988), pp. 315–34.
10 Ernst Mayr, *Systematics and the Origin of Species* (Cambridge: Cambridge University Press, 1942), p.120 for this original definition; Ann Gibbons, "The Species Problem," *Science* 331 (Jan. 28, 2011): 394 for its continued use. Mayr, *Toward a New Philosophy*, p. 318, pointed out that hybrids do occur, but that reproductive isolation "refers to the integrity of populations, even though an occasional individual may go astray." The reason species exist rather than a continuum of individuals, he says, is because it is "a protective device for well-integrated genotypes," p. 319.
11 Mayr, in Ereshefsky, *Units of Evolution*, p. 17.
12 Mayr, *Toward a New Philosophy*, pp. 337ff.
13 The exceptions involve intermediate objects such as Chiron, a "hybrid" between a comet and an asteroid. In such cases the classifier is faced with the practical question of whether to create a new Class, to employ dual classification in which the object is a member of two classes, or to designate it a subclass of each of the two classes. The answer may be dictated by utility or other factors, including social.
14 Mayr, "Species Concepts and Their Application," in Ereshefsky, *Units of Evolution*, pp. 15–25. Mayr discusses, and largely dismisses, other terms for "class" used by philosophers and others, including "natural kinds," "clusters," and "sets," *Toward a New Philosophy*, pp. 338ff.
15 Gibbons, "The Species Problem," p. 394.
16 Ereshefsky, *Units of Evolution*, p. 187.
17 Ernst Mayr, *Growth*, p. 185, and from species to taxa, p. 328. Of course, the purpose at hand for any particular scientist may change, so that one astronomer's useful classification may be of little use to another. Sandage's quote is open to challenge; "no idea" of how to choose key parameters seems too strong.
18 Lynn Margulis and Karlene V. Schwartz, *Five Kingdoms: An Illustrated Guide to the Phyla of Life on Earth* (New York: W. H. Freeman, 2nd ed., 1988).
19 On the rise of the Three Domain system see Jan Sapp, *The New Foundations of Evolution on the Tree of Life* (Oxford: Oxford University Press, 2009); on the claim for the Three Domains system as more natural, p. 265.
20 Lynn Margulis and Michael J. Chapman, *Kingdoms & Domains: An Illustrated Guide to the Phyla of Life on Earth* (Amsterdam: Elsevier, 2010), pp. 11–12. On the disputes between these two systems, and much else, see Jan Sapp, *New Foundations*, especially pp. 266–81, and on negotiation see M. Dietrich, "Paradox and Persuasion: Negotiating the Place of Molecular Evolution within Evolutionary Biology," *JHB* 31 (1998): 87–111.

21 Michael D. Gordin, *A Well-Ordered Thing: Dmitrii Mendeleev and the Shadow of the Periodic Table* (New York: Basic Books, 2004), pp. 3–4.
22 Ereshefsky, *Units of Evolution*, p. 187.
23 Eric Scerri, *The Periodic Table: Its Story and Significance* (Oxford: Oxford University Press, 2007), pp. 63–100.
24 Ibid., pp. 99–101.
25 Gordin, *A Well-Ordered Thing.*
26 Helge Kragh, *Quantum Generations: A History of Physics in the Twentieth Century* (Princeton: Princeton University Press, 1999), p. 321. Wolfgang Pauli is also reputed to have said, "Had I foreseen this, I would have gone into botany."
27 Andrew Pickering, *Constructing Quarks: A Sociological History of Particle Physics* (Edinburgh: Edinburgh University Press, 1984); Murray Gell-Mann, *The Quark and the Jaguar: Adventures in the Simple and the Complex* (New York: W. H. Freeman, 1994).
28 Mayr, *Toward a New Philosophy*, p. 314, and *Growth*, pp. 238ff.
29 On the other hand, there may be "genetic" elements in the uses of technology that detect new classes of astronomical objects, in terms of styles, design, and "survival of the fittest," as in detectors that utilize different sensors, resulting in transitions from visual to photographic, photoelectric, and electronic techniques, defining whole eras of astronomy. See Chapter 9.
30 John Hearnshaw, "Auguste Comte's Blunder: An Account of the First Century of Stellar Spectroscopy and How it Took One Hundred Years to Prove that Comte was Wrong!" *JAHH* 13 (2010): 90–104.
31 John Hearnshaw, *The Analysis of Starlight : One Hundred and Fifty Years of Astronomical Spectroscopy* (Cambridge: Cambridge University Press, 1986), pp. 24–29; Ralph H. Curtiss, "Classification and Description of Stellar Spectra," *Handbuch der Astrophysik*, vol. 5, part 1 (1932): 1–108: 1–3. On Fraunhofer in the context of the rise of precision optics in Germany see Myles Jackson, *Spectrum of Belief* (Cambridge, MA: MIT Press, 2000).
32 See especially David DeVorkin's work cited in Chapter 4.
33 Hearnshaw, *Analysis*, p. 81; also A. J. Meadows, "The Origins of Astrophysics," in Owen Gingerich, ed., *Astrophysics and 20th Century Astronomy to 1950*: Part A, (Cambridge: Cambridge University Press, 1984), pp. 4–15.
34 Hearnshaw, *Analysis*, pp. 89–94.
35 The discoverer of the true temperature sequence has been a subject of some dispute among astronomers over the years. In his MKK atlas published in 1943, Morgan wrote that "the idea of a temperature classification is based on the work of Miss Maury and Miss Cannon at Harvard and of Sir Norman Lockyer," W. W. Morgan, P. C. Keenan, and E. Kellman, *Atlas of Stellar Spectra, with an Outline of Spectral Classification* (Chicago: University of Chicago Press, 1943), Introduction. But in his history of Mt. Wilson Observatory, Sandage took Morgan to task for this attribution of "proof," arguing that it was Walter S. Adams and Hale who had used sunspot spectra and laboratory data to prove the temperature sequence in 1908, and, following Saha's ideas of ionization in stellar atmospheres in 1920, Fowler and Milne in 1923

and Cecilia Payne in 1925 who made it a "theoretical certainty"; Sandage, *Mt. Wilson Observatory*, pp. 251 and 255. The answer hinges on the question of what constitutes "proof." For more on this question see David DeVorkin and Ralph Kenat, "Quantum Physics and the Stars (I): Establishment of a Stellar Temperature Scale," *JHA* 14 (1983): 102–32.

36 David H. DeVorkin, "An Astronomical Symbiosis: Stellar Evolution and Spectral Classification (1860–1910)," PhD dissertation, University of Leicester (1978), and subsequent publications cited in this chapter and Chapter 4.

37 Hearnshaw, *Analysis*, p. 119. For more details on these systems, see pp. 104–42 in ibid. Cannon's revisions, including rearrangement of the spectral types, first appeared in print as Cannon and Pickering, "Spectra of Bright Southern Stars Photographed with the 13-inch Boyden Telescope as a Part of the Henry Draper Memorial, in *Annals of the HCO* 28 (1901, part II): 131–263. 1122 stars were classified in this report from plates taken at Harvard's Arequippa, Peru Southern station from November 1891 to December 1899.

38 Ralph H. Curtiss, "Classification and Description of Stellar Spectra," *Handbuch der Astrophysik*, vol. 5, part 1 (1932), pp. 1–108: p. 107. This important review of stellar spectroscopy was not quite complete before Curtiss's untimely death on Christmas Day, 1929, from pleurisy followed by a heart attack.

39 Curtiss, "Classification," p. 107; Hearnshaw, *Analysis*, pp. 283–93.

40 H. N. Russell, C. H. Payne-Gaposhkin, and D. H. Menzel, "The Classification of Stellar Spectra," *ApJ* 81 (1935): 107–18; quoted in Hearnshaw, *Analysis*, 284–85.

41 W. W. Morgan, "A Morphological Life," *ARAA* 26 (1988): 1–9: 3–4. On Morgan see *BEA*, vol. 2, and for the institutional context of his work see Donald Osterbrock, *Yerkes Observatory, 1892–1950: The Birth, Near Death and Resurrection of a Scientific Research Institution* (Chicago: University of Chicago Press, 1997), especially pp. 222ff.

42 W. W. Morgan, "On the Spectral Classification of the Stars of Types A to K," *ApJ* 85 (1937): 380–97: 384. The first log g-temperature plot is on page 387.

43 W. W. Morgan, "On the Determination of Color Indices of Stars from a Classification of Their Spectra," *ApJ* 87 (1938): 460–75. The Luminosity classes, including the division between the brightest (Ia) and fainter (Ib) supergiants, are listed on p. 466, table 3.

44 On the *Atlas,* its compilers and circumstances, see Donald Osterbrock, "Fifty Years Ago: Astronomy; Yerkes Observatory; Morgan, Keenan and Kellman," in C. J. Corbally, R. O. Gray, and R. F. Garrison, eds., *The MK Process at 50 Years* (San Francisco: Astronomical Society of the Pacific, 1994), 199–214.

45 Morgan, "A Morphological Life," p. 3. The similarity of this concept to Linnaeus's "the things themselves" is also notable (see opening epigraph).

46 M. W. Feast and A. D. Thackeray, "Red Supergiants in the Large Magellanic Cloud," *MNRAS* 116 (1956): 587–90.

47 Phillip Keenan, "Classification of Supergiants of types G, K, and M, Late-Type Stars," Proceedings of a Conference held in Tucson, AZ,

October 1970, G. Wesley Lockwood and H. Melvin Dyck, eds., *Kitt Peak National Observatory*, Contribution No. 554, 1971, pp. 35–9. W. W. Morgan and P. C. Keenan, "Spectral Classification," *ARAA* 11 (1973): 29–50: 44. The latter is a major article on classification.

48 Keenan, "Classification," p. 35.

49 Wolfgang Steinicke, *Observing and Cataloguing Nebulae and Star Clusters: From Herschel to Dreyer's New General Catalogue* (Cambridge: Cambridge University Press, 2010).

50 Morgan, "A Morphological Life," p. 6.

51 Allan Sandage, "Classification and Stellar Content of Galaxies Obtained from Direct Photography," in A. Sandage, M. Sandage and J. Kristian, eds., *Galaxies and the Universe*, 1975; online at http://nedwww.ipac.caltech.edu/level5/Sandage/frames.html; Sandage, "The Classification of Galaxies: Early History and Ongoing Developments," *ARAA* 43 (2005): 581–624.

52 Allan Sandage, *Mount Wilson Observatory*, pp. 230–31.

53 See H. D. Curtis's 1933 lengthy summary "The Nebulae," in *Handbuch der Astrophysik*, vol. 5, pt. 2, pp. 774–936, including appendix 5, "Systems of Nebular Classification," where he describes the nebular classification systems of William Herschel and John Herschel in the eighteenth and nineteenth centuries, S. I. Bailey (1908), Wolf (1908), Shapley (1928), Hubble (1921, 1926), and Lundmark (1927). All such systems prior to 1925 were necessarily constructed without knowledge of the distances of the extragalactic nebulae, and therefore consist of a mix of diffuse, planetary, and what turned out to be extragalactic "nebulae." Edwin Hubble, *The Realm of the Nebulae* (New Haven: Yale University Press, 1936); R. Hart and R. Berenzden, "Hubble, Lundmark and the Classification of Non-Galactic Nebulae," *JHA* 2 (1971): 200; R. Hart and R. Berendzen, "Hubble's Classification of Non-Galactic Nebulae, 1922–1926," *JHA* 2 (1971): 109–19; Sandage, "Classification," pp. 3–4.

54 Sandage, *Mt. Wilson Observatory*, 488–89. Lundmark's paper was published in *Arkiv fur Matematik, Astronomi och Fysik*, Band 19B, no. 8, 1926. On this controversy, see Pekka Teerikorpi, "Lundmark's Unpublished 1922 Nebula Classification," *JHA* 20 (1989): 165–70, where Teerikorpi puts the plagiarism charge to rest; also for more details, see Sandage, *Mt. Wilson Observatory*, pp. 485ff. Shapley's article was "On the Classification of Extra-Galactic nebulae," *HCO Bulletin*, no. 849 (1927): 1–4; see Robert Smith, *The Expanding Universe: Astronomy's Great Debate, 1900–1931* (Cambridge: Cambridge University Press, 1982), chap. 4, and Smith, "Beyond the Galaxy: The Development of Extragalactic Astronomy 1885–1965, Part 2," *JHA* 40 (2009): 78.

55 David L. Block and Kenneth C. Freeman, *Shrouds of the Night: Masks of the Milky Way and Our Awesome New View of Galaxies* (New York: Springer, 2008), p. 198.

56 The evidence is presented in Block and Freeman, *Shrouds*, pp. 183–203. In *The Realm of the Nebulae*, p. 44, Hubble mentions Reynolds's 1927 paper, "A Classification of Spiral Nebulae," *Observatory* 50, no. 185 (1927), and on p. 140 a paper by Reynolds on NGC 205, the companion to Andromeda, published in 1934. Reynolds's important 1920 paper is

"Photometric Measures of the Nuclei of some Typical Spiral Nebulae," *MNRAS* 80 (1920): 746–53.

57 Gérard de Vaucouleurs, "Classification and Morphology of External Galaxies," "*Handbuch der Physik* 53 (1959): 275ff.; Sidney van den Bergh, *Galaxy Morphology and Classification* (Cambridge: Cambridge University Press, 1998); Sandage, "Classification," p. 605; Ronald J. Buta, Harold G. Corwin, Jr., and Stephen C. Odewahn, *The de Vaucouleurs Atlas of Galaxies* (Cambridge: Cambridge University Press, 2007), p. 27.

58 Buta, Corwin, and Odewahn, *de Vaucouleurs Atlas*, p. 80.

59 Marsden, "On the Relationship between Comets and Minor Planets," *AJ* 75 (1970): 206–17. In classical Greek mythology Chiron was a centaur, the teacher of Aeneas, Jason, Achilles and many other ancient Greek cultural heroes.

60 Michael F. A'Hearn, "Pluto: A Planet or a Trans-Neptunian Object? in Hans Rickman, ed., *IAU Highlights of Astronomy*, 12 (San Francisco: ASP Conference series, 2002), 201–4; and see the entire section of articles in this volume. For the history of the archaeopteryx controversy see Sean B. Carroll, *Remarkable Creatures: Epic Adventures in the Search for the Origins of Species* (Boston: Houghton, Mifflin, Harcourt, 2009), pp. 161–79.

61 D. Latham et al., "Kepler-7b: A Transiting Planet with Unusually Low Density," *ApJ Letters* 713 (2010): L140–L144; D. R. Anderson et al., "Wasp-17b: An Ultra-Low Density Planet in a Probable Retrograde Orbit," *ApJ* 709 (2010): 159.

62 Raymond Harris, online at http://www.deepfly.org/TheNeighborhood/PlanetsInPowersOf10.html; J. J. Fortney et al., "Planetary Radii across Five Orders of Magnitude in Mass and Stellar Insolation: Application to Transits," *ApJ* 659 (2007): 1661–72; S. Marchi, "Extrasolar Planet Taxonomy: A New Statistical Approach," *ApJ* 666 (2007): 475–85.

63 The possibility of such a system was first broached in outline form in Steven J. Dick, "Extraterrestrial Life and Our World View at the Turn of the Millennium," Dibner Library Lecture delivered at the Smithsonian Institution on May 2, 2000, online at http://www.sil.si.edu/silpublications/dibner-library-lectures/extraterrestrial-life/etcopy-kr.htm. It was elaborated in more detail in my LeRoy E. Doggett Prize Lecture at the winter meeting of the American Astronomical Society in Washington, DC, in January 2006, "Astronomy's Three Kingdoms: Discovering, Classifying and Interpreting Astronomical Objects," *BAAS* 37 (2005): 1231, abstract at http://adsabs.harvard.edu/abs/2005AAS...207.4501D, and discussed again in the context of extended structure of discovery at the January 2012 meeting of the American Astronomical Society, "Discovery and Classification in Astronomy," abstract at http://adsabs.harvard.edu/abs/2012AAS...21911501D.

64 Martin Harwit, *Cosmic Discovery: The Search, Scope and Heritage of Astronomy* (New York: Basic Books, 1981), pp. 197–229: 197–98.

65 On the Three Domain versus Five Kingdom controversy in biology, see especially Jan Sapp, *The New Foundations of Evolution* (Oxford: Oxford University Press, 2009). On classification in physics and chemistry see Michael D. Gordin, *A Well-Ordered Thing: Dmitrii Mendeleev and the Shadow of the Periodic Table* (New York: Basic Books, 2004); Andrew Pickering,

Constructing Quarks: A Sociological History of Particle Physics (Edinburgh: Edinburgh University Press, 1984); and Murray Gell-Mann, *The Quark and the Jaguar: Adventures in the Simple and the Complex* (New York: W. H. Freeman, 1994).

66 Paul Davies, *Cosmic Jackpot: Why Our Universe Is Just Right for Life* (Boston and New York: Houghton-Mifflin, 2007), especially chapter 4. Isaac Asimov has made the same point in his popular books; for example, *Atom: Journey Across the Subatomic Cosmos* (New York: Penguin, 1992), p. 263.

67 Richard O. Gray and Christopher J. Corbally, *Stellar Spectral Classification* (Princeton and Oxford: Princeton University Press, 2009), p. 10.

68 Ibid., pp. 9–10, Morgan, "On the Spectral Classification of the Stars of Types A to K," *ApJ* 85 (1937): 380ff.

69 More details are given in a forthcoming book, *Astronomy's Three Kingdoms.*

70 E. O. Wilson, Foreword to Lynn Margulis and Michael Chapman, *Kingdoms and Domains: An Illustrated Guide to the Phyla on Earth* (Amsterdam: Elsevier, 2010), p. xi. In 2011, a group of biologists using a novel analysis estimated 8.7 million eukaryotic species exist, give or take a million. Eukaryotic species contain a nucleus, in contrast to prokaryotes. Daniel Strain, "8.7 Million: A New Estimate for All the Complex Species on Earth," *Science* 333 (2011): 1083.

71 Taxonomy has also evolved; see Mayr, *Growth*, p. 145, for stages in classification, and microtaxonomy versus macrotaxonomy.

72 Lisa Messeri, "The Problem with Pluto: Conflicting Cosmologies and the Classification of Planets," *Social Studies of Science* 40/2 (April 2010): 187–214: 189–90. Peter Galison, *Image and Logic: A Material Culture of Microphysics* (Chicago: University of Chicago Press, 1997).

73 Herschel to Piazzi, May 22, 1802, in Hughes and Marsden, *JAHH* 10 (2007): 22, citing C. J. Cunningham, *The First Asteroid: Ceres 1801–2001*, Historical Studies in Asteroid Research, volume 1 (Surfside, FL: Star Lab Press, 2002), pp. 251–52.

74 Hughes and Marsden, *JAHH* 10 (2007): 21–30: 23.

75 David DeVorkin's work, beginning with his 1978 dissertation, through many articles and his landmark biography of Henry Norris Russell, has analyzed this history and its relation to the Hertzsprung-Russell diagram early in the twentieth century, particularly with regard to Russell's achievements. David H. DeVorkin, "An Astronomical Symbiosis: Stellar Evolution and Spectral Classification (1860–1910)," PhD dissertation, University of Leicester (1978), distilled in "Community and Spectral Classification in Astrophysics: The Acceptance of E. C. Pickering's System in 1910," *Isis* 72 (1981): 29–49; "Stellar Evolution and the Origin of the Hertzsprung-Russell Diagram," *Astrophysics and Twentieth-Century Astronomy to 1950*, Part A, Owen Gingerich, ed. (Cambridge: Cambridge University Press, 1984), 90–108; *Henry Norris Russell: Dean of American Astronomers* (Princeton: Princeton University Press, 2000). Also A. V. Nielsen, "History of the Hertzsprung-Russell Diagram," *Centaurus* 9 (1963): 219–52.

76 In addition to DeVorkin, see Curtiss, "Classification," and Hearnshaw, *Analysis*, p. 123.

77 Curtiss, "Classification," p. 25.

78 A. H. Joy, Review of Morgan, Keenan and Kellman's *An Atlas of Stellar Spectra*, *ApJ* 98 (1943): 240; Hearnshaw, *Analysis*, pp. 288–90.
79 Sandage, *Mount Wilson Observatory*, pp. 253–55. "Victoria" referred to the work of Harry H. Plaskett on O stars and Wolf-Rayet stars at Victoria Observatory in Canada.
80 Sandage, *Mt. Wilson Observatory*, pp. 63 and 255.
81 Ibid., pp. 254, 256.
82 W. W. Morgan, "Remarks on the MK System," in M. F. McCarthy, A. G. D. Philip, and G. V. Coyne, eds., *Spectral Classification*, pp. 59–61.
83 Morgan, "Remarks," McCarthy, Philip, and Coyne, eds., *Spectral Classification*, pp. 59–61.
84 Sandage, *Mt. Wilson Observatory*, pp. 485–89 for details of this and rival systems.
85 J. H. Reynolds, "The Classification of the Spiral Nebulae," *The Observatory* 50 (1927): 185–89: 186–87; Sandage, *Mt. Wilson Observatory*, 489–91.
86 Morgan, "Remarks," McCarthy, Philip, and Coyne, eds., *Remarks*, pp. 59–61.
87 Sandage, "Classification," p. 1; online at http://abyss.uoregon.edu/~js/ast123/lectures/lec11.html.

9. Technology and Theory as Drivers of Discovery

1 R. D. Ekers and K. I. Kellermann, "Discoveries in Astronomy," *Proceedings of the American Philosophical Society* 155 (2011): 129–33.
2 Robert Smith, "Beyond the Galaxy: The Development of Extragalactic Astronomy 1885–1965, part 2 ," *JHA* 40 (2009): 72.
3 Albert van Helden and Thomas Hankins, "Introduction: Instruments in the History of Science," *Osiris*, 2nd series, ix (1994): 1–6: 4.
4 Kenneth I. Kellermann, J. M. Cordes, R. D. Ekers, J. Lazio, and Peter Wilkinson, "The Exploration of the Unknown," delivered at the International Union special session "Accelerating the Rate of Astronomical Discovery," August 11–14, 2009, Rio de Janeiro, Brazil, online at http://pos.sissa.it/cgi-bin/reader/conf.cgi?confid=99#session-0, p. 2. An earlier version appeared in P. N. Wilkinson, K. I. Kellermann, R. D. Ekers, J. M. Cordes, and T. J. Lazio, "The Exploration of the Unknown," *New Astronomy Reviews* 48 (2004): 1551–63, as part of an argument for funding the new Square Kilometer Array (SKA) radio telescope.
5 See also the papers from the Symposium "Discoveries in Astronomy," held at the American Philosophical Society on the occasion of the 400th anniversary of the telescope, *Proceedings of the American Philosophical Society*, 155 (2011): 129–57, online at http://pos.sissa.it/cgi-bin/reader/conf.cgi?confid=99#session-0.
6 Kellermann et al., "Exploration of the Unknown," p. 2.
7 On elementary particle physics see Andrew Sessler and Edmund Wilson, *Engines of Discovery: A Century of Particle Accelerators* (Hackensack, NJ: World Scientific Publishing, 2007); Peter Galison, *How Experiments End* (Chicago: University of Chicago Press, 1987), and Galison, *Image and Logic: A Material Culture of Microphysics* (Chicago: University of Chicago Press, 1997). On astronomy see Robert Smith, "Engines of Discovery: Scientific

Instruments and the History of Astronomy and Planetary Science in the United States in the Twentieth Century," *JHA* 32 (1997): 49–77, and Toby E. Huff, "Inventing the Discovery Machine," in *Intellectual Curiosity and the Scientific Revolution: A Global Perspective* (Cambridge: Cambridge University Press, 2011), pp. 22–47.

8 Van Helden, "Telescopes and Authority from Galileo to Cassini," *Instruments, Osiris*, second series, vol. 9 (1994): 8–29. Van Helden points out that, unlike Galileo's telescopes, Fontana's used a convex instead of a concave ocular, giving a much greater field of view and supporting higher magnification. Despite the fact that "Fontana launched a telescope race that lasted for half a century" (p. 16), no new classes of objects were discovered.

9 On the intrepid Bevis (who died at age 76 after falling from his telescope) see *BEA* 1 (2007): 118–19. Lord Rosse was the first to refer to it as the "Crab Nebula." Only in 1939 did Mayall argue based on its expansion that a "reasonable working hypothesis" was that it was associated with the supernova of 1054.

10 J. A. Bennett, "On the Power of Penetrating into Space: The Telescopes of William Herschel," 7 (1976): 75–108; also Wolfgang Steinicke, *Observing Nebulae and Cataloguing Nebulae and Star Clusters: From Herschel to Dreyer's New General Catalogue* (Cambridge: Cambridge University Press, 2010), pp. 18–24. Herschel discovered his first planetary nebulae, today known as the Saturn Nebulae, NGC 7009 in Aquarius, using his "small 20-foot" with 12-inch aperture mirror, observing from Datchet. His large 20-foot with 18.7-inch mirror was ready on October 23, 1783, and became his standard telescope for nebulae searches.

11 As John Lankford has pointed out, only toward the end of the 1880s did photography become a legitimate research tool. John Lankford, "The Impact of Photography on Astronomy," in Owen Gingerich, ed., *Astrophysics and Twentieth Century Astronomy to 1950* (Cambridge: Cambridge University Press), pp. 16–39.

12 Charles C. Steidel et al., "Lyα Imaging of a Proto-Cluster Region at <z>=3.09," *ApJ* 532 (2000): 170–82: 172, 181–82.

13 On astronomical discovery metrics in the context of Hubble see David S. Leckrone, "The Secrets of Hubble's Success," forthcoming, based on *Science News* statistics as reported in C. Christian; G. Davidson, "The Science News Metrics," in *Organizations and Strategies in Astronomy Volume 6 - Astrophysics and Space Science Library*, vol. 335, pp. 145–56. On the science and politics of the SDSS see Ann Finkbeiner, *A Grand and Bold Thing* (New York and London: Free Press, 2010).

14 Allan Sandage, "The Ability of the 200-inch Telescope to Discriminate Between Selected World Models," *ApJ* 133 (1961): 355–92.

15 Angular resolution is inversely proportional to the diameter of the telescope aperture.

16 On what Harwit and others have called "phase space filters" such as sensitivity and so on, see Harwit, *Cosmic Discovery*, pp. 158–93: 189. For an example of how polarization was responsible for an important discovery, interstellar magnetic fields as indicated by interstellar polarization, see Steven Dick, *Sky and Ocean Joined: The U.S. Naval Observatory, 1830–2000* (Cambridge: Cambridge University Press,

2003), pp. 409–11. In this case the discovery was in the form of a new phenomenon rather than a new class of objects.

17 David Jewitt and Jane Luu, "Discovery of the Candidate Kuiper Belt Object 1992 QB1," *Nature* 362 (1993): 730–2. On the telescopes of Messier, see http://seds.org/messier/xtra/history/m-scopes.html.

18 Smith, "Engines of Discovery," p. 62.

19 C. R. O'Dell, private communication.

20 No overall nuanced history exists of these empirical discoveries, but on Ritter and the ultraviolet, see Jan Frercks, Heiko Weber, and Gerhard Wiesenfeldt, "Reception and Discovery: The Nature of Johann Wilhelm Ritter's Invisible Rays," *Studies in History and Philosophy of Science Part A* 40, no. 2 (2009): 143–56, a sophisticated study of the nature and reception of this particular discovery. On Hertz's work see Jed Z. Buchwald, *The Creation of Scientific Effects* (Chicago: University of Chicago Press, 1994). Maxwell's theory was published in James Clerk Maxwell, "A Dynamical Theory of the Electromagnetic Field," *Philosophical Transactions of the Royal Society of London* 155 (1865): 459–512. For the context of Maxwell's work see, for example, Basil Mahon, *The Man Who Changed Everything: The Life of James Clerk Maxwell* (Chichester, UK: Wiley, 2003).

21 George H. Rieke, "History of Infrared Telescopes and Astronomy," *Experimental Astronomy* 25 (2009): 125–41.

22 On this early history as written by some of the pioneers, see Lyman Spitzer, Jr., "Ultraviolet Spectra of the Stars," in Paul Hanle and von del Chamberlain, eds., *Space Science Comes of Age: Perspectives in the History of the Space Sciences* (Washington, DC: Smithsonian Institution Press, 1981), pp. 2–13, Leo Goldberg, "Solar Physics," ibid., pp. 15–30, and Herbert Friedman, "Rocket Astronomy – An Overview," ibid., pp. 31–44. For an overview of these events to 1990, see Herbert Friedman, *The Astronomer's Universe: Stars, Galaxies and Cosmos* (New York: W. W. Norton, 1990). On the early history of radio astronomy, see particularly Woodruff T. Sullivan, III, *Cosmic Noise: A History of Early Radio Astronomy* (Cambridge: Cambridge University Press, 2009), and on the early history of X-ray astronomy, Richard F. Hirsh, *Glimpsing an Invisible Universe: The Emergence of X-ray Astronomy* (Cambridge: Cambridge University Press, 1983). On the ultraviolet, see Martin A. Barstow and Jay B. Holberg, *Extreme Ultraviolet Astronomy* (Cambridge: Cambridge University Press, 2003. On V-2s and sounding rockets, see David H. DeVorkin, *Science with a Vengeance: How the Military Created the US Space Sciences After World War II* (New York: Springer, 1992).

23 S. Weinreb, A. H. Barrett, M. L. Meeks, and J. C. Henry, "Radio Observations of OH in the Interstellar Medium," *Nature* 200 (1963): 829–31. For the competitive environment, see Alan H. Barrett's personal account, "The Beginnings of Molecular Radio Astronomy," in K. Kellermann and B. Sheets, *Serendipitous Discoveries in Radio Astronomy* (Green Bank: NRAO, 1983), pp. 280–90.

24 Alan Barrett, in Kellermann and Sheets, *Serendipitous Discoveries*, p. 286. In 1970, George Carruthers found direct evidence of molecular hydrogen in dust clouds, using an extreme ultraviolet detector on an Aerobee rocket flight. Unlike gaseous molecular clouds, dust clouds had long been known from the work of E. E. Barnard and others. On

Carruthers's ultraviolet work see Herbert Friedmann, *The Astronomer's Universe*, p. 56, and George Carruthers, "Rocket Observation of Interstellar Molecular Hydrogen," *ApJ* 161 (1970): L81–L85; for a review see Carruthers, "Atomic and Molecular Hydrogen in Interstellar Space," *SSR* 10 (1970): 459–82, especially 476–80 for a history of how molecular hydrogen was deduced in dark dust clouds.

25 N. Z. Scoville and P. M. Solomon, "Molecular Clouds in the Galaxy," *ApJ* 199 (1975): L105–L109; Lewis Snyder, "The Search for Biomolecules in Space," in K. I. Kellermann and G. A. Seielstad, eds., *The Search for Extraterrestrial Intelligence*" (Green Bank, WVA: National Radio Astronomy Observatory, 1986), 39–50.

26 Scoville and Solomon, "Molecular Clouds in the Galaxy," *ApJ* 199 (1975): L105–L109; P. M. Solomon, D. B. Sanders, and N. Z. Scoville, "Giant Molecular Clouds in the Galaxy: A Survey of the Distribution and Physical Properties of GMC's," *BAAS* 9 (1977): 554; full article in Solomon, Sanders, and Scoville, "Giant Molecular Clouds in the Galaxy - The Distribution of ^{13}CO Emission in the Galactic Plane," *ApJ* 232 (1979): L89–L93. The 1977 article may have been the first use of the term "Giant Molecular Cloud," extending the earlier use of "Molecular Cloud." Molecular clouds are dominated by molecular hydrogen, followed by carbon monoxide, which is much easier to observe. The most famous molecular clouds in our Galaxy are the Orion Molecular Cloud and the Sagittarius B2 cloud near the center of our Galaxy. Like H I and H II regions, molecular clouds are found primarily in spiral arms, but molecular hydrogen is concentrated much more toward the center of the galaxy.

27 See Kellermann and Sheets, *Serendipitous Discoveries*.

28 F. Curtis Michel, "Pulsar Planetary Systems," *ApJ* 159 (Jan. 1970): L25–L28; J. G. Hills, "Planetary Companions to Pulsars," *Nature* 226 (1970): 730–31; M. J. Rees, V. L. Trimble, and J. M. Cohen, "Planet, Pulsar, 'Glitch' and Wisp," *Nature* 229 (1971): 395–96.

29 Perhaps no telescope is a better case study on the technical, social, and institutional difficulties in constructing a new technology; see Lovell's own account of this history in Lovell, *Out of the Zenith: Jodrell Bank, 1957–1970* (New York: Harper & Row, 1974).

30 A. Wolszczan, and D. A. Frail, "A Planetary System around the Millisecond Pulsar PSR 1257+12," *Nature* 355 (1992): 145–47; A. Wolszczan, "Confirmation of Earth - Mass Planets Orbiting the Millisecond Pulsar PSR B1257+12," *Science* 264 (1994): 538.

31 D. C. Backer et al., "A Second Companion of the Millisecond Pulsar 1620–26," *Nature* 365 (1993): 817; S. E. Thorsett et al., "PSR B1620–26: A Binary Radio Pulsar with a Planetary Companion?" *ApJ Letters* 412 (1993): L33; the Hubble confirmation of the white dwarf is S. Sigurdsson et al., "A Young White Dwarf Companion to Pulsar B1620–26: Evidence for Early Planet Formation," *Science* 301 (2003): 193–96.

32 G. H. Rieke, "History of Infrared Telescopes and Astronomy," in Bernhard Brandl, Remko Stuik and Jeanette Katgert-Merkelijn, eds., *400 Years of Astronomical Telescopes* pp. 125–41. See also George Rieke, *The Last of the Great Observatories: Spitzer and the Era of Faster, Better, Cheaper at NASA* (Tucson: University of Arizona Press, 2006).

33 H. H. Aumann, F. C. Gillett et al., "Discovery of a Shell around Alpha Lyrae," *ApJ Letters* 278 (1984): L23–L27: L23; front page of *Washington Post* for August 10, 1983, "Satellite Discovers Possible Second Solar System."
34 "Protoplanetary Systems," *Science* 225 (July 6, 1984): 39; "Infrared Evidence for Protoplanetary Rings around Seven Stars," *Physics Today* (May 1984): 17–20.
35 B. A. Smith and R. J. Terrile, "A Circumstellar Disk around Beta Pictoris," *Science* 226 (1984): 1421–24.
36 Karl Popper, *Logik der Forschung*, 1934 (The Logic of Scientific Discovery, 1959); Norwood Russell Hanson, *Patterns of Discovery: An Inquiry into the Conceptual Foundations of Science* (Cambridge: Cambridge University Press, 1969); Martin Curd and J. A. Cover, *Philosophy of Science: The Central Issues* (New York: W.W. Norton, 1998); Alex Rosenberg, *Philosophy of Science: A Contemporary Introduction* (London and New York: Routledge, 2000), pp. 137ff.
37 Buchdahl, *Creation*, p. 217.
38 Martin O. Harwit, "Observational Discovery vs Theoretical Discovery," in Kellermann, *Serendipitous Discoveries*, pp. 197–210.
39 Joshua Gilder and Anne-Lee Gilder, *Heavenly Intrigue: Johannes Kepler, Tycho Brahe, and the Murder Behind One of History's Greatest Scientific Discoveries* (New York: Doubleday, 2004) This theory is now disproved.
40 Max Caspar, *Kepler*, trans. C. Doris Hellman (New York: Dover Publications, 1993), p. 197; Toby E. Huff, *Intellectual Curiosity and the Scientific Revolution: A Global Perspective* (Cambridge: Cambridge University Press, 2011), pp. 57–59. As noted in Chapter 2, Kepler published his observations of Jupiter in the *Narratio de observatis a se quatuor Jovis satellitibus erronibus* (Narration on the Four Wandering Satellites of Jupiter Observed by Him). It appeared in late October 1610, with a 1611 imprint, and was only twelve pages in length; Westman, *The Copernican Question*, p. 480. Ironically, Kepler's *Dioptrice,* published in 1611, explicated the formation of images by a combination of convex lenses, inventing on paper what became known as the Keplerian telescope. But Kepler himself never built one, this being left for Gascoigne, Huygens and others. See also *Kepler's Conversation with Galileo's Sidereal Messenger*, trans. Edward Rosen (New York and London, 1965).
41 Eric G. Forbes, "Gauss and the Discovery of Ceres," *JHA* 2 (1971): 195–99.
42 On the interplay of theory and observation with the discovery of white dwarfs, see Section 4.3; see also J. B. Holberg and F. Wesemael, "The Discovery of the Companion of Sirius and Its Aftermath," *JHA* 38 (2007): 161–74, and Holberg, "The Discovery of the Existence of White Dwarf Stars: 1860 to 1930," *JHA* 30 (2009): 137–54.
43 J. C. Adams, "On the Orbit of the November Meteors," *MNRAS* 27 (1867): 247–52; G. V. Schiaparelli, "Intorno al corso ed all'origine probabile delle telle meteoriche," *Bulletin meteorologico dell'osservatorio del Collegio omano* 5 (1867): 8–12 and vol. 6, 2; C. F. W. Peters, Letter dated January 29, *Astronomische Nachrichten* 68 (1867): 287–8; Steven J. Dick, "Observation and Interpretation of the Leonid Meteors Over the Last Millennium," *JAHH* 1 (1998): 1–20: 5–6.

44 Alexander von Humboldt, *Cosmos: A Sketch of a Physical Description of the Universe*, vol. 1 (New York: Harper & Brothers (1850), p. 44.
45 E. W. Brown, "On the Predictions of Trans-Neptunian Planets from the Perturbations of Uranus," *PNAS* 16 (1930): 364–71.
46 Gibson Reaves, "The Prediction and Discoveries of Pluto and Charon," in S. Alan Stern and David J. Tholen, eds., *Pluto and Charon* (Tucson: University of Arizona Press, 1997), p. 18; Clyde W. Tombaugh and Patrick Moore, *Out of the Darkness: The Planet Pluto* (New York: New American Library, 1980), pp. 6–7.
47 K. E. Edgeworth, "The Evolution of our Planetary System," *JBAA* 53 (1943): 181–88: 186; Edgeworth, "The Origin and Evolution of the Solar System," *MNRAS* 109 (1949): 600–9; and Gerard Kuiper, "On the Origin of the Solar System," in J. A. Hynek, ed., *Astrophysics: A Topical Symposium* (New York: McGraw-Hill, 1951), p. 402.
48 Thomas Gold, "Rotating Neutron Stars as the Origin of the Pulsating Radio Sources," *Nature* 218 (1968): 731–32; reprinted in Bartusiak, *Archives*, 518–21. The Baade and Zwicky prediction was Walter Baade and Fritz Zwicky, "Remarks on Super-Novae and Cosmic Rays," *Physical Review* 46 (1934): 76–77. The 1966 paper was David Melzer and Kip Thorne, "Normal Modes of Radial Pulsation of Stars at the End Point of Thermonuclear Evolution," *ApJ* 145 (1966): 514–43.
49 J. R. Oppenheimer and G. M. Volkoff, "On Massive Neutron Cores," *Physical Review* 55 (1939): 374–81; Oppenheimer and Hartland Snyder, "On Continued Gravitational Contraction," *Physical Review* 56 (1939): 455–59. Caltech physicist Kip Thorne tells the story in detail in *Black Holes and Time Warps: Einstein's Outrageous Legacy* (New York: W. W. Norton, 1994), as does Arthur I. Miller in more popular fashion in *Empire of the Stars: Obsession, Friendship, and Betrayal in the Quest for Black Holes* (Boston: Houghton Mifflin, 2005).
50 Charles A. Whitney, *The Discovery of Our Galaxy* (New York: Alfred A. Knopf, 1971), pp. 77–86. The passage is here quoted as in Whitney, p. 84.
51 According to the U.S. National Academy of Sciences a theory "refers to a comprehensive explanation of some aspect of nature that is supported by a vast body of evidence"; online at http://en.wikipedia.org/wiki/Scientific_theory. This definition is in general supported by philosophers of science; see, for example, Alex Rosenberg, *Philosophy of Science: A Contemporary Introduction* (London and New York: Routledge, 2000), p. 69. The question of when a hypothesis becomes a theory is crucial, but even a theory is always provisional.
52 R. Hart and R. Berenzden, "Hubble's Classification of Non-Galactic Nebulae," *JHA* 2 (1971): 109–19: 113. The Jeans work is *Problems of Cosmogony and Stellar Dynamics* (Cambridge: Cambridge University Press, 1919), updated as *Astronomy and Cosmogony* (Cambridge: Cambridge University Press, 1928). Hubble's 1926 paper is "Extra-Galactic Nebulae," *ApJ* 64 (1926): 321–72.
53 H. I. Ewen and E. M. Purcell, "Observation of a Line in the Galactic Radio Spectrum: Radiation from Galactic Hydrogen at 1,420 Mc/sec," *Nature* 168 (1951): 356, followed by C. A. Muller and J. H. Oort, "Observation of a Line in the Galactic Radio Spectrum: The Interstellar Hydrogen Line at 1,420 Mc./sec., and an Estimate of Galactic Rotation," 357. See Section 3.3.

54 See Chapter 4.
55 D. H. Menzel, "The Planetary Nebulae," *PASP* 38 (1926): 295–312; H. Zanstra, *Physical Review* 27 (1926): 644. See Section 3.2.

10. Luxuriant Gardens and the Master Narrative

1 William Herschel, "Catalogue of a second Thousand Nebulæ and Clusters of Stars; with a few introductory Remarks on the Construction of the Heavens," *Philosophical Transactions of the Royal Society* 79 (1789): 212–55.
2 Harlow Shapley, *Beyond the Observatory* (New York: Charles Scribner's Sons, 1967), pp. 15–16.
3 Andrew Sessler and Edmund Wilson, *Engines of Discovery: A Century of Particle Accelerators* (Hackensack, NJ: World Scientific Publishing, 2007).
4 Parts of what follows are based on Steven J. Dick, "Cosmic Evolution: History, Culture, and Human Destiny," in Steven J. Dick and Mark L. Lupisella, *Cosmos and Culture: Cultural Evolution in a Cosmic Context* (Washington, DC: NASA SP 4802, 2009), pp. 25–62.
5 John Greene, *Science, Ideology, and World View: Essays in the History of Evolutionary Ideas* (Berkeley: University of California Press, 1981), p. 130.
6 James Secord, *Victorian Sensation: The Extraordinary Publication, Reception and Secret Authorship of Vestiges of the Natural History of Creation* (Chicago: University of Chicago Press, 2000).
7 The quotation is from Simon Schaffer, "The Nebular Hypothesis and the Science of Progress," in J. R. Moore, ed., *History, Humanity and Evolution: Essays in Honor of John C. Greene* (Cambridge: Cambridge University Press, 1989), pp. 131–64: 132. On popular writings see Bernard Lightman, "The Evolution of the Evolutionary Epic," in *Victorian Popularizers of Science: Designing Nature for New Audiences* (Chicago: University of Chicago Press, 2007), pp. 219–94.
8 Helge Kragh, *Cosmology and Controversy: The Historical Development of Two Theories of the Universe* (Princeton: Princeton University Press, 1996), p. 4. See also Stephen Toulmin and June Goodfield, *The Discovery of Time* (Chicago: University of Chicago Press, 1982).
9 Kragh, *Cosmology and Controversy*, p. 4.
10 Joe D. Burchfield, *Lord Kelvin and the Age of the Earth* (Chicago: University of Chicago Press, 1990); Martin J. S. Rudwick, *Bursting the Limits of Time: The Reconstruction of Geohistory in the Age of Revolution* (Chicago and London: University of Chicago Press, 2005).
11 Allan Sandage, "Current Problems in the Extragalactic Distance Scale," *ApJ* 127 (1958): 513–26; David A. Weintraub, *How Old Is the Universe?* (Princeton: Princeton University Press, 2011).
12 John North, *The Measure of the Universe; Helge Kragh, Cosmology and Controversy*, pp. 3–38.
13 Helge Kragh and Robert Smith, "Who Discovered the Expanding Universe?" *History of Science* 41 (2003): 141–62.
14 Kragh, *Cosmology and Controversy*, pp. 28–59.
15 Allan Sandage, *The Mount Wilson Observatory: Breaking the Code of Cosmic Evolution* (Cambridge: Cambridge University Press, 2004).
16 Stephen G. Brush, *Nebulous Earth: The Origin of the Solar System and the Core of the Earth from Laplace to Jeffreys* (Cambridge: Cambridge University Press, 1996); Ronald L. Numbers, *Creation by Natural Law:*

Laplace's Nebular Hypothesis in American Thought (Seattle: University of Washington Press, 1977).

17 Steven J. Dick, *The Biological Universe: The Twentieth Century Extraterrestrial Life Debate and the Limits of Science* (Cambridge: Cambridge University Press, 1996), pp. 160–221.

18 H. D. Curtis, "The Nebulae," in *Handbuch der Astrophysik*, vol. 5, pt. 2, pp. 774–936: 831.

19 Although as early as 1956 the Soviet astronomer Joseph Shklovskii suggested that planetary nebulae are formed from the outer layers of some red giants, the American astronomers George Abell and Peter Goldreich presented the first coherent evidence that planetary nebulae indeed evolve from red giants. In their 1966 paper, they gave due credit to Joseph Shklovskii, Robert O'Dell, and others who had found pieces of the puzzle. G. O. Abell and P. Goldreich, "On the Origin of Planetary Nebulae," *PASP* 78 (1966): 232–41; Abell and Goldreich cite I. S. Shklovskii, *Astr. Zhurnal USSR* 33 (1956): 315, and C. R. O'Dell, "The Evolution of the Central Stars of Planetary Nebulae," *ApJ* 138 (1963): 67–78.

20 Kragh, *Cosmology and Controversy*, pp. 142–388.

21 E. Margaret Burbidge, Geoffrey R. Burbidge, William A. Fowler, and Fred Hoyle, "Synthesis of the Elements in Stars," *Reviews of Modern Physics* 29 (1957): 547–650.

22 Peter J. Bowler, *Evolution: The History of an Idea* (Berkeley: University of California Press, 1989); Catherine Baker, *The Evolution Dialogues: Science, Christianity, and the Quest for Understanding* (Washington, DC: AAAS), 2006. The latter is notable for having been published by the American Association for the Advancement of Science.

23 George Gaylord Simpson, *The Meaning of Evolution: A Study of the History of Life and of Its Significance for Man* (New Haven: Yale University Press, 1950), pp. 281–82.

24 Julian Huxley, *Evolutionary Humanism* (Buffalo, NY: Prometheus Books, 1992; first ed., 1964); Theodosius Dobzhansky, *Mankind Evolving: The Evolution of the Human Species* (New Haven and London: Yale University Press, 1969; first ed., 1962). See also Vassily Betty Smocovitis, *Unifying Biology: The Evolutionary Synthesis and Evolutionary Biology* (Princeton: Princeton University Press, 1996).

25 Harlow Shapley, *The View from a Distant Star: Man's Future in the Universe* (New York: Basic Books, 1963), p. 5. On Shapley's cosmic evangelism see Joann Palmeri, "Bringing Cosmos to Culture: Harlow Shapley and the Uses of Cosmic Evolution," in Steven J. Dick and Mark L. Lupisella, *Cosmos and Culture: Cultural Evolution in a Cosmic Context* (Washington, DC: NASA SP 4802, 2009), pp. 489–521.

26 Stuart Kauffman, *At Home in the Universe: The Search for the Laws of Self-Organization and Complexity* (Oxford: Oxford University Press, 1996); John A. Wheeler, *At Home in the Universe* (New York: American Institute of Physics, 1997).

27 Ursula Goodenough, *The Sacred Depths of Nature* (Oxford: Oxford University Press, 2000).

28 Michael Dowd, *Thank God for Evolution: How the Marriage of Science and Religion will Transform Your Mind and Our World* (New York: Viking, 2008).

29 John C. Greene, *Science, Ideology and World View: Essays in the History of Evolutionary Ideas* (Berkeley: University of California Press, 1981), p. 2; John Durant, "Evolution, Ideology and World View: Darwinian Religion in the Twentieth Century," in James R. Moore, ed., *History, Humanity and Evolution: Essays for John C. Greene* (Cambridge: Cambridge University Press, 1989), pp. 355–73: 368, and Greene in the "Afterword" to this volume, 403–13: 404.
30 Steven J. Dick, "Cosmology and Biology," Proceedings of the 2008 Conference on the Society of Amateur Radio Astronomers, June 29–July 2, 2008 (NRAO: Green Bank, WV), pp. 1–15. This theme is embodied in the debate over the multiverse and the anthropic principle, e.g., Brandon Carter, "Large Number Coincidences and the Anthropic Principle in Cosmology," in M. S. Longair, ed., *Confrontation of Cosmological Theories with Observational Data* (Dordrecht: Reidel Publishing, 1974), pp. 291–98; John D. Barrow and Frank J. Tipler, *The Anthropic Cosmological Principle* (Oxford: Oxford University Press, 1986); Bernard Carr, editor, *Universe or Multiverse?* (Cambridge: Cambridge University Press, 2007).

11. The Meaning of Discovery

1 Yudhijit Bhattacharjee, "A Week in Stockholm," *Science* 336 (2012): 26–31.
2 Robert Williams, "Discovery and the Culture of Astronomy," Accelerating the Rate of Astronomical Discovery, *Proceedings of Science*, online at http://pos.sissa.it/archive/conferences/099/001/sps5_001.pdf.
3 H. N. Russell quoting Pickering, in Arthur Miller, *Empire of the Stars* (Boston: Houghton-Mifflin, 2005), p. 45. Russell is said to have made the remark with regard to white dwarfs at the Paris Colloquium on the subject in 1939.
4 Adrian Cho, "Last Hurrah: Final Tevatron Data Show Hints of Higgs Boson," *Science* 335 (2012): 1159. It is notable that the uncertainties of discovery also extend to theory. No less than six theorists could claim to have a role in predicting the Higgs mechanism. Adrian Cho, "Who Invented the Higgs Boson?" *Science* 337 (2012): 1286–89.
5 Brian Vastag and Joel Achenbach, "Scientists' Search for Higgs Boson Yields New Subatomic Particle," *Washington Post*, July 4, 2012; Adrian Cho, "Higgs Boson Makes Its Debut After Decades-Long Search," *Science* 337 (2012): 141–43; Johanna Miller, "The Higgs Particle, or Something Much Like It, Has Been Spotted," *Physics Today* 65 (2012): 12–15.
6 Bhattacharjee, "A Week in Stockholm," pp. 26–31. The editors at *Scientific American* recently made the same plea: "Solve the Nobel Prize Dilemma," *SciAm* 307 (October 2012): 12.
7 Jane Gregory, *Fred Hoyle's Universe* (Oxford: Oxford University Press, 2005), pp. 271–79, especially p. 275.
8 David Lindley, *The End of Physics* (New York: Basic Books, 1994); John Horgan, *The End of Science* (New York: Addison Wesley, 1996).
9 Martin Harwit, *Cosmic Discovery: The Search, Scope and Heritage of Astronomy* (New York: Basic Books, 1981), pp. 39–44, 219–24, and 291–97.
10 John Hearnshaw, "The 25 Greatest Discoveries in Astronomy and Astrophysics of the 20th Century," online at http://cosmicdiary.

org/blogs/john_hearnshaw/?p=96, and "The Top 51 Discoveries in Astronomy and Astrophysics of the 20th Century," online at http://cosmicdiary.org/blogs/john_hearnshaw/?p=637, accessed Feb. 21, 2012.

11 Michael Nielsen, *Reinventing Discovery: The New Era of Networked Science* (Princeton: Princeton University Press, 2012).

12 The Virtual Observatory is funded by NASA and the NSF. See the Virtual Observatory site at http://www.us-vo.org/ and http://virtualobservatory.org/. The National Virtual Observatory is collaborating with the International Virtual Observatory Alliance for worldwide coordination of ground-based and space-based astronomical data.

Select Bibliographical Essay

The Smithsonian Astrophysical Observatory/NASA Astrophysics Data System (http://adsabs.harvard.edu/) is a tremendous resource for much of the scientific literature used and cited in the notes. The reader can access most of that literature by accessing the ADS Web site and entering author and date. I have also found very useful the following compilations and commentaries: Marcia Bartusiak, ed., *Archives of the Universe: A Treasury of Astronomy's Historic Works of Discovery* (New York: Pantheon Books, 2004); Kenneth Lang and Owen Gingerich, eds., *A Source Book in Astronomy and Astrophysics, 1900–1975* (Cambridge, MA: Harvard University Press, 1979; Helmut Abt, ed., *American Astronomical Society Centennial Issue: Selected Fundamental Papers Published This Century in the Astronomical Journal and the Astrophysical Journal*, *Astrophysical Journal* 525 (1999). The *Journal for the History of Astronomy*, edited for forty years by Michael Hoskin at Churchill College, Cambridge, is a vast treasure trove of information that I have also used extensively. It is also available via ADS.

Part I. Entrée

The discovery of Pluto has been the subject of a good deal of historical and scientific literature, beginning with William G. Hoyt, *Planets X and Pluto* (University of Arizona Press, 1980), continuing with Alan Stern and Jacqueline Mitton, *Pluto and Charon: Ice Worlds on the Ragged Edge of the Solar System* (New York: Wiley, 1999), and most recently David H. DeVorkin, "Pluto: The Problem Planet and Its Scientists," in Roger Launius, ed., *Exploring the Solar System: The History and Science of Planetary Exploration* (New York: Palgrave MacMillan, 2013). On Tombaugh, see David H. Levy, *Clyde Tombaugh: Discoverer of the Planet Pluto* (Tucson: University of Arizona Press, 1991). Tombaugh's own recollections are Clyde W. Tombaugh and Patrick Moore, *Out of the Darkness: The Planet Pluto* (New York: New American Library, 1980).

James W. Christy describes his discovery of Charon, the first moon of Pluto, in J. W. Christy, "The Discovery of Pluto's Moon, Charon, in 1978," in S. A. Stern and D. J. Tholen, eds., *Pluto and Charon* (Tucson: University Arizona Press, 1997), pp. xvii–xxi. See also R. M. Marcialis, "The Discovery of Charon: Happy Accident or Timely Find?" *JBAA* 99 (1989): 27–29.

The situation regarding Trans-Neptunian objects on the eve of the IAU action "demoting" Pluto is well documented in David Weintraub, *Is Pluto a Planet? A Historical Journey through the Solar System* (Princeton: Princeton University Press, 2007). For a popular account of the discovery of Trans-Neptunian Objects, see Govert Schilling, *The Hunt for Planet X: New Worlds and the Fate of Pluto* (New York: Copernicus Books, 2009). Michael Brown, a pioneer in the field, has given his thoughts in *How I Killed Pluto and Why It Had It Coming* (New York: Spiegel & Grau, 2010).

The debate over Pluto as a planet is described in Neil deGrasse Tyson's *The Pluto Files: The Rise and Fall of America's Favorite Planet* (New York: W. W. Norton, 2009). Owen Gingerich, the chairman of the planet definition committee of the IAU at the time of the IAU decision, has given his insights in "The Inside Story of Pluto's Demotion," *Sky and Telescope* 112 (November 2006): 34–39; also Gingerich, "Planetary Perils in Prague," *Daedalus* 136, no.1 (2007): 137–40, preceded in the same issue by David Jewitt and Jane X. Luu, "Pluto, Perception and Planetary Politics," pp. 132–36. A scholarly historical study is Lisa Messeri, "The Problem with Pluto: Conflicting Cosmologies and the Classification of Planets," *Social Studies of Science* 40/2 (April 2010): 187–214.

Among many other articles, the general problems associated with planet definition are discussed in Steven Soter, "What Is a Planet?" *AJ* 132 (2006): 2513–19, and its popular version, "What Is a Planet?" *Scientific American* (January 2007); also S. Allen Stern and Harold F. Levinson, "Regarding the Criteria for Planethood and Proposed Planetary Classification Schemes," Presented at the 34th General Assembly of the IAU, Manchester, UK, August 7–18, 2000, in *Highlights in Astronomy* 12 (2002): 205–13, online at http://www.boulder.swri.edu/~hal/PDF/planet_def.pdf.

Part II. Narratives of Discovery

For Galileo in context see, most recently, John Heilbron, *Galileo* (Oxford: Oxford University Press, 2010), one of the best of many biographies. Galileo's firsthand account of his discovery of the moons of Jupiter is given in *Sidereus Nuncius, or The Sidereal Messenger*, translated with introduction, conclusion, and notes by Albert van Helden (Chicago: University of Chicago Press, 1989), pp. 64–86. Stillman Drake discusses the events surrounding the discovery of the moons of Jupiter in *Galileo at Work: His Scientific Biography* (Chicago and London: University of Chicago Press, 1978), pp. 142–54. Van Helden has analyzed the discovery of Saturn's rings in two penetrating articles: "Saturn and his Anses," *Journal for the History of Astronomy* 5 (1974): 105–21, and "Annulo Cingitur: the Solution of the Problem of Saturn," *Journal for the History of Astronomy* 5 (1974): 155–74.

The most comprehensive work dealing with the discovery of the minor planets, including much primary source material, is Clifford J. Cunningham, *The First Asteroid: Ceres, 1801–2001* (Surfside, FL: Star Lab Press, 2001). On the separation of the planets into classes by size in the seventeenth century, see Albert van Helden, *Measuring the Universe: Cosmic Dimensions from Aristarchus to Halley* (Chicago: University of Chicago Press, 1985).

Michael Hoskin has been the leading scholar on William Herschel over the last two generations. His *William Herschel and the Construction of the Heavens* (London: Oldbourne, 1963), based solely on Herschel's published work, has now been superseded by his volume, *The Construction of the Heavens: William Herschel's Cosmology* (Cambridge: Cambridge University Press, 2012), which also makes use of the Herschel manuscripts. Hoskin's latest research, aside from the extensive introduction to his *Construction* volume, is found in *Discoverers of the Universe: William and Caroline Herschel* (Princeton: Princeton University Press, 2011).

On Herschel and the nebulae see also Simon Schaffer, "Herschel in Bedlam: Natural History and Stellar Astronomy," *BJHS* 13 (1980): 211–39; Kenneth Glyn Jones, *The Search for the Nebulae* (Chalfont St. Giles: Science History Publications, 1975); and Wolfgang Steinicke's recent volume, *Observing and Cataloguing Nebulae and Star Clusters: From Herschel to Dreyer's New General Catalogue* (Cambridge: Cambridge University Press, 2010), a treasury of information. Owen Gingerich, "Messier and His Catalogue," in John H. Mallas and Evered Kreimer, *The Messier Album* (Cambridge, MA: Sky Publishing, 1978), includes the original French of the 1781 third edition [actually a reprint of this edition published in 1784]. Barbara Becker places Huggins's discovery of nebulosity in context in her biography of Huggins, *Unravelling Starlight: William and Margaret Huggins and the Rise of the New Astronomy* (Cambridge: Cambridge University Press, 2011). Kenneth Glyn Jones, *Messier's Nebulae and Star Clusters* (Cambridge: Cambridge University Press, 2nd ed., 1991) remains a useful work. William Sheehan, *The Immortal Fire Within: The Life and Work of Edward Emerson Barnard* (Cambridge: Cambridge University Press, 1995), especially chapter 22, discusses the discovery of dark nebulae. On the discovery of hydrogen regions via the 22-cm line, see *Serendipitous Discoveries in Radio Astronomy*, Ken Kellermann and B. Sheets, eds. (Green Bank: NRAO, 1983).

On the history of astronomical spectroscopy, see John B. Hearnshaw, *The Analysis of Starlight: One Hundred and Fifty Years of Astronomical Spectroscopy* (Cambridge: Cambridge University Press, 1986), as well as other works cited in the notes. On the discovery of giant and dwarf stars, and much else, see David DeVorkin, *Henry Norris Russell: Dean of American Astronomers* (Princeton: Princeton University Press, 2000). Allan Sandage provides much insight into the discovery of later classes of stars, from the perspective of a practitioner, in *The Mount Wilson Observatory*, volume 1 of the Centennial History of the Carnegie Institution of Washington (Cambridge: Cambridge University Press, 2004). On the MK system of stellar classification, see Robert F. Garrison, ed., *The MK Process and Stellar Classification*. Proceedings of the Workshop in Honor of W. W. Morgan and P. C. Keenan, held at the University of Toronto, Canada, June

1983 (University of Toronto, 1984), as well as Chrisopher Corbally, Robert O. Gray, and Robert F. Garrison, eds., *The MK Process at 50 years: A Powerful Tool for Astrophysical Insight. A Workshop of the Vatican Observatory, Tucson, Arizona, U.S.A., September 1993* (Astronomical Society of the Pacific, San Francisco, CA, 1994). The continued utility and expansion of the MK system is detailed in Richard O. Gray and Christopher J. Corbally, *Stellar Spectral Classification* (Princeton and Oxford: Princeton University Press, 2009).

On the discovery of the more exotic stars, the endpoints of stellar evolution, see especially J. B. Holberg, "The Discovery of the Existence of White Dwarf Stars: 1860 to 1930," *JHA* 30 (2009): 137–54; Arthur Miller, *Empire of the Stars* (Boston: Houghton-Mifflin, 2005), especially pp. 102, 132–33; and Matthew Stanley, *Practical Mystic: Religion, Science and A. S. Eddington* (Chicago: University of Chicago Press, 2007). The literature on black holes, once considered a phantom phenomenon, is now too large to list; see, however, references cited in the notes. A good starting point is Caltech physicist Kip Thorne's book, *Black Holes and Time Warps: Einstein's Outrageous Legacy* (New York: W. W. Norton, 1994).

On the discovery of the galaxies see Robert Smith, *The Expanding Universe: Astronomy's 'Great Debate,' 1900–1931* (Cambridge: Cambridge University Press, 1982), as well as Robert Smith, "Beyond the Galaxy: The Development of Extragalactic Astronomy 1885–1965, Part 1," *JHA* 39 (2008): 91–119; and Part 2, *JHA* 40 (2009): 71–107. Richard Berendzen, Richard Hart, and Daniel Seeley, *Man Discovers the Galaxies* (New York: Science History Publications, 1976) also remains useful, as do their earlier articles, R. Hart and R. Berendzen, "Hubble's Classification of Non-Galactic Nebulae, 1922–1926," *JHA* 2 (1971): 109–19; and "Hubble, Lundmark and the Classification of Non-Galactic Nebulae," *JHA* 2 (1971): 200. From the point of view of a practitioner, Allan Sandage's article remains useful: "Classification and Stellar Content of Galaxies," in Allan Sandage, Mary Sandage, and Jerome Kristian, eds., *Galaxies and the Universe* (Chicago and London: University of Chicago Press, 1975). The best biography of Hubble is Gale E. Christianson, *Edwin Hubble: Mariner of the Nebulae* (New York: Farrar, Straus and Giroux, 1995). Later galaxy classification schemes are covered in Ronald J. Buta, Harold G. Corwin, Jr., and Stephen C. Odewahn, *The de Vaucouleurs Atlas of Galaxies* (Cambridge: Cambridge University Press, 2007). Most recently, David L. Block and Kenneth C. Freeman, *Shrouds of the Night: Masks of the Milky Way and Our Awesome New View of Galaxies* (New York: Springer, 2008), contains new information on galaxies and their classification.

The early radio discoveries of new classes of objects are covered in Woodruff T. Sullivan III, *Cosmic Noise: A History of Early Radio Astronomy* (Cambridge: Cambridge University Press, 2009). The discovery of quasars is detailed in Maarten Schmidt, "Discovery of Quasars," in K. Kellermann and B. Sheets, *Serendipitous Discoveries in Radio Astronomy* (Green Bank: NRAO, 1983), pp. 171–74; also Maarten Schmidt, "The Discovery of Quasars," *Proceedings of the American Philosophical Society* 155 (2011): 142–46, and Richard Preston, "Beacons in Time – Maarten Schmidt and the Discovery of Quasars," *Mercury* 17 (1988): 2–11.

Part III. Patterns of Discovery

The literature on the nature of discovery (in contrast to the reports of specific discoveries) is not large. In astronomy, the pioneering work is Martin Harwit, *Cosmic Discovery: The Search, Scope & Heritage of Astronomy* (New York: Basic Books, 1981). The American Philosophical Society symposium on "Discoveries in Astronomy" is found online at http://www.amphilsoc.org/publications/proceedings/v/155/n/2, and the IAU Symposium "Accelerating the Rate of Astronomical Discovery" is online at "http://pos.sissa.it/cgi-bin/reader/conf.cgi?confid=99#session-0" http://pos.sissa.it/cgi-bin/reader/conf.cgi?confid=99#session-0. In a broader context, Thomas S. Kuhn, *The Structure of Scientific Revolutions* (Chicago: University of Chicago Press, 1962, expanded edition, 1970), discusses the extended nature of discovery, but this discussion is largely lost in his emphasis on paradigms and scientific revolutions. Norwood Russell Hanson, "An Anatomy of Discovery," *Journal of Philosophy* 64 (June 8, 1967): 321–52, and Theodore Arabatzis, "On the Inextricability of the Context of Discovery and the Context of Justification," in J. Schickoreand F. Steinle, eds., *Revisiting Discovery and Justification* (Dordrecht: Springer, 2006), pp. 215–30, both parse the varieties of discovery.

Early work on discovery as a social event is given in Ludwik Fleck, *Genesis and Development of a Scientific Fact*, Thaddeus J. Trenn and Robert K. Merton, eds. (Chicago: University of Chicago Press, 1979, translated from the 1935 original); Robert K. Merton, *Science, Technology and Society in Seventeenth Century England*, *Osiris*, 4, pt. 2 (1938): pp. 360–632; Michael Polanyi, *Personal Knowledge: Towards a Post-Critical Philosophy* (Chicago: University of Chicago Press, 1958) and *The Tacit Dimension* (Garden City, NY: Doubleday, 1966). Augustine Brannigan, *The Social Basis of Scientific Discoveries* (Cambridge: Cambridge University Press, 1981) remains important. Among the most recent works in what is now a very large literature on social influences on the practice and content of science, see Steven Shapin, *Never Pure: Historical Studies of Science as If It Was Produced by People with Bodies, Situated in Time, Space, Culture, and Society, and Struggling for Credibility and Authority* (Baltimore: Johns Hopkins University Press, 2010).

Ken Caneva, "'Discovery' as a Site for the Collective Construction of Scientific Knowledge," *HSPS* 35 (2005): 175–291 is the most comprehensive treatment of the idea of collective discovery, and reviews much of the literature on the subject. On the problematic nature of observation as related to discovery, see Norwood Russell Hanson, *Patterns of Discovery: An Inquiry into the Conceptual Foundations of Science* (Cambridge: Cambridge University Press, 1958), and Daston and Lunbeck, *Histories of Scientific Observation*; also Augustine Brannigan, *The Social Basis of Scientific Discoveries* (Cambridge: Cambridge University Press, 1981).

The literature on classification is large in biology, but not in astronomy. Among the most important works relevant to biology are Paul L. Farber, *Finding Order in Nature: The Naturalist Tradition from Linnaeus to E. O. Wilson* (Baltimore and London: Johns Hopkins University Press, 2000); Ernst Mayr, *The Growth of Biological*

Thought: Diversity, Evolution and Inheritance (Cambridge, MA: Harvard University Press, 1982); Ernst Mayr, *Toward a New Philosophy of Biology*, (Cambridge, MA: Harvard University Press, 1988); Jan Sapp, *The New Foundations of Evolution on the Tree of Life* (Oxford: Oxford University Press, 2009); David L. Hull, *Science as a Process: An Evolutionary Account of the Social and Conceptual Development of Science* (Chicago: University of Chicago Press, 1988); Marc Ereshefsky, *The Units of Evolution: Essays on the Nature of Species* (Cambridge, MA: MIT Press, 1992); and Lynn Margulis and Michael J. Chapman, *Kingdoms & Domains: An Illustrated Guide to the Phyla of Life on Earth* (Amsterdam: Elsevier, 2010),

For the development of classification in chemistry, in particular the Periodic Table, see Michael D. Gordin, *A Well-Ordered Thing: Dmitrii Mendeleev and the Shadow of the Periodic Table* (New York: Basic Books, 2004), and Eric Scerri, *The Periodic Table: Its Story and Significance* (Oxford: Oxford University Press, 2007). For classification in physics in relation to the Standard Model, see Andrew Pickering, *Constructing Quarks: A Sociological History of Particle Physics* (Edinburgh: Edinburgh University Press, 1984); Murray Gell-Mann, *The Quark and the Jaguar: Adventures in the Simple and the Complex* (New York: W. H. Freeman, 1994).

For broader treatments of the importance of classification in science, see Carol Kaesuk Yoon, *Naming Nature: The Clash Between Instinct and Science* (New York: W. W. Norton, 2009) and Andrew Hamilton and Quentin D. Wheeler, "Taxonomy and Why History of Science Matters for Science," *Isis* 99 (June 2008): 331–40. On classification as it relates to daily human life, see Geoffrey C. Bowker and Susan Leigh Star, *Sorting Things Out: Classification and Its Consequences* (Cambridge, MA: MIT Press, 1999).

Part IV. Drivers of Discovery

On the role of instruments in the history of science, a place to start is the special *Osiris* volume introduced by Albert van Helden and Thomas Hankins, "Introduction: Instruments in the History of Science," *Osiris*, 2nd series, ix (1994): 1–6: 4. On technology as an engine of discovery in elementary particle physics, see Andrew Sessler and Edmund Wilson, *Engines of Discovery: A Century of Particle Accelerators* (Hackensack, NJ: World Scientific Publishing, 2007); Peter Galison, *How Experiments End* (Chicago: University of Chicago Press, 1987), and Galison, *Image and Logic: A Material Culture of Microphysics* (Chicago: University of Chicago Press, 1997).

On the circumstances of the invention of the telescope and Galileo's role, see the excellent set of articles in Albert van Helden, Sven Dupré, Rob van Gent, and Huib Zuidervaart, eds., *The Origins of the Telescope* (Amsterdam: KNAW Press, 2010); also Toby E. Huff, "Inventing the Discovery Machine," in *Intellectual Curiosity and the Scientific Revolution: A Global Perspective* (Cambridge: Cambridge University Press, 2011), pp. 22–47.

For twentieth-century astronomy see Robert Smith, "Engines of Discovery: Scientific Instruments and the History of Astronomy and Planetary Science in the United States in the Twentieth Century," *JHA* 32 (1997): 49–77. For the effect of opening the electromagnetic spectrum for astronomy in the radio wavelength regime, see Woodruff T. Sullivan III, *Cosmic Noise: A History of Early Radio Astronomy* (Cambridge: Cambridge University Press, 2009). On the early history of X-ray astronomy, Richard F. Hirsh, *Glimpsing an Invisible Universe: The Emergence of X-ray Astronomy* (Cambridge: Cambridge University Press, 1983). On the infrared, G. H. Rieke, "History of Infrared Telescopes and Astronomy," in Bernhard Brandl, Remko Stuik, and Jeanette Katgert-Merkelijn, eds., *400 Years of Astronomical Telescopes* New York: Springer, 2010) pp. 125–41. On the ultraviolet, see Martin A. Barstow and Jay B. Holberg, *Extreme Ultraviolet Astronomy* (Cambridge: Cambridge University Press, 2003). On V-2s and sounding rockets, David H. DeVorkin, *Science with a Vengeance: How the Military Created the US Space Sciences After World War II* (New York: Springer, 1992).

Part V. The Synthesis of Discovery

On the history and uses of the cosmic evolution idea, see Steven J. Dick, "Cosmic Evolution: History, Culture, and Human Destiny," in Steven J. Dick and Mark L. Lupisella, *Cosmos and Culture: Cultural Evolution in a Cosmic Context* (Washington, DC: NASA SP 4802, 2009), pp. 25–62. Elements of the idea of cosmic evolution are discussed in Stephen G. Brush's much underrated *Nebulous Earth: The Origin of the Solar System and the Core of the Earth From Laplace to Jeffreys* (Cambridge: Cambridge University Press, 1996). On Shapley's cosmic evangelism, see Joann Palmeri, "Bringing Cosmos to Culture: Harlow Shapley and the Uses of Cosmic Evolution," in Dick and Lupisella, *Cosmos and Culture,* pp. 489–521. On how discovery is changing in the modern era, see Michael Nielsen, *Reinventing Discovery: The New Era of Networked Science* (Princeton: Princeton University Press, 2012).

Spectacular imagery of the objects discussed in this volume is now widely available, not only from sites such as the Hubble Space Telescope (http://hubblesite.org/gallery/album/) and other spacecraft Web sites, but also from ground-based projects such as the Sloan Digital Sky Survey and the Web sites of particular observatories. In addition, programs such as WikiSky's Sky Map (http://www.sky-map.org/), Microsoft's World Wide Telescope (http://www.worldwidetelescope.org/Home.aspx), and Google Sky (http://www.google.com/sky/) provide imagery, information, and associated sky positions of specific objects.

Glossary of Concepts Related to Discovery

We have emphasized throughout this volume the necessity of precision of language. This glossary is a summary of terms and concepts as defined in this volume.

Class. A category, or *taxonomic level*, in a classification system. Arguably the optimal way to define "class" most precisely is through such a system, exemplified in Appendix 1. In astronomy the majority of new classes have been discovered by detection (planetary satellites and rings), some by extended inference (gas giant planets, dwarf, and giant stars), and only a few by declaration (dwarf planets, lenticular galaxies) when class status is otherwise ambiguous.

Class, Evolution of. The meaning of a particular class may sometimes evolve, e.g., double stars to binary stars (physical systems), terrestrial and giant planets based on size to those based on composition, spiral and elliptical nebulae to spiral and elliptical galaxies. The vast majority of classes, however, are stable and do not evolve in terms of their fundamental meaning. At the same time, the boundaries of a particular class may expand over time, as in the case of those active galaxies known as blazars.

Classification System. A system that both defines and orders classes, either based on phenomena (an artificial classification system) or on deeper physical meaning (natural classification system), sometimes referred to as "the thing itself." A classification system may be hierarchical, as in biology, or flat, as in chemistry's periodic table. Modern classification systems for biology began with Linnaeus in the eighteenth century. Influenced by Linnaeus, William Herschel was the first to produce a classification system for a set of astronomical objects, the nebulae.

Discovery. A complex process of uncovering novelty, characterized by an extended structure consisting of detection, interpretation, and understanding, each with its own technological, conceptual, and social components. In this volume detection refers to the detection of a new class of astronomical object, a stage sometimes replaced by inference or declaration. Discovery thus defined is often preceded by pre-discovery (see Appendix 2) and always by a post-discovery. The extended structure of discovery leads directly to the idea of collective discovery.

Discovery, Collective. The idea that, due to the extended structure of discovery, many individuals contribute to any particular discovery, with implications for community consensus, acceptance, and credit.

Discovery, End of. Discoveries may be defined as ending with a basic understanding of the fundamental properties of a class, but before mature understanding as defined by knowing an object's place in an evolutionary scheme. The end of any particular discovery involves "persuasive evidence," a necessarily nebulous concept that depends on a scientific community's collective decision.

Pre-discovery. A phase preceding discovery, consisting of casual sightings of astronomical objects with the naked eye, or telescopic observations that go unreported, unrecognized, or undistinguished as new classes of objects. The concept applies not only to early observations of meteors, comets, and novae, but also to planets, stars, and nebulae before they were separated as classes and their true nature was understood. In the case of nebulae that proved to be galaxies, the definition of the class evolved. Pre-discovery inference is based on data or theory before the discovery is made.

Post-discovery. A phase during which the detailed properties of the object, rather than the object itself and its fundamental properties, are discovered. Post-discovery also encompasses the reception of the discovery among scientists and the public, its place in any evolutionary scheme, and issues of credit and reward.

Index